CHEMISTRY FOR THE PROTECTION OF THE ENVIRONMENT 3

ENVIRONMENTAL SCIENCE RESEARCH

Recent Volumes in this Series

Volume 45 — GLOBAL CLIMATE CHANGE: Linking Energy, Environment, Economy, and Equity
Edited by James C. White

Volume 46 — PRODUCED WATER: Technological/Environmental Issues and Solutions
Edited by James P. Ray and F. Rainer Engelhardt

Volume 47 — GLOBAL ENERGY STRATEGIES: Living with Restricted Greenhouse Gas Emissions
Edited by James C. White

Volume 48 — GLOBAL ATMOSPHERIC–BIOSPHERIC CHEMISTRY
Ronald G. Prinn

Volume 49 — BIODEGRADATION OF NITROAROMATIC COMPOUNDS
Edited by Jim C. Spain

Volume 50 — BIOMONITORS AND BIOMARKERS AS INDICATORS OF ENVIRONMENTAL CHANGE: A Handbook
Edited by Frank M. Butterworth, Lynda D. Corkum, and Judith Guzmán-Rincón

Volume 51 — CHEMISTRY FOR THE PROTECTION OF THE ENVIRONMENT 2
Edited by Lucjan Pawłowski, William J. Lacy, Christopher G. Uchrin, and Marzenna R. Dudzińska

Volume 52 — PRODUCED WATER 2: Environmental Issues and Mitigation Technologies
Edited by Mark Reed and Ståle Johnsen

Volume 53 — EVALUATING CLIMATE CHANGE ACTION PLANS: National Actions for International Commitment
Edited by James C. White

Volume 54 — BIOTECHNOLOGY IN THE SUSTAINABLE ENVIRONMENT
Edited by Gary S. Sayler, John Sanseverino, and Kimberly L. Davis

Volume 55 — CHEMISTRY FOR THE PROTECTION OF THE ENVIRONMENT 3
Edited by Lucjan Pawłowski, Marjorie A. Gonzales, Marzenna R. Dudzińska, and William J. Lacy

---

CHEMISTRY FOR THE PROTECTION OF THE ENVIRONMENT 3

Edited by

Lucjan Pawłowski
Technical University of Lublin
Lublin, Poland

Marjorie A. Gonzales
Lawrence Livermore National Laboratory
Livermore, California

Marzenna R. Dudzińska
Technical University of Lublin
Lublin, Poland

and

William J. Lacy
Lacy and Associates
Alexandria, Virginia

Springer Science+Business Media, LLC

Library of Congress Cataloging-in-Publication Data

Chemistry for the protection of the environment 3 / edited by Lucjan Pawłowski ... [et al.].
p. cm. -- (Environmental science research ; v. 55)
"Proceedings of the 11th International Conference on Chemistry for the Protection of the Environment, held September 10-17, 1997, in Cairo, Assuan and Luxor, Egypt"--T.p. verso.
Includes bibliographical references and index.
ISBN 978-1-4757-9666-7
1. Environmental chemistry--Congresses. 2. Environmental protection--Congresses. 3. Environmental engineering--Congresses. I. Pawłowski, Łucjan. II. International Conference on Chemistry for the Protection of the Environment (11th : 1997 : Cairo, Egypt, etc.) III. Series.
TD193.C4718 1998
628--dc21 98-41248
CIP

Proceedings of the 11th International Conference on Chemistry for Protection of the Environment, held September 10 – 17, 1997, in Cairo, Assuan, and Luxor, Egypt

ISBN 978-1-4757-9666-7 ISBN 978-1-4757-9664-3 (eBook)
DOI 10.1007/978-1-4757-9664-3

Originally published by Plenum Press, New York in 1998
Softcover reprint of the hardcover 1st edition 1998

http://www.plenum.com

10 9 8 7 6 5 4 3 2 1

PREFACE

The first meeting in this series was organized by Prof. Pawłowski and Dr. Lacy in 1976 at the Marie Curie-Sklodowska University in Lublin, Poland. The conference dealt with various physicochemical methodologies for water and wastewater treatment research projects that were jointly sponsored by US EPA and Poland.

The great interest expressed by the participants led the organizers to expand the scope of the second conference, which was also held in Poland in September 1979. The third and enlarged symposium was again successfully held in 1981 in Lublin, Poland. At that time the participating scientists and engineers expressed their desire to broaden the coverage as well as the title of the conference series. The International Committee, approved the title "Chemistry for the Protection of the Environment" and designated that date of the fourth conference, CPE IV, which was convened in September 1983 at the Paul Sabatier University in Toulouse, France, and was hosted and arranged by Prof. A. Verdier. This conference series included participants from various government agencies, academia, and the private sector, representing industrialized countries as well as emerging nations, both the East and West in an independent, non political forum.

The central goals of CPE are to improve technology transfer and scientific dialogue, thereby leading to a better comprehension of and solution to a broad spectrum of environmentally related problems. The fifth conference was held in September 1985 at the Catholic University in Leuven, Belgium. It was hosted by Profs. A. Van Haute and G. Alaerts. CPE V covered topics dealing with treatment technologies and phenomena related to hazardous waste and the utilization of fossil fuels. It provided an opportunity for interdisciplinary discussions and encouraged the exchange of ideas among international specialists from diverse fields and backgrounds.

Under the leadership of Profs. Mentasti and Sarzanini and with the assistance of Dr. Gennero, CPE VI, was held in 1987 at the University of Turin in Italy. Over 150 selected scientific papers and posters were presented to an audience of specialists from 32 nations. This assemblage comprised in equal measure scientists from Europe, the New World, and developing nations.

CPE VII, was convened at the Catholic University in Lublin, Poland in 1989. The exchange of information by approximately 200 scientists and engineers made this a memorable scientific occurrence. The scientific committee selected presenters of high intellectual and technical merit. The distinguished participants of CPE VII included Poland's Minister and Deputy Minister for Environmental Protection, U.S. Scientific Council, Israel's Deputy Minister of the Environment, presidents and vice presidents of five universities, representatives of the Academies of Sciences for Czechoslovakia, France, Italy, Poland, and the U.S.S.R., as well as many department heads and acclaimed scientists.

CPE VIII was scheduled to convene in Budapest, Hungary in September 1991. But due to international administrative difficulties it was moved to Lublin. Despite this last minute change of venue, the scientific meeting was voted an outstanding success by the participants. One of the key note speakers was Dr.Gilbert S. Jackson, Senior Environmental Engineer for Latin America and The Caribbean of The U.S. Agency for International Develpment. Another was Debra A.Jacobson, Counsel to The Committee on Energy and Commerce of The U.S.Houce of Representatives, Washington, D.C. The technical presentation were original and informative, the major topics included chemical/physical/biological/treatment technologies, monitoring modeling and risk assessment. There were over 120 attendees and some 90 scientific oral and poster presentations. Through CPE International Committee, Dr. Hartstein of U.S. Dept. of Energy, had the proceedings published in the USA

CPE IX held in September 1993 in Alexandria, Cairo/Luxor Egypt and included a joint conference with Dr. Ahmed Hamza and Dr. James Gallup, EPA/U.S.AID's Fourth International Symposium on Industry in the Developing World. The issues covered were an extremely successful workshop on industrial pollution prevention and clean technologies plus cooperation and institutional issues. It too was deemed a great success, highly informative by the attendees. The multi-disciplined technical group from 27 countries and international (organization affiliated with the global environmental movement) were a captive like audience aboard a Nile River Crusier. This atmosphere promoted free, open exchanges and dialogues between all the attendees. Selected papers were published in a volume by the High Institute of Public Health, Alexandria, Egypt.

CPE X the 20th anniversary meeting was held in the city of its birth, Lublin, Poland. The spirit of this 20th anniversary not only permeated the program and the international group of experts but was reflected in the various folk music festival, folk dancing and social/cultural programs. Some of the major benefits for the participants were technology transfer and exchange of novel, innovative and alternative treatment methods and information about activities in other countries related to environmental problems. The meeting was enhanced by the participation of large delegations from both The Peoples Republic of China and the Taiwan Chinese Republic, with papers published under title „Chemistry for the Protection of the Environment, 2" in Environmental Science Research Series, by Plenum Press.

The XI CPE was held in Cairo, Egypt and on the Nile between Luxor and Assuan, including a site visit on the Assuan Dam. Interesting and informative papers and posters were presented on the following topics: adsorption, analytical methods, chemical/ biological/ treatment, groundwater studies, ion exchange, modeling, risk assessment, sludge treatment, waste minimization, innovative technology, acid rain, and for the first time during CPE conferences on ISO 14001 - environment management and quality systems. Selected papers are published in this volume.

The next CPE XII is scheduled to be held in China just before the end of this millennium - in autumn 1999. Prof. Cao Zhihong of Nanjing Institute of Soil Sciences will be a host of that conference. We hope to attract even more attention of scientists from the Asia countries.

L. Pawłowski, M. A. Gonzalez, M. R. Dudzińska, W. J. Lacy

CONTENTS

Section 3: New Technologies in Wastwater Treatment

Section 4: Solid Waste Utilization

CHEMISTRY FOR THE PROTECTION OF THE ENVIRONMENT 3

DRINKING WATER PRODUCTION WITH A DUAL FLOATING MEDIUM–SAND FILTER SYSTEM

B. A. Bolto,[1] H. H. Ngo,[2] and S. Vigneswaran[2]

[1]CSIRO Division of Molecular Science
Private Bag 10, Clayton South MDC, Victoria 3169, Australia
[2]School of Civil Engineering
University of Technology
PO Box 123, Broadway, NSW 2007, Australia

ABSTRACT

Low quality water supplies have been treated in a combined downflow filter column comprising a floating medium and a coarse sand layer at the bottom, the focus being on a comparison of inorganic coagulants and cationic polyelectrolytes as primary coagulants. In one study a commercially available organic polyelectrolyte was not as effective as polyaluminium chloride, but an experimental organic polymer having some aromatic and hydrophobic character was comparable with the inorganic coagulant for particulates uptake, and was almost as good for organics removal. In another comparison ferric chloride was less effective than highly charged organic polymers for turbidity removal. In both cases the dose of organic polymer required was much less than that of the inorganic coagulant. This poses operational disadvantages of filter run length and sludge volume for the inorganic additives.

Keywords: Cationic polyelectrolyte; primary coagulation; clay; humic substances; floating filter medium; dual media; filtration.

1. INTRODUCTION

A downflow buoyant medium packed bed filtration system with in-line flocculation arrangement has been successfully developed for water and wastewater treatment.[1–5] The system is a direct filter that operates under constant head pressure, and incorporates in-line

Chemistry for the Protection of the Environment 3, edited by Pawłowski *et al.* Plenum Press, New York, 1998

flocculation. After addition of coagulant, the feed water enters the filter at the top where it comes into contact with the floating medium. Flocculation then occurs in the floating medium due to the promotion of interparticle contacts by the water flow around individual grains of media. This is followed by the separation of particles and flocs on the floating filter medium. Thus it has a dual function of flocculation and solid-liquid separation. A detailed study with a short depth (400 mm) of polypropylene medium showed that the system has good pollutant removal capacity at a filtration velocity of 5–10 m^3/m^2h. It also demonstrated very low headloss development and produced uniform, microflocs of 26–40 μm size.[1–4] Detailed studies have shown that the system can treat water at high loading rates of up to 60 m^3/m^2h with a suitable depth (>1200 mm) of polystyrene medium. At high filtration rates, although the operation of frequent but short duration backwashes is necessary, the backwash required is less than 1% of water production and the energy requirement is minimal.[5] To maximize pollutant removal efficiency, a dual floating medium-sand filter system was applied with the concept of using a floating medium filter as a prefilter with the sand filter as a subsequent polishing filter.[3]

The present paper looks at the removal of humic substances and clay. Humic substances in natural waters are but one component of natural organic matter (NOM), which in the form of dissolved organic carbon is defined as that which will pass through a membrane having pores of 0.45 μm size. It is present in water sources at levels in the range 0.1–115 mg/L, with 5.75 mg/L being a world average for streams.[6] Normal drinking water sources contain 2–10 mg/L of NOM, of which only 10–30% has been identified. There are several reasons why it poses a problem for the water treatment industry: apart from causing colour, taste and odour, especially after disinfection with chemicals, it can form potentially carcinogenic chlorinated hydrocarbons when disinfection is carried out with chlorine, and aldehydes and carboxylic acids when ozone is employed. As well, NOM or its degradation products interfere with the oxidation of dissolved iron and manganese to insoluble, easily removed forms; it fouls membranes and ion-exchange resins; and it exacerbates corrosion and the deterioration of water quality in distribution systems because of biological growths.[7,8]

There is a whole range of organic compounds involved: amino acids, fatty acids, phenols, sterols, sugars, hydrocarbons, urea, porphyrins and polymers. The polymers that are present include polypeptides, lipids, polysaccharides and humic substances. The proportions of the various compounds can range from 1% for non-functionalised hydrocarbons to 10% for humic acid and 40% for fulvic acid; this varies seasonally as well as regionally.[6,9] The higher the colour, the higher the MW of the NOM; deciduous forest areas give more colour than coniferous forest areas.[9] Significant structural differences in NOM, which are known to vary wildly with the raw water source and the season, call for caution in the use of synthetic NOM to simulate the natural one. It is important to know the performance variation for process planning purposes.

The present study also compares the performance of cationic polyelectrolytes with that of inorganic coagulants. That organic polymers may be used as primary coagulants as well as in the more traditional flocculation step of binding already formed small flocs into larger particles is well documented.[10–18] The polymer acts as a destabilising agent via a charge neutralisation/precipitation mechanism, and as well is an agent for floc growth.[19]

In summary, the literature indicates that polymers have a considerable number of advantages, mostly arising from the lack of additional solids in the form of metal hydroxide which adds to the burden of the separation process. Thus in utilising cationic polymers in lieu of inorganic coagulants the performance is not dependent on pH, and a lower dose of coagulant is applied, in proportion to the turbidity. There is a lower level of dissolved ions in the product water, there are no soluble residues from added metal ions such as Al or Mn, and the alkalinity is maintained. Because of the lower content of insoluble solids in the reaction mixture, processing is faster, filter runs are longer, the head loss is significantly lower, there is a much smaller sludge volume—usually half—and the sludge has a lower water content. Costs are competitive if the dose is not extreme, but costing needs to be holistic, taking into account sludge disposal benefits and the other advantages mentioned. However, health and environmental issues require attention, and a much clearer understanding of the reaction mechanism is required to optimise the choice of polymer.

2. EXPERIMENTAL

The filter consisted of a perspex column (90 mm inside diameter, 1600 mm height). Polypropylene beads (diameter 3.8 mm, density 0.87 g/cm^3) were packed in the column to a depth of 400 mm and restrained by a grid of stainless steel coarse mesh at the top. A layer of coarse sand (diameter 1.7 mm) was placed at the bottom of the filter column to a depth of 400 mm. A simple rapid mixing arrangement was provided to mix the flocculant with suspension prior to its entry into the floating medium. There was a 400 mm space between the two media. The layout is illustrated in Fig. 1.

During experiments, the filtration velocity in the filter was maintained constant at a known value (5, 7.5 and 10 m^3/m^{2}h). In order to evaluate the filtration performance, headloss measurement as well as water sampling for filtrate quality analysis were carried out periodically. The headloss through the bed was directly recorded from a manometer reading. The filter column was cleaned after each filter run using a combination of air (35–70 Kpa) and water (6 L/m^{2}s). Typical cleaning of the column involved air scouring (30 s) followed by a water backwash (30 s). This procedure was repeated twice and followed by a

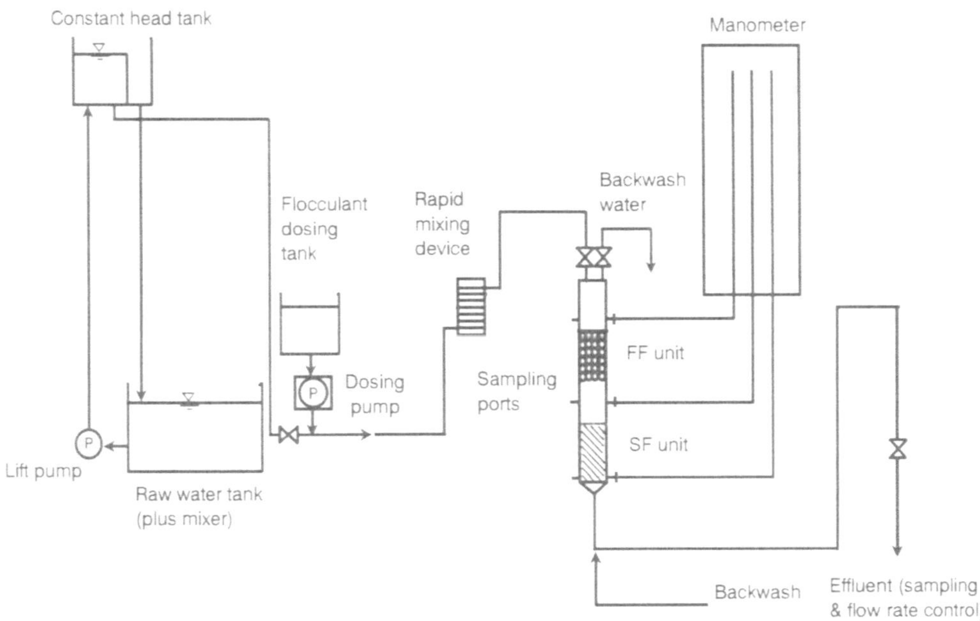

Figure 1. Experimental set-up, with FF denoting the floating filter and SF the sand filter.

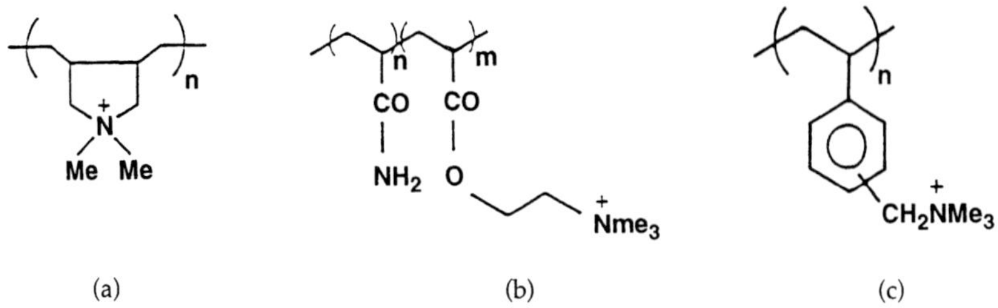

Figure 2. Structures of (a) DADMAC; (b) CPAM; (c) CPS.

final backwash with water for four minutes. This was done to clean the media thoroughly between runs, and did not represent the optimum backwashing which would apply in an operational situation. The use of a buoyant medium significantly reduces the backwash water requirement.

The feed consisted of a synthetic water containing 50 mg/L of kaolin and 0.5 mg/L of fulvic acid from an International Humic Substances Standard Suwannee River sample. The coagulants tested were polyaluminium chloride (PACl), from Atochem, France, polydiallyldimethylammonium chloride (DADMAC) of charge density (CD) 100 wt% and medium MW and a range of high MW cationic polyacrylamides (CPAM) of CD 10, 50, 63, and 80 wt% from Allied Colloids, Australia, and cationic polystyrene (CPS) of CD 85 wt% and low molecular weight (MW) from Dow Chemical, USA. The weakly basic polyallylamine hydrochloride (PAll) was obtained as a low MW polymer from Bio-Scientific Pty Ltd, Australia. The structures of the cationic polymers are shown in Fig. 2, with the anionic counter ions omitted.

The optimum polymer dose was determined by standard jar tests, and the filtrate quality was measured in terms of turbidity, suspended solids (SS) and total organic carbon (TOC). The SS were measured by the American Water Works Association standard method, with drying at 103–105°C; TOC was measured with a Dohrman DC80 analyser; UV absorption was measured at a wavelength of 195 nm using a Unicam 5625 UV/VIS spectrometer.

3. RESULTS AND DISCUSSION

3.1. Relative Performance in Jar Tests

In standard jar tests to determine the optimum dose of polymer at pH 5–6, it was observed that the turbidity of the waters increased with polymer dose, to peak at 0.1–0.2 mg/L of polymer. This was ascribed to the precipitation of the fulvate anions by the cationic polymers, the effectiveness of which was found to be better for the very high MW polymers which have superior floc bridging ability; CD was less important in a series of CPAMs, which gave peak turbidity readings as in Table 1.

On the other hand, measurement of the organics remaining after filtration through a 0.45μm membrane, carried out by UV absorption at a wavelength of 195 nm, showed the CD to be the dominating factor, as depicted in Fig. 3. The best removal at pH 6 was in the order CD 80 > 63 > 50%. Polyallylamine performed much as for the CPAM of CD 80%.

Table 1. Residual turbidity after optimum dose of CPAM as a function of polymer CD and MW

CD %	MW M	Maximum turbidity NTU
80	7	30
10	9	27
63	3	17
50	3	14

Protonation of the primary amino groups in this polymer at pH 6 can be assumed to be essentially complete.[20]

3.2. Dual floating/Settled Filter Media Runs

A striking feature on variation of coagulant is the greatly reduced headloss arising from the use of organic polymers, which require much lower doses than PACl. For example, after a run time of three hours at a filtration rate of 5 m^3/m^2h, the headloss through the dual system was about 40 mm of water for CPS and 20 mm for DADMAC, versus 190 mm for PACl.[10] This parallels earlier findings with direct filtration, where it took five to six hours to achieve a head loss of 1000 mm of water when a dose of 14 mg/L of alum was used; when 10 mg/L of a cationic polyelectrolyte was the coagulant instead of alum, the time required was extended to 24 hours.[16]

The effective run time is a function of filtration rate, the three hour headloss being 60 mm of water at 2.5 m^3/m^2h for PACl, but 265 mm at 7.5 m/h. For CPS the losses are 20 and 120 mm respectively. At a filtration rate of 5 m^3/m^2h the relative performances of PACl, DADMAC and CPS are shown in Table 2. The removal of SS is good with all, while the removal of finer dispersed material as turbidity is superior with PACl and CPS. PACl is best for TOC removal. At a filtration rate of 7.5 m^3/m^2h the story is similar, with PACl and CPS being on a par in their response to SS and turbidity and organics removal (Table 3).

Some idea of the contrasting sludge volumes produced can be estimated from the total loads on the filter. At a filtration rate of 5 m^3/m^2h after a three hour filter run the solids

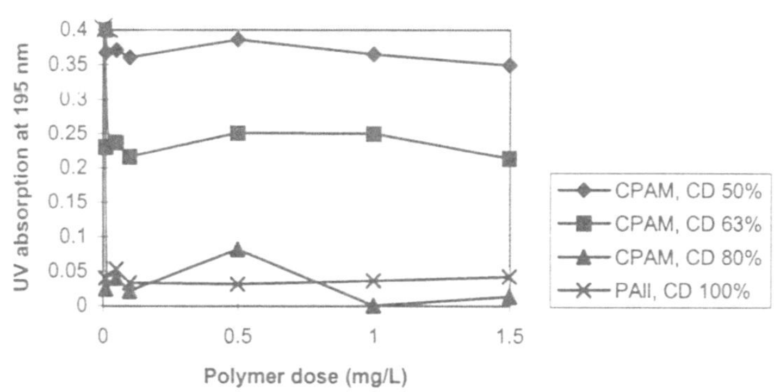

Figure 3. Removal of UV absorbing organic compounds as a function of polymer dose.

Table 2. Percentage removals after dual filtration at 5 m^3/m^2h^a

Coagulant	Dose mg/L	SS % removal	Turbidity % removal	TOC % removal
PACl	88.5	97	88	58
DADMAC	1	94	80	38
CPS	2	97	90	44

[a]Initial SS 50 mg/L, turbidity 30-46 NTU and TOC 6-7 mg/L

Table 3. Percentage removals after dual filtration at 7.5 m^3/m^2h^a

Coagulant	Dose mg/L	SS % removal	Turbidity % removal	TOC % removal
PACl	88.5	98	93	54
CPS	2	98	94	56

[a] Initial SS 50 mg/L, turbidity 29-43 NTU and TOC 5-8 mg/L

Table 4. Estimates of filter load as a function of coagulant after 3 h at 5 m^3/m^2h

Coagulant	SS and TOC load kg/m^3	Total solids load kg/m^3
PACl	0.68	1.78
CPS	0.64	0.65
DADMAC	0.61	0.61

Table 5. The performance of floating medium filtration with different flocculants[a]

Coagulant	Dose of coagulant mg/L	Average turbidity removal %	Headloss after 4 h mm water
$FeCl_3$	35	84	490
PAll, CD 100%	1	91	1830
CPAM, CD 63%	1	92	510
CPAM, CD 50%	1	48	100

[a]Influent turbidity 54 NTU; filter depth 1210 mm; filtration velocity 30 m^3/m^2h; filtration time 4 h

loading on the filter, including the coagulant itself, is more than doubled when PACl is the additive compared to when organic polymers are used, as shown in Table 4. Also, because of the more gelatinous nature of the metal hydroxide sludge, for the same total solids loading on the filter the head loss is significantly greater; at 0.6 kg/m^3 it is more than five times that of the organic polymer sludge.[10]

Experiments were also conducted at a high filtration rate of 30 m^3/m^2h with a single polystyrene medium bed of 1210 mm depth, to give the results shown in Table 5. When polymers such as CPAM of CD 63% or PAll were used as a single flocculant instead of ferric chloride, the filter performance improved further in terms of turbidity removal (91 and 92% respectively, versus 84%). However, the headloss in the case of PAll was high at 1830 mm after filtration for four hours, versus 510 mm for the CPAM. This is contrary to what would be expected from the chain lengths of the two polymers, which have low and very high MWs respectively, and may be a reflection of the role of molecular structure on performance. In the case of the CPAM of CD 50%, removal efficiency and headloss were significantly lower, with only 48% turbidity removal accompanied by an insignificant headloss. The lower solids removal clearly results in less resistance to flow through the filter.

4. CONCLUSIONS

In both the comparisons outlined here the dose of organic polymer required was much less than that of the inorganic coagulant. This poses operational disadvantages of filter run length and sludge volume for the inorganic additives. Highly charged organic polymers are generally as effective as their inorganic counterparts for turbidity removal, and nearly as effective for the removal of the dissolved organics in the raw water studied here, in the form of low MW fulvic acids. Health and environmental issues need to be considered; these are well documented except for CPS and PAll.

ACKNOWLEDGMENTS

The work was supported by an Australian Research Council Large Research Grant, 1997-98.

REFERENCES

1. Ngo, H. H. and S. Vigneswaran, Application of floating medium filter in organic removal, *J. Indian Assocn. for Environmental Management,* 1994, *21* (3), 55–62.
2. Ngo, H. H. and S. Vigneswaran, Floating medium downflow flocculator with coarse sand filter—a system for a small community water supply, *Water,* 1995, *22* (3), 34–37.
3. Ngo, H. H. and S. Vigneswaran, Application of floating medium filter in water and wastewater treatment with contact-flocculation filtration arrangement, *Water Research,* 1995, *29,* 2211–2213.
4. Ngo, H. H. and S. Vigneswaran, Application of downflow floating medium flocculator/prefilter—coarse sand filter in nutrient removal, *Water Science and Technology,* 1996, *33,* 63–70.
5. Vigneswaran, S. and H. H. Ngo, A high rate flocculation-filtration system in water treatment, *Indian Journal of Engineering & Material Sciences* , 1997 (in press).
6. Boggs, S., D. G. Livermore and M. G. Seitz, Humic macromolecules in natural waters, *Rev. Macromol. Chem. Phys.* 1985, *C25,* 599–657.
7. Huang, W.-J. and H.-H. Yeh, Organic fractionation for water treatment processes evaluation, *Proc. Water Technology Conf.,* Pt. I,, Amer. Water Works Assoc., Denver, 1993, p. 257.
8. Duguet, J.-P. and J. Mallevialle, Influence of NOM on water treatment, *Proc. 21st Congress International Water Services Association,* Madrid, 20–26 September 1997, p. SS13–8.
9. Thorsen, T., A. Harz and H. Ødegaard, Influence of raw water characteristics and membrane pore size on the performance of ultra-filters for NOM removal, *Proc. 21st Congress International Water Services Association,* Madrid, 20–26 September 1997, p. SS13–5.
10. Kancharla, V., H. H. Ngo, S. Vigneswaran and B. A. Bolto, The use of polyelectrolyte in downflow filtration in a dual system of floating medium and sand, *Proceedings of the AWWA 17th Federal Convention,* Vol. 1, Australian Water and Wastewater Assoc., Artarmon, 1997, p. 506.
11. Glaser, H. T. and J. K. Edzwald, Coagulation and direct filtration of humic substances with polyethyleneimine, *Environ. Sci. Technol.,* 1979, *13,* 299–305.
12. Jackson, G. E., Granular media filtration in water and wasteweater treatment—Part 2, *CRC Critical Reviews in Environmental Control,* Vol. 11, CRC Press, Boca Raton, Florida, 1980, p. 1.
13. Schlauch, R. M., Coagulation for gravity type clarification and thickening, in: *Polyelectrolytes for Water and Wastewater Treatment,* ed. W. K. Schwoyer, CRC Press, Boca Raton, Florida, 1981, p. 91.

14. Amy, G. L. and P. A. Chadik, Cationic polyelectrolytes as primary coagulants for removing trihalomethane precursors, *J. Am. Water Works Assoc.*, 1983, *75*, 527–531.
15. Rebhun, M., Z. Fuhrer and A. Adin, Contact flocculation-filtration of humic substances, *Water Research*, 1984, *18*, 963–970.
16. Edzwald, J. K., Conventional water treatment and direct filtration: treatment and removal of total organic carbon and trihalomethane precursors, in: *Organic Carcinogens in Drinking Water*, ed. N. M. Ram, E. J. Calabrese and R. F. Christman, Wiley, New York, 1986, p. 208.
17. Vik, E. A. and B. Eikebrokk, Coagulation process for removal of humic substances from drinking water, in: *Aquatic Humic Substances*, ed. I. H. Suffet and P. MacCarthy, Adv. Chem. Seies, Vol. 219, Am. Chem. Soc., Washington, 1989, p. 385.
18. Coccagna, L., Direct filtration, in: *Water, Wastewater, and Sludge Filtration*, ed. S. Vigneswaran and R. Ben Aim, CRC Press, Boca Raton, Florida, 1989, p. 57.
19. Bolto, B. A., Soluble polymers in water purification, *Prog. Polym. Sci.*, 1995, *20*, 987–1041.
20. Chen, W. and T. J. McCarthy, Layer-by-layer deposition: a tool for polymer surface modification, *Macromolecules*, 1997, *30*, 78–86.

2

DETERMINATION OF REDUCED SULFUR COMPOUNDS IN THE AQUATIC ENVIRONMENT BY HIGH-PERFORMANCE LIQUID CHROMATOGRAPHY AND CAPILLARY ELECTROPHORESIS

Elżbieta Kaniowska, Rafał Głowacki, Grażyna Chwatko, Paweł Kubalczyk, and Edward Bald

Department of Environmental Chemistry
University of Lodz
163 Pomorska Str.
90-236 Lodz, Poland

1. ABSTRACT

A sensitive and reproducible method for the determination of the mixture of hydrophilic thiols and hydrogen sulfide in water has been developed. The essential steps in the assay include conversion of the thiols and hydrogen sulfide to stable derivatives with the use of 2-chloro-1-methylpyridinium iodide (CMPI), separation of the derivatives by reversed-phase high-performance liquid chromatography or capillary electrophoresis (HPCE), and detection and quantification by UV-spectrophotometry. Even in the case of a large excess of hydrogen sulfide the CMPI-HPLC method has a sensitivity higher than 2 pmol and coefficients of variation from 0.47% for 100 nmol/ml level of thioglycolic acid to 4.58% for 5 nmol/ml level of homocysteine. The linear calibration graphs were obtained for concentration of the thiols between 5 and 100 nmol/ml and for hydrogen sulfide from 5 to 600 nmol/ml.The CMPI-HPCE method proved sensitivity higher than 1 pmol and better resolution as compared with CMPI-HPLC. The CMPI-HPLC method was applied to sediment porewater samples.

Key words: HPLC, HPCE, enviromental reduced sulfur compounds, N-acetylcysteine, cysteine, glutathione, homocysteine, hydrogen sulfide, N-(2-mercaptopropionyl)glycine, thioglycolic acid, thiomalic acid

2. INTRODUCTION

Reduced sulfur compounds are ubiquitous in aqueous and atmospheric systems. Natural sources of reduced sulfur species in the environment result from biological reduction of sulfate, anaerobic microbial processes in sewage systems, putrefaction of biogenic matter, oxidative decomposition of pyrite[1], and activities of marine organisms in the upper layers of the ocean[2]. In some studies it was estimated that the amount of hydrogen sulfide emission from higher plants constitutes 50% of the biogenic sulfur in the atmosphere[3].

Among reduced sulfur compounds hydrogen sulfide and thiols play particular role. They are chemically and biochemically very active components of the sulfur cycle of the natural environment. Hydrogen sulfide and simple low molecular mass thiols are volatile compounds sparingly soluble in water. Their oxidation forms are even less soluble. For instance, when propylthiol is oxidatively coupled at ambient temperature, the aqueous solubility drops from 9.0 to 0.03 g/l[4]. This large solubility reduction provides a simple means of removing mercapto compounds from water. In contrast, polyfunctional thiols like cysteine, which appear in natural waters mainly as a result of breakdown of organic matter, are hydrophilic. These thiols are of particular importance because many have large complex-formation constants with certain transition and heavy metals and therefore have environmental implication for trace metal speciation[5]. Another study proves the responsibility of thiols for the abiotic dehalogenation of haloorganic compounds in natural waters[6].

Mercaptans (thiols) and hydrogen sulfide are autoxidized in the presence of oxygen in alkaline medium. In general, the oxidation is slow in the absence of catalysts because of unfavourable spin state symmetries that result from differences in the electronic configuration of the reactants. However, the reaction proceeds rapidly in the presence of traces of metal ions. The catalysts tend to alter the electronic structure of either the reductant or oxygen so as to surmount the activation energy barrier imposed on the reaction by spin-state symmetry restriction. The following equation shows the acid-base equilibrium of the thiol group and the formation of a radical group, which represents the high reactivity of the thiol group to oxygen and its easy transformation into disulfide.

$$RSH \xrightleftharpoons{-H^+} RS^- \xrightarrow{-e} RS\cdot \xrightarrow{RS\cdot} RSSR$$

In order to slow this process addition of a chelating agents such as ethylenediamine tetraacetic acid and babbling nitrogen through the medium are recommended. The great susceptibility of reduced sulfur compounds to oxidation and, in general, nucleophilic reactions very advantageous in terms of the environmental pollution control, cause problems in their analysis. Besides, environmental thiols lack the structural properties necessary for the production of signals compatible with common detectors which are in use with modern analytical systems. To effectively measure such mixtures, a protocol should accomplish two things: allow for the fixation of a variety of reduced sulfur compounds to prevent their reactions during storage and sample preparation, and separate and quantify as many compounds as possible with a single assay.

We propose a method developed in our laboratory, the 2-halopyridinium salt method of reduced sulfur compounds derivatization with subsequent separation by high-performance liquid chromatography (HPLC) or high-performance capillary electrophoresis

(HPCE) and ultraviolet detection. In this report, we describe the determination of hydrogen sulfide and hydrophilic thiols in the presence of each other with a single assay.

3. BACKGROUND

One of the major disadvantages of HPLC and HPCE, despite all of the progress these methods made in the past, is a lack of detectors, particularly universal detectors for a reasonable price which can match the sensitivity of gas chromatography (GC) detectors. The best and most popular detectors coupled with these methods currently available are spectrophotometric and fluorimetric detectors. It seems logical, therefore, to solve the immediate detection problems by derivatization. Most of the procedures which are in use by environmental chemist for determination of reduced sulfur compounds involve precolumn derivatization followed by separation via reversed-phase HPLC. Obviously, at this time, there does not appear to be one ideal method, and the choice will depend on the needs of the investigator, which in turn are determined by the problem to be solved and laboratory equipment. Such techniques include the monobromobimane (bimane–HPLC) method[7], the halogenosulfonylbenzofurazans (ABD-F or SBD-F-HPLC) method[8], and the o-phthalaldehyde (OPA-HPLC) method[9,10].

The following reaction equation demonstrates the chemical concept of the HPLC method for determination of reduced sulfur compounds developed in our laboratory[11,14].

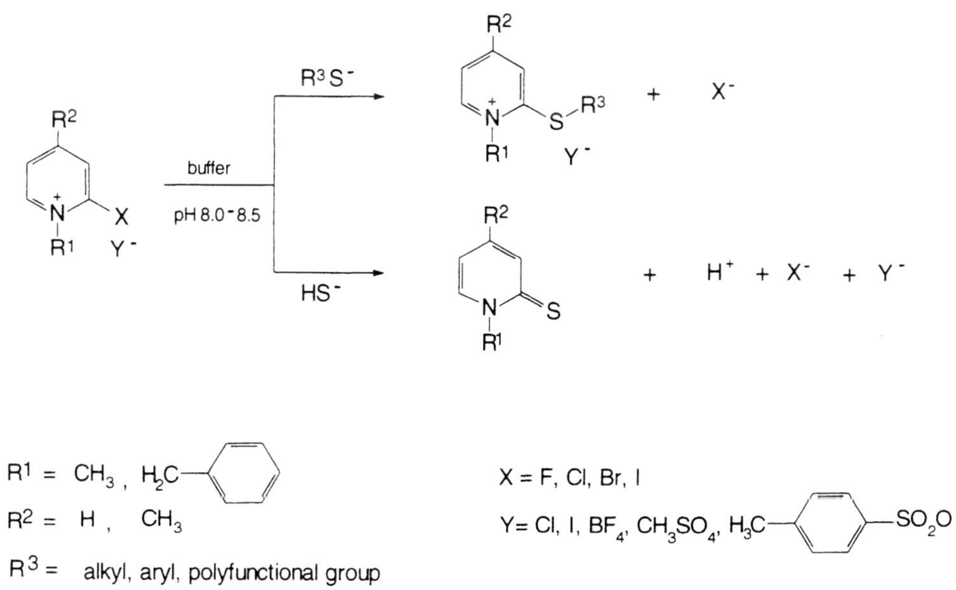

This derivatization scheme also goes very well with electromigration methods [15,16] because its products—S-pyridinium derivatives—are electrically charged. As in majority of thiol derivatization reactions, and in this case too, we are taking advantage of high nucleophilicity of the sulfhydryl function. In slightly alkaline water solution thiols react rapidly with 2-halopyridinium salts to form stable thioethers. These derivatives exhibit a well defined absorption maximum at 308–316 nm region in the UV spectrum as a consequence of the batochromic shift from the reagent maximum. Of different functionalities (e.g. -COOH, -NH_2, -SH) of hydrophilic thiols potentially able to undergo nucleopilic attack at the 2-position of pyridine ring in aqueous solution, unlike in anhydrous conditions[17] only the sulfhydryl group reacts. This means that no multiple derivatives are formed. Hydrogen sulfide gives under above conditions corresponding 2-thiopyridone.

4. MATERIALS AND METHODS

4.1. Chemicals and Solutions

2-Chloro-1-methylpyridinium iodide (CMPI), -the derivatization reagent- , was prepared as described previously[18]. For reduced sulfur compounds derivatization prior to HPLC and HPCE analysis, a 0.1 mol/l water solution of CMPI was used. Reduced glutatione (GSH) and cysteine (CSH) were purchased from Reanal (Budapest, Hungary). Homocysteine (HSH), N-acetylcysteine (ACSH), N-(2-mercaptopropionyl)glycine (MPG), thioglycolic acid (TGA) and thiomalic acid (TMA) were from Fluka (Buchs, Switzerland).Ion-pairing reagent—1-octanesulphonic acid sodium salt—was from Sigma (St. Louis, MO, USA). All other chemicals including sodium sulfide (HS) and solvents were HPLC-grade and were supplied by Baker (Deventer, Netherlands).

A stock standard solutions of 10 μmol/ml sodium sulfide and thiols in water containing respectively 0.02 mol/l sodium hydroxide and 0.02 mol/l hydrogen chloride were prepared followed by standardization with HMB method[19]. These solutions could be kept at 4°C for several days without noticeable change of analytes content. The working solutions were prepared by appropriate dilutions as needed. Calibration curves were obtained by assaying standard solutions of thiols and sodium sulfide at ten and nine concentration levels, respectively.

4.2. Instrumentation

HPLC analysis were performed with a Hewlett Packard 1100 Series system equipped with quaternary pump, an autosampler, thermostateol column compartment, vacuum degasser and diode array detector. For HPCE analysis Hewlett Packard HP^{3D} Capillary Electrophoresis instrument comprising an automatic injection device, an autosampler and a diode array detector was used. For instruments control, data acquisition and data analysis, a HP ChemStation for LC 3D system and HP^{3D} CE ChemStation software were used, respectively. For pH measurement, a Hach One pH-Meter was used. Water was purified using Millipore Milli-QRG system.

4.3. Derivatization, Chromatography and Electrophoresis

Derivatization. In a 5 ml calibrated flask, a water sample, 1ml of 0.1 mol/l pH 8.1 phosphate derivatization buffer and an appropriate amount of 0.1 M solution of CPMI were placed. The flask was stopped, mixed by inversion and put aside for 30 min. The reaction mixture was quenched with 4M phosphoric acid to pH 2.5 (indicator paper), made to a volume of 5 ml, then an aliquot was injected into the HPLC or HPCE system.

Chromatography. Samples, usually 20 μl, were injected using an autosampler into a 125x3 mm column packed with 3 μm particles of ODS-Hypersil, equipped with a 4x4 mm guard column containing 5 μm particles of ODS-Hypersil. The mobile phase consisted of isopropanol, methanol and 0.175 mol/l pH 2.19 citric buffer containing 12 mmol/l of octane sulfonic acid sodium salt (2:10:88, v/v). The temperature was 50°C, the flow rate 0.7 ml/min and the peaks were monitored using diode array detector at 312 nm and 340 nm for thiols and hydrogen sulfide, respectivelly.

Electrophoresis. An aliquot of final analytical mixture was hydrodynamically injected into standard fused-silica capillary having an effective length 40 cm, total length 48.5 cm and an internal diameter of 100 μm. The separation was performed using 0.12 mol/l pH 7.2 phosphate buffer at 24°C. The applied voltage was 9 KV, and the peaks were monitored at 312 nm.

4.4. Sediment Porewater Sample Preparation

Sediment cores were collected with a 50 mm diameter polyvinylchloride (PVC) corer from a small marsh pond located in central part of Poland on August 30, 1997. Cores were immediately transfered to a N_2-filled polyethylene jar that was placed in a refrigerator. Upon return to the laboratory the next day, the samples were centrifuged (14.000g), and supernatants were assayed according to the recommended procedure by the CMPI-HPLC method.

5. RESULTS AND DISCUSSION

5.1. Derivatization and Separation

The lipophilic thiols and hydrogen sulfide in water are determined in the presence of each other. The analytical procedure involves derivatization of analytes with 2-chloro-1-methylpyridinium iodide in slightly alkaline water solution and subsequent separation by HPLC or HPCE and ultraviolet detection. The derivatives—UV-sensitive compounds—are formed as a consequence of nucleophilic attack of thiolate (RS^-) and bisulfide (HS^-) anions at position 2 of pyridine ring in CMPI. S-pyridinium thioethers are the derivatization products of thiol component of the mixture, and 1-methyl-2-thiopyridone for hydrogen sulfide. In both cases, we are dealing with stable molecules. Experiments were carried out to determine the reaction time and derivatization reagent excess necessary for completion of the reaction. The results are shown in Figure 1.

It was established that the reaction occurs immediately and proceeds to a maximum in about 25 min. Based on this, for routine assay, a derivatization reaction time of 30 min and seven fold molar excess of the CMPI is recommended before the reaction mixture is injected to the final analytical system.

The S-pyridinium derivatives of all investigated compounds were found to be stable at room temperature for a reasonable time, which allows for long, unattended runs. No significant change was noted when 7 thiols and hydrogen sulfide CMPI-derivatives (Figure 2.) were kept at 4°C for nine days.

Different molar ratios of thiols and hydrogen sulfide do not influence significantly the results of derivatization and separation, which is demonstrated in Figure 3.

Under the experimental conditions used in this study, seven thiols and hydrogen sulfide are separated by CMPI-HPLC method. Results are shown in Table and in Figures 4 and 5. Ionic compounds, such as CMPI-thiol derivatives, show poor retention in standard reversed-phase systems. The retention may be enhanced by the addition of oppositely charged, hydrophobic pairing ions to the mobile phase. Three basic mechanisms have

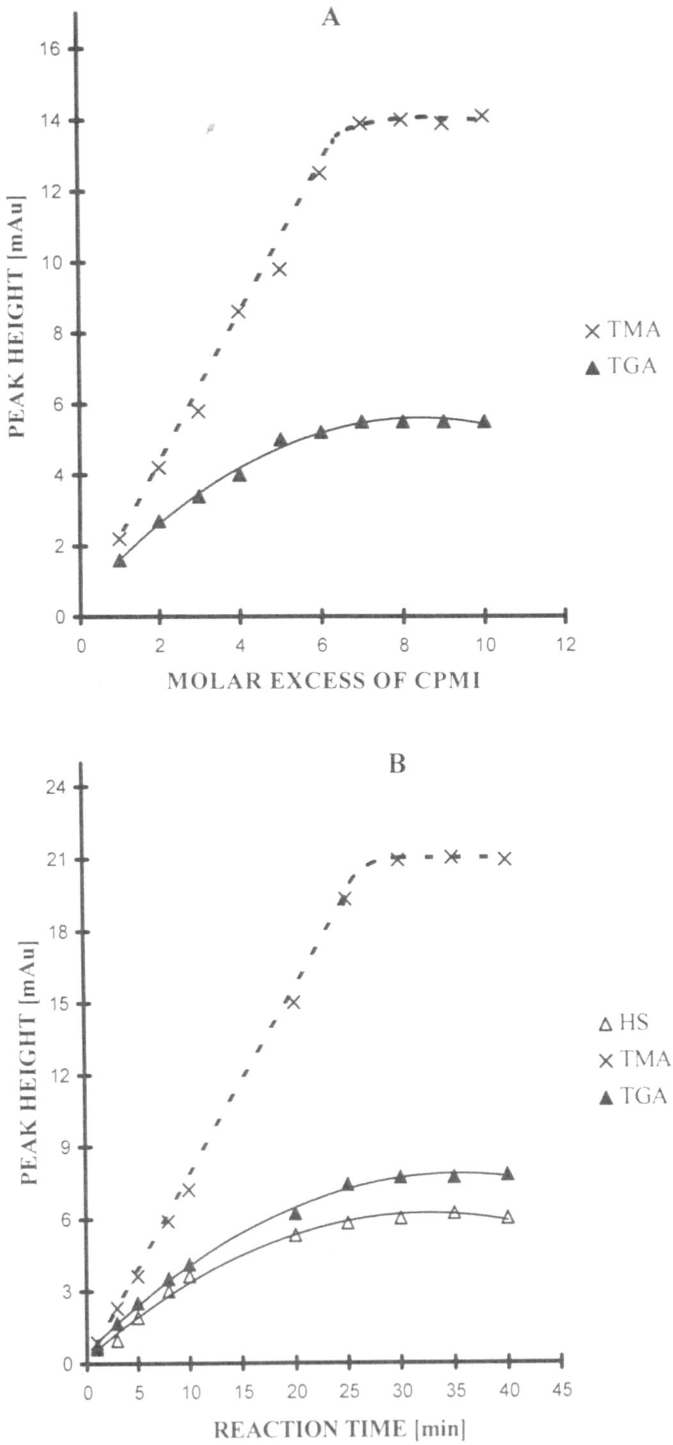

Figure 1. Derivatization reaction yield as a function of: A—excess of the reagent, reaction time 30 min; B—time, seven fold molar excess of CMPI in respect to each analyte.

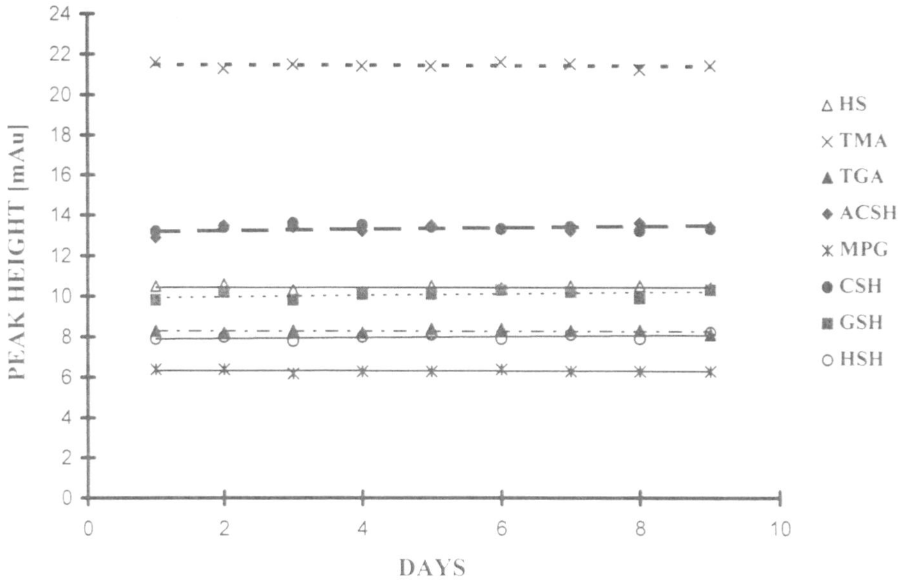

Figure 2. Stability of the S-pyridinium derivatives in the derivatization reaction mixture acidified to pH 2.5 and kept at 4°C.

been proposed[20–23] to model the observed increase in retention: (i) an ion pair forms in the mobile phase, increasing the lipophilicity of the ion followed by partitioning into the stationary phase, (ii) the stationary phase is modified by the adsorbed pairing agent to form a kind of dynamic ion exchanger and (iii) modification of the stationary–mobile phase interface with non-stoichiometric interaction of the pairing ion and analyte. It is difficult to prove the dominance of one of the three mechanisms for any given separation system as the chromatographic retention factor k is the thermodynamically derived quantity, and its measurement does not directly shed any light on the kinetics or mechanism of the process. In this case, the increased retention of cationic thiol CMPI-derivatives was achieved by the use of 1-octane sulphonic acid sodium salt. As expected—1-methyl-2-thiopyridone-CMPI derivative of hydrogen sulfide, which is neutral under chromatographic conditions, shows no dependence of retention on mobile phase addition of this ion-pairing agent. As is shown in Figure 4 eight analytes are separated from each other and CMPI excess during about twelve minutes.

Figure 5 shows that, as a result of the bathochromic shift from the CMPI, absorption maximum of the S-pyridinium derivatives (analytical wawelength of the detector) falls far away from the derivatization reagent maximum enabling application of its large excess without interference with the chromatogram.

5.2. Detection Limit, Reproducibility and Linearity

The performance of the CMPI-HPLC method was evaluated with a mixture of seven thiols and hydrogen sulfide that were well separated by reversed-phase ion-pair with isocratic elution. Results of the evaluation are summerized in Table 1 and in Figure 6 and 7.

The detector response was linear over the range of 5–100 nmol/ml for thiols and 5–600 nmol/ml for hydrogen sulfide.

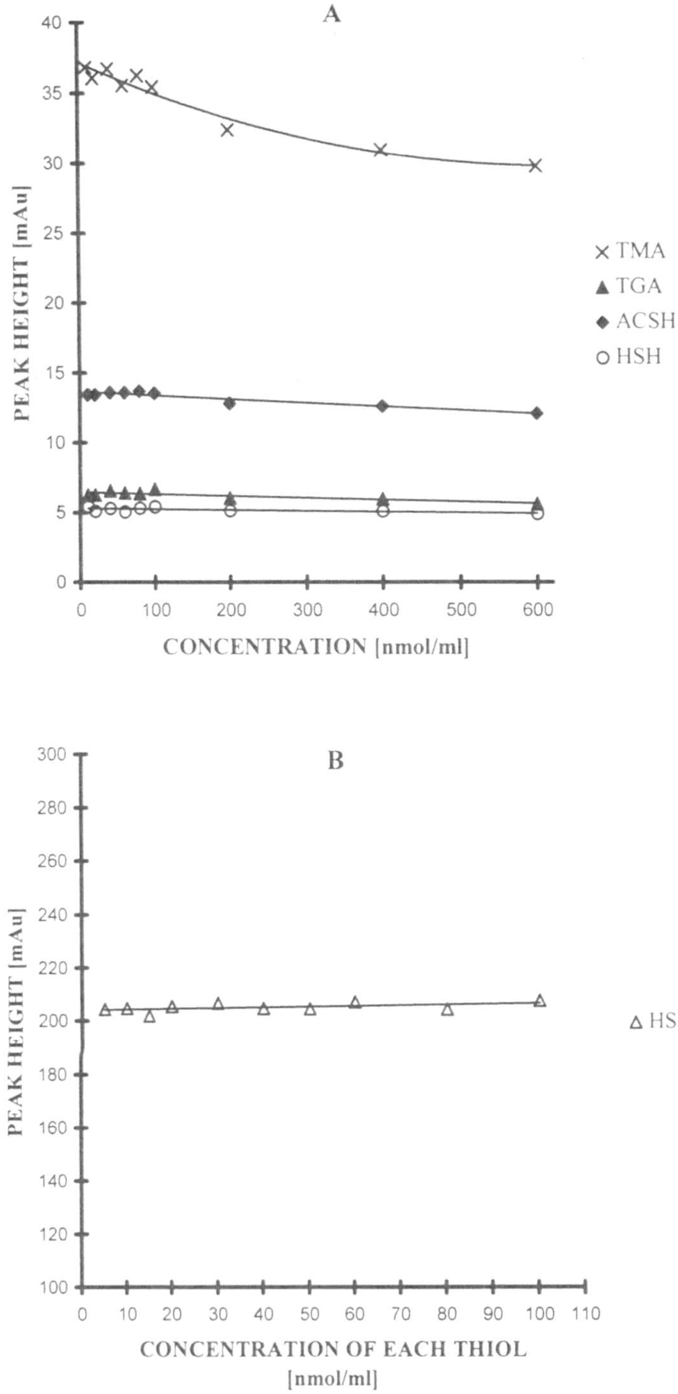

Figure 3. Determination of: A—thiols in the presence of various amounts of hydrogen sulfide, concentration of each thiol in the mixture 20 nmol/ml; B—hydrogen sulfide in the presence of various amounts of thiols, hydrogen sulfide content 200 nmol/ml.

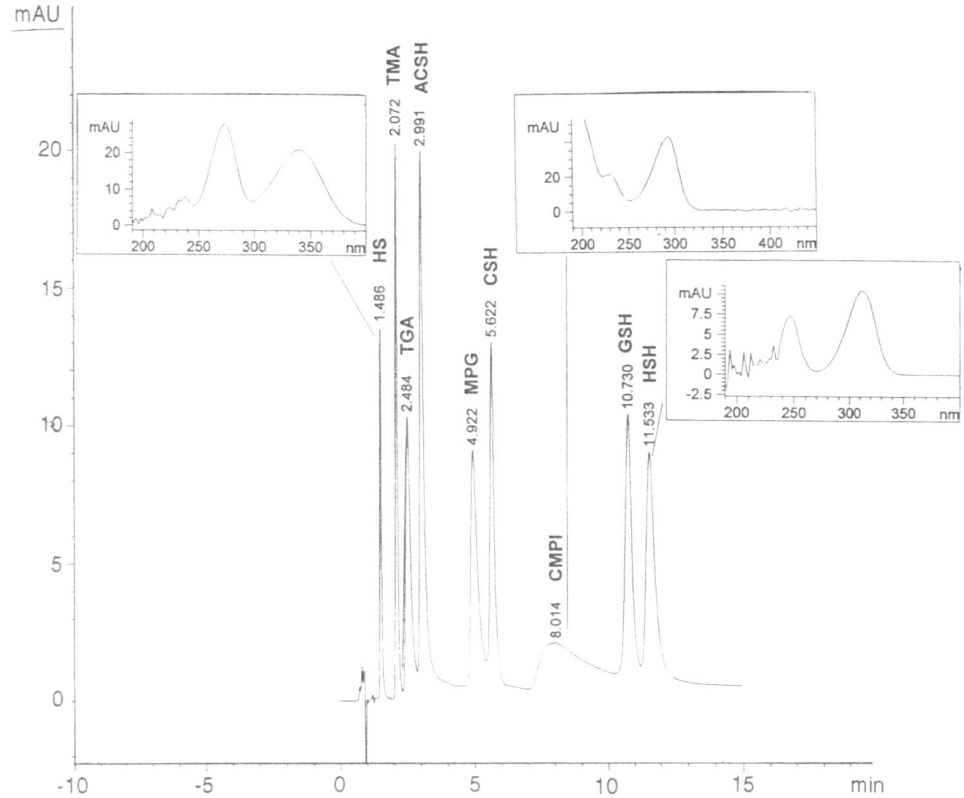

Figure 4. HPLC profile of standard reduced sulfur compounds mixture derivatized with CMPI. Upper pictures : characteristic absorption spectra of the S-pyridinium derivative of hydrogen sulfide, homocysteine and CMPI taken with the use of diode array detector.

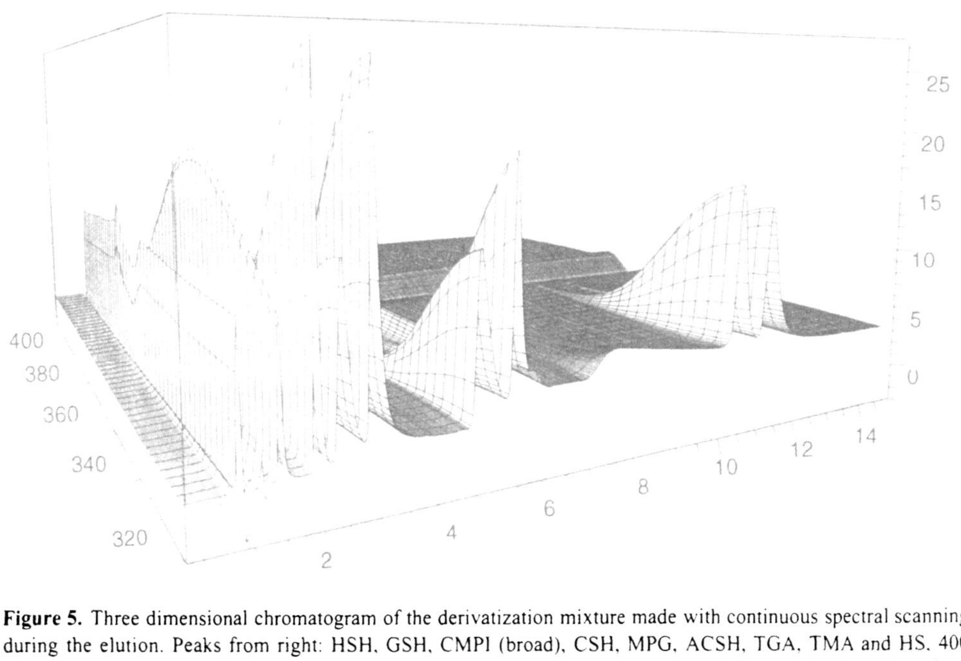

Figure 5. Three dimensional chromatogram of the derivatization mixture made with continuous spectral scanning during the elution. Peaks from right: HSH, GSH, CMPI (broad), CSH, MPG, ACSH, TGA, TMA and HS. 400 pmol of sulfur compound in each peak. Initial molar reagent to each analyte ratio 7:1. Axis titles: X—time, Y—wavelength and Z—absorbance.

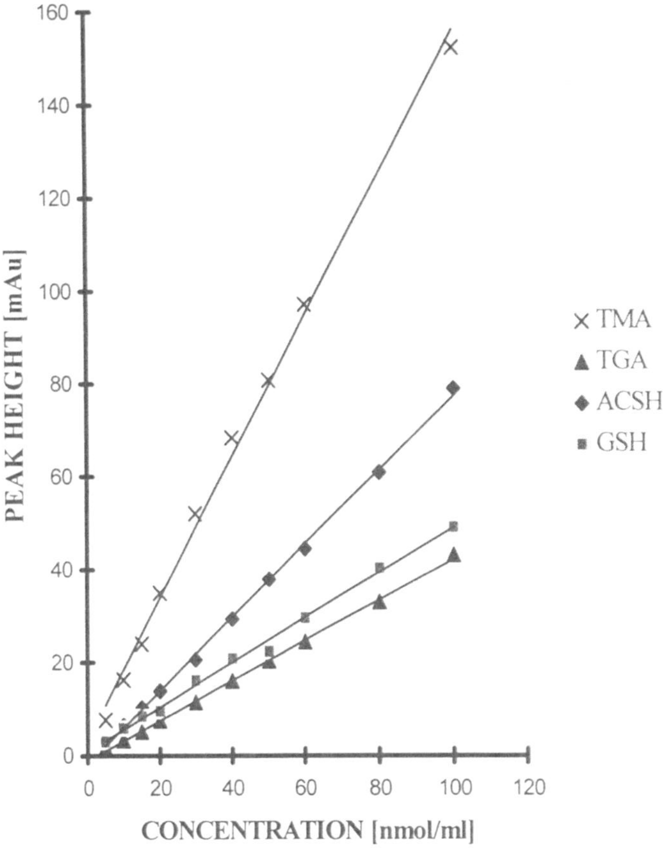

Figure 6. Calibration curves for CMPI-HPLC determination of reduced sulfur compounds.

Table 1. CMPI-HPLC analysis validation data

Sulfur compound[a]	Retention time [min]	Regression equation	Correlation coefficient[b]	Coefficient of variation [%][c]	Detection limit [pmol]
HS	1.486	y = 2.0576x – 2.8951	0.9984	4.32–0.65	0.4
TMA	2.072	y = 1.5309x + 3.1402	0.9961	3.44–0.47	2
TGA	2.484	y = 0.4340x – 1.2210	0.9987	3.13–0.75	2
ACSH	2.991	y = 0.7939x – 1.9841	0.9987	3.60–0.57	2
MPG	4.922	y = 0.4398x – 0.2342	0.9766	3.80–1.20	2
CSH	5.622	y = 0.7674x – 2.2219	0.9978	3.95–1.04	2
GSH	10.730	y = 0.4841x + 0.6856	0.9954	3.33–0.86	2
HSH	11.533	y = 0.3283x + 0.6679	0.9929	4.58–1.94	2

[a]Samples for standard thiol mixtures were from 5 to 500 nmol/ml; for HS, from 5 to 600 nmol/ml
[b]n = 10 ; for HS n = 9
[c]For bottom and top of the calibration range; n = 3

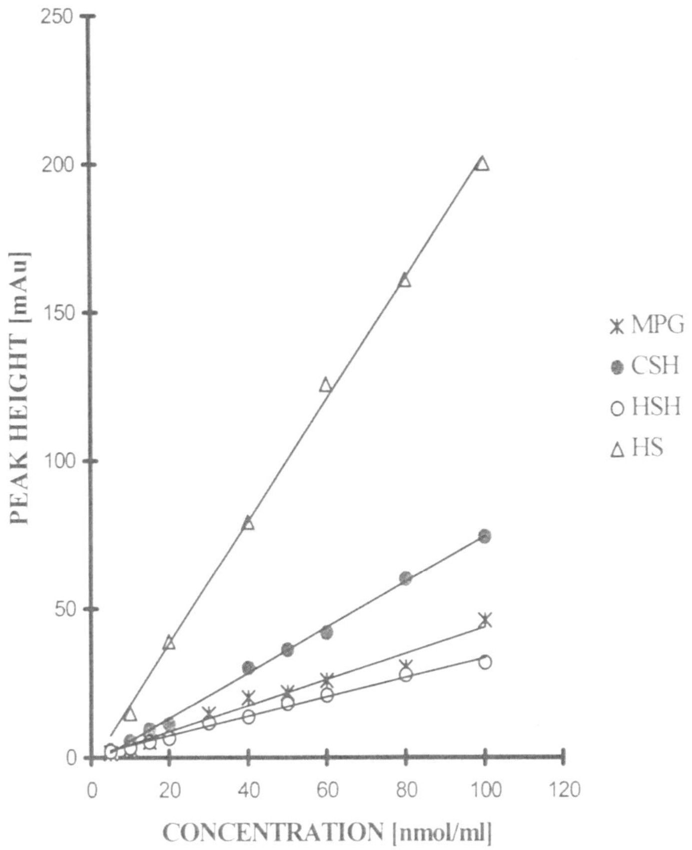

Figure 6. (*continued*)

Detection limits for these two methods were far below 2 pmol and 1 pmol respectively (Figure 6 A and B). Coefficients of variation (Table 1) were from 0.47% for 100 nmol/ml level of TMA to 4.58% for 5 nmol/ml level of HSH.

The CMPI-HPCE method was evaluated with a mixture of four thiols. As can be seen from Figure 7 detection limits and resolutions are much better with CMPI-HPCE.

5.3. Application of the Method to Sediment Porewater Analysis

The CMPI-HPLC method was applied to the sediment porewater analysis and appropriate chromatogram is shown in Figure 8.

6. CONCLUSION

We have shown that the coupling derivatization with the use of 2-chloro-1-methylpyridinium iodide with HPLC or HPCE yields sufficient sensitivity and selectivity to determine hydrophilic thiols and hydrogen sulfide in different quantitative ratios. The simplicity, portability and stability of the CMPI-derivatives make this method suitable for preserving the samples in the field by derivatization and assaying them later in the labora-

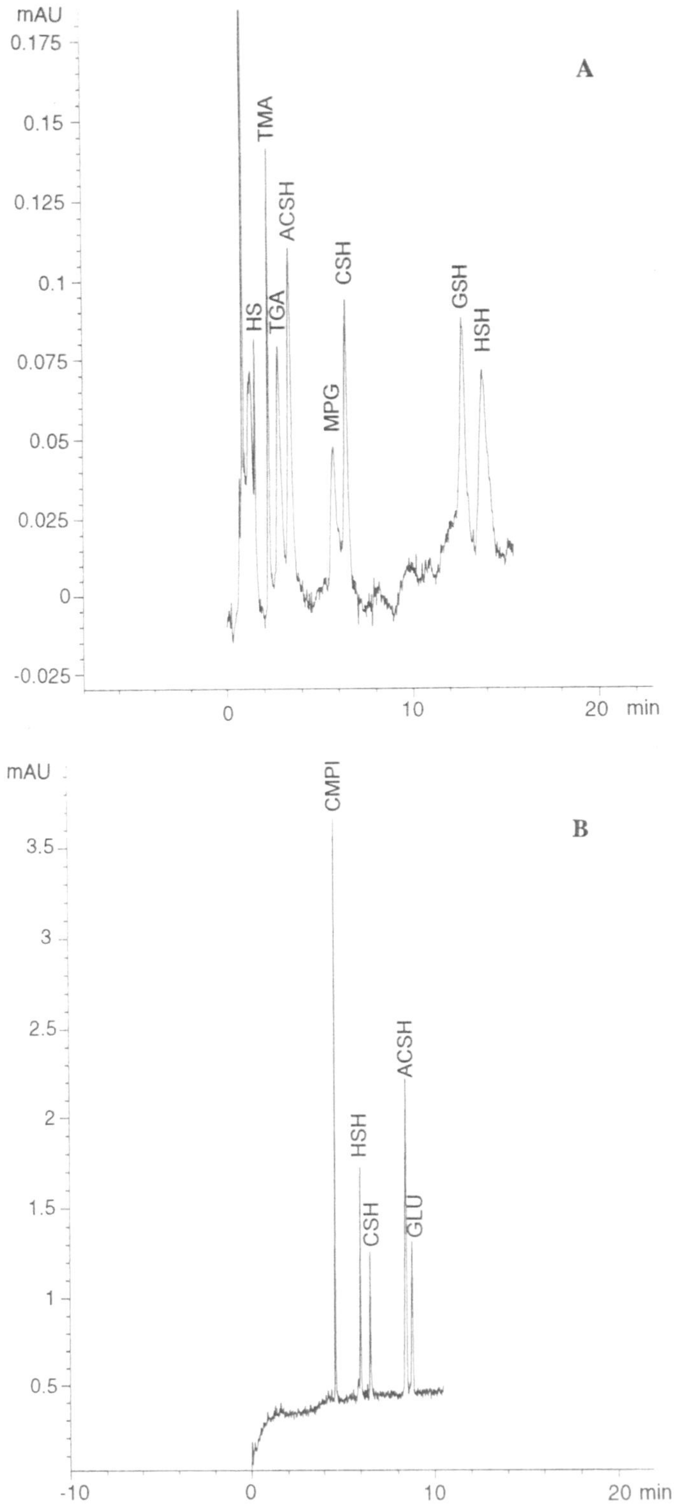

Figure 7. Detection limits for analysis of reduced sulfur compounds: A—CMPI-HPLC method, 2 pmol in peak: B—CMPI-HPCE method, 1 pmol in peak. In both cases peaks were monitored at 312 nm.

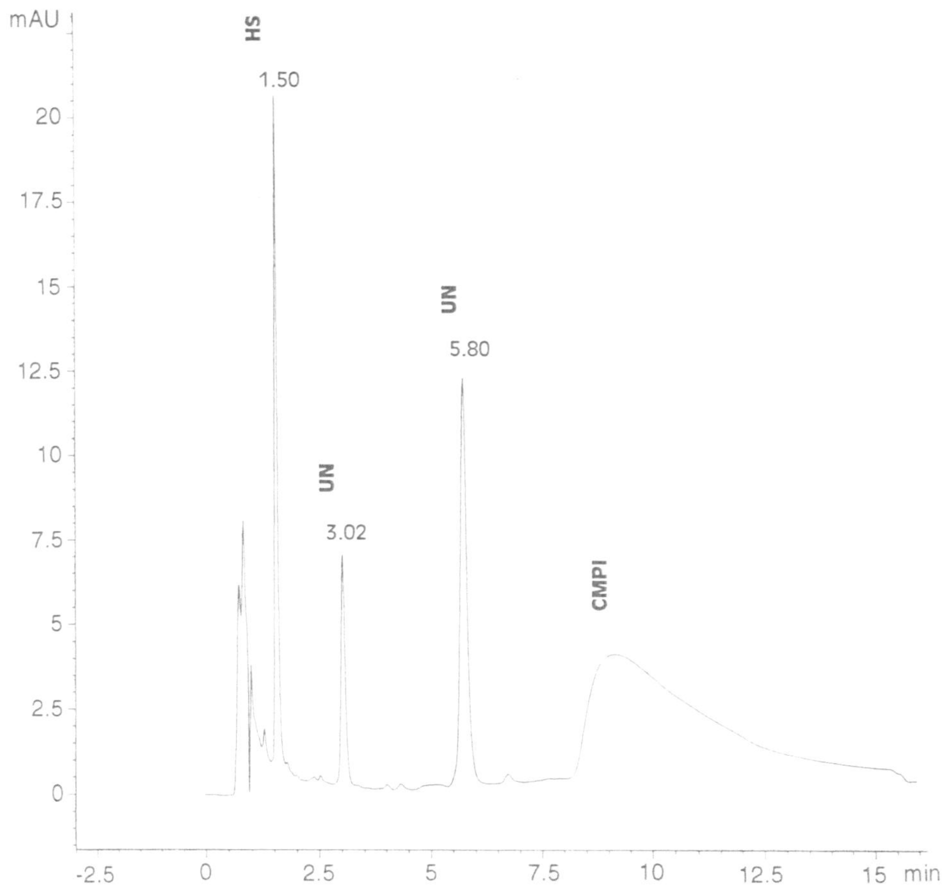

Figure 8. Chromatogram for porewater sample. Peaks: HS—hydrogen sulfide; UN—unknown. Three reduced sulfur compounds were detected as was proved by UV-spectrum taken with the use of a diode array detector (data not shown) but only one of which—hydrogen sulfide—was identified. Its concentration in the analysed sample was 37.5 nmol/ml.

tory in unattended runs. Further work on electrophoresis of reduced sulfur compounds, particularly by means of micellar electrokinetic chromatography, is now in progress.

ACKNOWLEDGMENT

This work was supported in part by Grant No 505/600 from the University of Lodz.

REFERENCES

1. Stedman D.H., Geophy. Res. Lett., 1984, **1**, 858.
2. Cline J.D. and Bates T.S., Geophy. Res. Lett., 1983, **19**, 949.
3. Winner W.E., Smith C.L., Voch G.W., Mooney H.A., Bewley J.D. and Krouse H.R., Nature, 1981, **289**, 672.
4. Giles D.W., Cha J.A. and Lim P.K., Chem. Eng. Sci., 1986, **41**, 3129.
5. Shea D. and MacCrehan W.A., Sci. Total Environ., 1989, **73**, 135.

6. Barbash J.E. and Reinhard M., in Biogenic Sulfur in the Environment, Ed. Saltzman E.S. and Cooper W.J., ACS Symposium Series, Washington 1989, p. 101.
7. Fahey R.C. and Newton G.L., Methods Enzymol., 1987, **143**, 76.
8. Toyo'oka T. and Imai K., J. Chromatogr., 1983, **282**, 495.
9. Roth M., Anal. Chem., 1971, **42**, 880.
10. Mopper K. and Delmas D., Anal. Chem., 1984, **56**, 2557.
11. Bald E., and Sypniewski S., J. Chromatogr., 1993, **641**, 184.
12. Sypniewski S. and Bald E., J. Chromatogr. A, 1994, **676**, 321.
13. Sypniewski S. and Bald E., J. Chromatogr. A, 1996, **729**, 335.
14. Bald E., Sypniewski S., Drzewoski J. and Stêpieñ M., J. Chromatogr. B, 1996, **681**, 283.
15. Bald E., J. Chromatogr. 1979, **174**, 483.
16. Bald E., Chromatographia, 1982, **15**, 525.
17. Bald E., Saigo K. and Mukaiyama T., Chem. Lett., 1975, 1163.
18. Bald E., Talanta, 1980, **27**, 281.
19. Wronski M. Talanta, 1977, **24**, 347.
20. Melander W.R. and Horvath C., in Ion-Pair Chromatography, Ed. Hearn M.T.W., Marcel Dekker, New York, 1985, p.42.
21. Shea D. and MacCrehan W.A., Anal. Chem., 1988, **60**, 1449.
22. Bidlingmeyer B.A., Deming S.N., Price W.P., Sachok B. and Petrusek M., J. Chromatogr., 1979, **186**, 419.
23. Knox J.H. and Hartwick R.A., J. Chromatogr., 1981, **204**, 3.

3

METAL SPECIATION IN OVERFLOW AND LEACHATE FROM A THERMAL POWER PLANT ASH POND

Impact on Receiving Waters

D. K. Banerjee and Balaram Pani

School of Environmental Sciences
Jawaharlal Nehru University
New Delhi—110 067, India

ABSTRACT

The speciation of six environmentally important heavy metals, viz., Fe, Mn, Cd, Cu, Zn and Ni in terms of free ions, labile complexes, slowly labile complexes and stable complexes, has been investigated in the overflow and leachates from an ash pond of a 720 mw thermal power plant in Delhi. They find their way through a canal to the receiving waters of the river Yamuna and depending on the relative concentration of the different forms of these metals in water and various sizes of suspended particulates, may have a deleterious effect on water quality. Lysimeter studies simulating the generation of ash pond leachates have also been conducted to ascertain the speciation status of metals at two different depths. It is found that the speciation is more prominent in the smallest size fraction of the particulates and therefore, ecotoxicologically more significant. There is a seasonal variation of leachability. Also, the free metal ions plus their labile complexes are more in the receiving waters than the stable complexes. This may impart metal toxicity to the water body.

Key Words: Metal speciation, thermal power plant, ash pond, overflow, leachate, receiving waters.

1. INTRODUCTION

The increasing use of coal for power generation results in an increasing potential for adverse environmental impacts. One of these emanates from the disposal of ash from thermal power plants through settling ash ponds. Overflow and leachates from these ponds are discharged to adjacent receiving water bodies. An environmentally important component of these discharges is toxic metals. Although they are generally present at relatively low concentrations in coal, significant mobilisation still occurs because of the very large quantities of coal consumed. Therefore, it is imperative to assess their impact, particularly on the receiving waters.

The fate and transport of heavy metals in land fill leachates depend upon their variable complexing abilities, the relative concentrations of the other constituents and upon the environmental conditions viz., pH, Eh, temp. etc. encountered by the leachate.

Landfill leachates contain many organic and inorganic ligands, with different ability to form complexes. These complexes will impact on the mobility and possible toxic effects of the metals in the soil. It is therefore important to determine the speciation of the elements in the leachates.

1.1. Objectives

The present work is concerned about the speciation of some selected heavy metals in the ash pond overflow and leachates from a thermal power plant. For this purpose the ash pond of Badarpur Thermal Power Station (BTPS) in Delhi has been chosen as the source. BTPS is a 720 mw power station situated at the southern most fringe of metropolitan Delhi. In this plant 7500 tonnes bituminous coal is consumed per day. Just behind this power plant, three ash ponds are situated. However, the first and second ash ponds have been filled up and the third is now being used. Ash slurries are discharged into this ash pond, the effluents from which are carried through a canal ultimately to the river Yamuna which flows at distance of 3 km. from the power plant. The effluent is discharged to the Yamuna river on continuous basis, i.e., there is a constant flow from the power plant. Thus the quality of surface water is obviously affected continuously. The pollution potential of coal fired power plant depends on the quality and quantity of the coal used by the power plant. The BTPS uses coal from different coal mines which differs in quality. Six environmentally important heavy metals viz., Fe, Mn, Cu, Cd, Zn and Ni have been selected for the present study because these metals are of common environmental concern vis-à-vis their ecotoxicology and health hazards to the human beings as well as to aquatic biota. The speciation or different chemical forms of these heavy metals are very closely correlated to each other. These metals are more likely to affect the water quality parameters depending on the nature of their different chemical forms, distribution patterns, order of the stabilities of their complexes and so on.

Laboratory attenuation studies also have been conducted by generating ash leachate in the laboratory. Settled ash from BTPS ash ponds has been subjected to infiltration by the sluicing water in soil columns and the soil columns effluents at particular depths have been analysed for the metal constituents and speciation.

2. METHODOLOGY

For physico-chemical characterisation of ash pond overflow and the receiving waters, the following sites (Fig.1) were chosen for collecting samples.

1. Inlet of ash pond.
2. Outlet of ash pond.

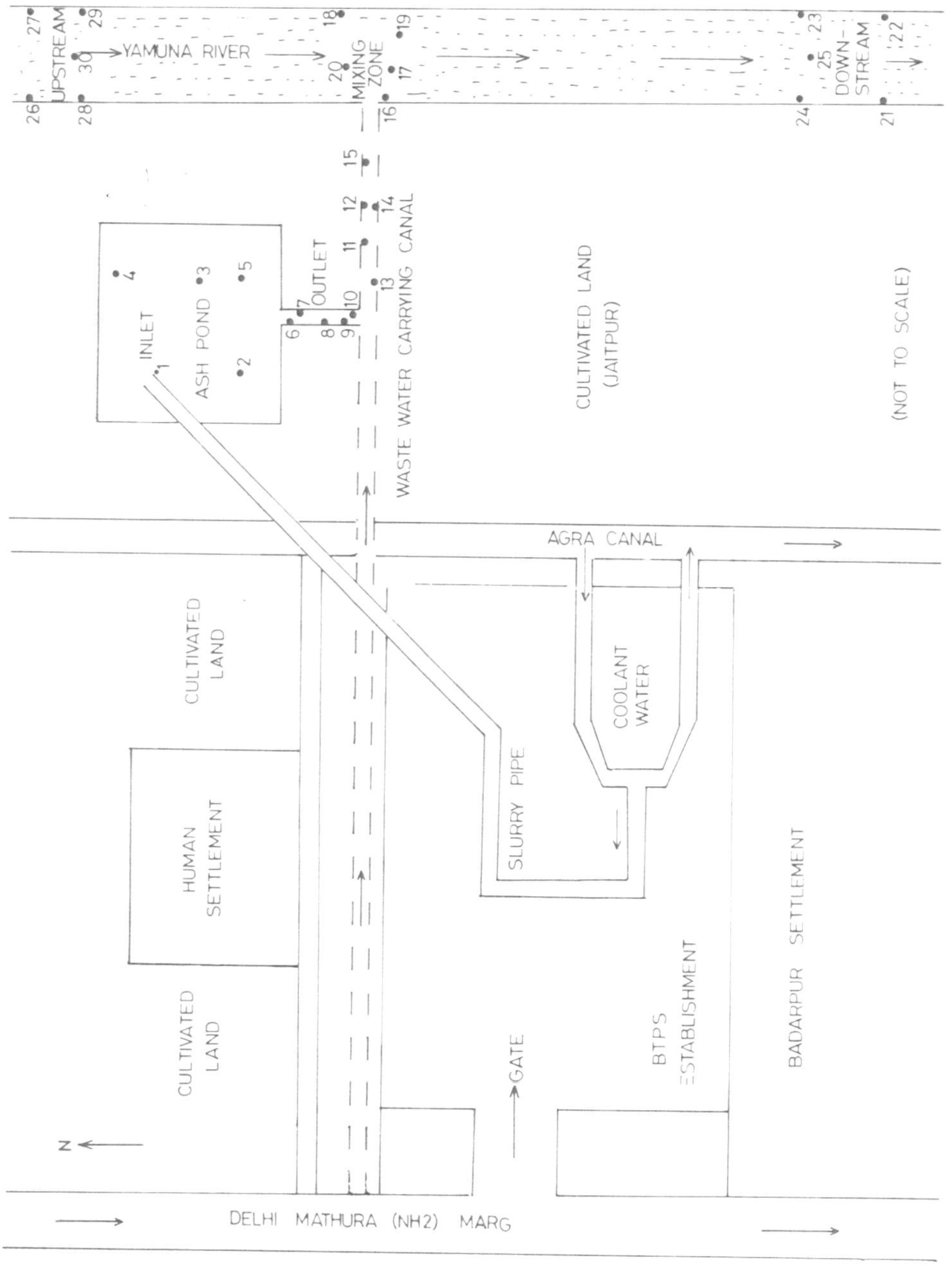

Figure 1. Schematic diagram of Badarpur Thermal Power Station area and sampling points

Table 1. Different chemical parameters of the overflow and river water

	pH	EC mho	TDS ppm	Alk ppm	Eh mv	DO ppm	COD ppm	Cl^- ppm	SO_4^{--} ppm	PO_4^{---} ppm	NO_3^- ppm
Inlet											
PRE	6.08	448.12	453.67	390.00	65.00	1.24	176.00	88.92	36.38	5.91	7.13
MON	6.81	617.84	529.21	309.00	148.00	1.93	262.00	169.08	46.99	4.28	7.91
POS	6.79	584.13	493.82	323.00	110.00	1.82	238.00	148.12	44.06	4.84	7.67
WIN	6.63	566.42	457.79	350.00	87.00	1.41	204.00	126.90	42.21	3.96	7.25
Outlet											
PRE	6.42	520.87	401.14	335.00	79.00	1.79	173.00	90.58	28.08	6.53	5.63
MON	6.96	623.36	485.14	297.00	184.00	2.48	284.00	133.21	47.41	5.21	6.75
POS	6.87	567.30	422.81	318.00	125.00	2.17	247.00	124.54	43.60	5.87	6.38
WIN	6.96	552.57	451.83	306.00	90.00	1.86	212.00	101.58	34.73	5.44	5.89
CANAL											
PRE	6.73	596.34	268.31	310.00	41.00	2.91	156.00	68.01	28.74	5.47	3.98
MON	7.42	692.31	342.56	237.00	136.00	3.72	232.00	111.68	44.63	4.09	5.20
POS	6.91	661.74	327.97	254.00	114.00	3.53	207.00	104.03	39.96	4.92	4.71
WIN	7.22	637.83	299.71	285.00	64.00	3.24	183.00	94.13	35.53	4.49	4.35
Mixing zone											
PRE	7.15	632.84	135.87	275.00	92.00	3.76	151.00	94.13	21.67	4.74	3.76
MON	7.64	756.20	247.39	233.00	170.00	4.97	246.00	108.66	38.74	3.97	4.60
POS	7.23	734.98	209.94	221.00	115.00	4.45	212.00	95.89	31.39	4.28	4.15
WIN	7.59	682.45	167.45	247.00	147.00	4.16	184.00	83.32	26.61	3.31	3.92
Down stream											
PRE	7.38	655.85	87.81	186.00	186.00	6.83	204.00	40.82	21.29	3.81	2.31
MON	7.91	776.25	184.76	150.00	334.00	8.54	332.00	59.06	49.16	2.72	2.98
POS	7.43	738.21	147.19	171.00	278.00	7.65	298.00	51.93	33.97	3.33	2.52
WIN	7.82	702.98	111.16	169.00	220.00	7.26	256.00	47.87	24.82	2.17	2.39
Up stream											
PRE	7.49	691.28	63.98	160.00	310.00	9.95	296.00	52.27	15.92	2.84	2.25
MON	8.13	789.68	117.56	115.00	395.00	12.76	485.00	71.26	36.61	1.62	3.31
POS	7.97	736.00	93.50	146.00	364.00	10.76	432.00	64.09	27.74	2.08	2.92
WIN	8.04	717.23	76.43	128.00	378.00	10.98	367.00	54.71	20.77	1.21	2.64

3. Canal carrying the ash pond effluents.
4. Effluent out fall (effluents meeting the river)
5. Downstream (≈ 500m) from out fall.
6. Upstream (≈500m) from out fall.

Overflow and river water samples were collected at five different points at each site (Fig.1) at monthly intervals for a period of one year. The average seasonal values of some parameters are given in Tables 1 and 2.

2.1. Laboratory Leachates

In the laboratory, leachates were generated from the settled ash collected at 3ft depth from the inlet and outlet of ash pond no. 2 which had recently filled up. The sluicing water used for this purpose was the same as being used at the power station.

For the laboratory leachate experiments, the lysimeter columns were fabricated from perspex sheet (length 8", width 8" and height 36"), at the bottom of each column there was a perforated plate for water out flow. There were 3 sampling points (0.5" holes) on both sides of the columns at 8" intervals for soil collection, and were closed with corks. The

Table 2. Different chemical parameters of the leachate

	pH	EC mho	TDS ppm	Alk ppm	Eh mv	DO ppm	COD ppm	Cl^- ppm	SO_4^{--} ppm	PO_4^{---} ppm	NO_3^- ppm
2.5 ft.											
Inlet											
PRE	5.43	454.92	585.21	486.00	13.00	0.43	96.00	226.90	39.22	7.84	7.65
MON	6.01	503.17	655.92	426.00	70.00	0.93	162.00	288.92	72.41	6.12	9.73
POS	6.32	483.74	627.84	441.00	50.00	0.72	141.00	269.08	54.84	6.71	8.92
WIN	6.54	469.38	616.42	465.00	40.00	0.51	112.00	248.12	47.28	5.46	8.30
Outlet											
PRE	5.89	431.74	539.14	421.00	33.00	0.83	113.00	194.58	37.08	7.21	7.78
MON	6.47	537.40	622.91	380.00	65.00	1.14	194.00	253.02	65.73	6.49	8.76
POS	6.42	501.29	571.55	404.00	60.00	1.07	167.00	234.13	52.91	5.92	8.25
WIN	6.71	480.90	596.02	364.00	49.00	0.91	142.00	211.68	45.52	5.37	7.94
1.5 ft.											
Inlet											
PRE	6.10	466.74	517.87	394.00	20.00	1.11	136.00	154.75	34.41	6.32	7.29
MON	6.52	608.11	591.41	325.00	71.00	1.62	208.00	215.59	56.17	5.36	8.30
POS	6.62	587.24	566.21	360.00	52.00	1.43	186.00	209.11	51.25	5.89	7.92
WIN	6.85	510.48	542.93	315.00	35.00	1.34	153.00	183.32	42.87	4.98	7.65
Outlet											
PRE	6.24	491.00	492.02	331.00	41.00	1.25	124.00	151.92	29.55	6.07	7.10
MON	6.84	587.29	554.92	287.00	90.00	1.86	238.00	195.89	49.38	5.43	7.96
POS	6.30	568.75	532.79	305.00	60.00	1.67	197.00	152.11	43.44	4.91	7.51
WIN	6.72	528.88	509.48	293.00	54.00	1.78	164.00	157.12	34.72	5.15	7.27

lysimeter was packed with treated gravel, sand and the ash pond samples following usual procedures. Sluicing water used at the power plant and collected from the site was used for ponding the sample and equilibriate or two weeks.

After two weeks the sluicing water was added slowly (drop by drop) every day in each column for a period of 5 to 6 hrs. The water layer about 2" thick, which was formed at the top of the soil layer took 12 hrs. to reach the bottom layer of the column, and it took nearly 20 to 22 hrs. for all the water to percolate through the sample soil. After the leachate sample was collected, the column was allowed to dry before new run was made. Replicate leachate samples were collected at intervals of 48 hrs. for speciation analysis.

2.2. Analytical Methods

The important physico-chemical parameters which have a bearing on speciation like pH. E.C., Eh, alkalinity, COD, etc., were determined using standard methods (1). For speciation studies, the procedures were followed as per the schemes (2,3,4). Basically these procedures employ the calcium saturated cation exchange resin chelex-100 to determine the fractions of free metal ions and complexed metals. Total metal concentrations were determined by flame atomic absorption method, using a Philips PU9200X instrument after preconcentrating the samples through standard procedures (1,5.) Characterisation of the metal complexes involves a resin column with a short leachate retention time retaining metal bound as "labile complexes" followed by a batch with a high amount of resin and a long equilibriation time to retain metal bound as "slowly labile complexes". Metal staying in solution is characterised as "stable complexes" (4).

Total concentration of metals in the solution before passing through the resin columns (chelex-100) is taken as initial concentration. The concentration of the metals in the solution after passing through the resin column for 24 hours is taken as equilibrium concentration and that in the solution immediately after passing through the resin column is taken as effluent concentration. These can be correlated as:

$$\text{Free ions} + \text{Labile complex (C+F)} = \frac{\text{Initial conc.}-\text{effluent conc.}}{\text{Initial conc.}} \times 100$$

$$\text{Slowly labile complex (CS1)} = \frac{\text{Effluent conc.}-\text{equilibrium conc.}}{\text{Initial conc.}} \times 100$$

Stable complex (CS) = 100 – [(C+F) + (CS1)]

The leachate and overflow samples were successively filtered through standard pore-size filter papers (Whatman and Millipore) to yield filtrates containing four size ranges (>12 um., >1–12um, > 0.45–1um, and > 0.25–4.5um) of suspended particulates. These filtrates were analysed for speciation, following the same procedure as mentioned earlier.

3. RESULTS AND DISCUSSION

The speciation of heavy metals has been assessed in terms of free ions and labile complex (C+F), slowly labile complex (CS1) and stable complex (CS) as percentage of total metal concentration. The speciation of any heavy metal is influenced by the physico-chemical parameters of the water column like pH, Eh, DO, COD, etc., and also the presence of inorganic and organic ligands. In the present study, the different forms of chemical species (C+F, CS1 and CS) and their distribution in suspended particulates of four different sizes as mentioned above at all the sampling stations, for both overflow and leachate, have been assessed for individual metals (Table 3 and 4). In addition to that, the seasonal variations of the metals in the water body and leachates have also been studied. For the laboratory leachates study, the leachates were investigated at two different depths (2.5 ft. and 1.5 ft.) to find out the influence of different soil depths on the characteristics and behaviours of the metals such as, different chemical species, size distribution, and leachability. From the experimental results it is found that the leachability is more at the depth 2.5 ft than that at 1.5 ft. (Table 5).

From Table 3 it is found that among the different chemical forms of iron, free plus labile complex ions are more than the stable complex ions at all the sampling points. The slowly labile complex ions percentage is always the highest. Manganese also has a similar pattern of distribution. The presence of different chemical forms of a metal depends on the solubility of that metal which is influenced by the factors, pH and redox potential (Eh). Their behaviour is also influenced by their ability to form complex ions with inorganic ligands, like sulphate, phosphate, chloride, nitrate, oxides, hydroxides etc., and organic ligands like humic, fulvic acids etc. In case of iron, under the oxidising conditions and in the pH range 5 to 8, it is precipitated as highly insoluble ferric hydroxide. From Table 1 it is found that pH of all the samples is in this range and Eh has a positive value which is favourable for oxidation. So under these conditions the ferric state of iron may get precipitated out and settled at the bottom sediments. The ferric form, Fe(III), generally forms

Table 3. Percentage of metal forms associated with different particle sizes (Overflow)

	>12 μm			>1–12 μm			>0.45–1 μm			>0.25–0.45 μm		
	C+F	CS1	CS	C+F	CS1	CS	C+F	CS1	CS	C+F	CS1	CS
Iron												
Intel	27.20	57.15	15.65	26.22	56.15	17.63	27.05	57.03	15.92	25.35	55.07	19.58
Outlet	26.07	58.38	13.55	27.17	57.27	15.56	27.90	57.90	14.20	25.90	55.52	18.58
Canal	29.00	59.33	11.67	27.95	58.60	13.45	27.90	58.18	13.92	25.95	55.95	18.10
Mixingzone	29.05	59.75	11.20	27.70	58.25	14.05	28.75	59.08	12.17	26.00	56.05	17.95
Upstream	29.60	59.62	10.78	28.92	59.00	12.08	29.87	60.05	10.08	28.47	58.60	12.93
Downstream	30.00	59.80	10.20	28.12	58.85	13.03	29.95	60.50	9.55	27.50	57.57	14.93
Manganese												
Inlet	24.87	57.33	17.80	23.45	56.40	20.15	24.37	57.50	18.13	23.27	55.55	21.18
Outlet	25.45	58.55	16.00	23.95	56.47	19.58	24.62	58.05	17.33	23.87	55.83	20.30
Canal	25.40	58.80	15.80	25.00	56.70	18.30	25.17	58.17	16.66	24.42	56.55	19.03
Mixing zone	26.35	58.60	15.05	25.02	57.60	17.38	25.85	58.60	15.55	25.62	56.25	18.13
Upstream	27.18	59.62	13.20	26.05	57.65	16.30	26.80	59.02	14.18	26.85	57.58	15.03
Downstream	26.50	58.90	14.60	25.92	57.35	16.73	26.47	58.82	14.71	26.42	57.55	16.03
Copper												
Inlet	13.50	56.47	30.03	13.32	55.57	31.11	15.50	57.37	27.13	14.45	57.80	27.75
Outlet	14.20	56.82	28.98	13.55	56.15	30.30	16.52	58.85	24.63	15.45	59.22	25.33
Canal	15.92	57.55	26.53	14.12	56.42	29.46	17.82	59.82	22.63	16.17	60.15	23.68
Mixing zone	16.37	58.85	24.78	14.52	56.80	28.68	17.85	60.42	21.73	16.97	59.80	23.23
Upstream	17.07	60.90	22.03	15.57	58.67	25.76	18.57	60.10	20.33	18.07	60.35	21.58
Downstream	16.82	59.45	23.73	15.52	57.77	26.71	17.50	60.57	21.93	17.55	60.32	22.13
Cadmium												
Inlet	11.90	55.85	32.25	11.42	55.57	33.01	14.60	56.22	29.18	12.90	56.15	30.95
Outlet	12.60	56.95	30.45	12.15	56.22	31.63	15.00	57.67	27.33	14.00	57.37	28.63
Canal	13.05	57.47	29.38	12.42	56.47	31.11	15.52	58.92	25.56	14.77	58.02	27.21
Mixing zone	13.32	57.68	29.00	12.60	56.57	30.83	14.90	58.85	26.25	14.85	58.15	27.00
Upstream	14.42	59.75	25.83	14.73	57.55	28.08	16.52	60.22	23.26	16.82	60.10	23.08
Downstream	13.60	59.00	27.40	13.15	57.00	29.85	16.05	59.47	24.48	15.57	59.00	25.43
Zinc												
Inlet	12.25	54.62	33.13	11.77	54.45	33.78	12.95	55.80	31.25	12.47	55.02	32.51
Outlet	13.10	55.67	31.23	12.77	55.50	31.73	14.37	56.47	29.16	12.85	55.42	31.73
Canal	13.92	56.97	29.11	13.72	57.22	29.06	15.05	57.30	27.65	13.72	57.35	28.93
Mixing zone	15.87	57.72	26.41	14.52	56.95	28.53	15.47	57.62	26.91	14.07	57.37	28.56
Upstream	16.85	58.30	24.85	16.35	58.10	25.55	15.95	58.02	26.03	15.42	58.55	26.03
Downstream	16.52	58.72	24.76	14.95	57.47	27.58	15.77	57.95	26.28	14.85	58.05	27.10
Nickel												
Inlet	16.20	56.32	27.48	14.52	55.70	29.77	14.15	55.77	30.08	13.27	56.42	30.31
Outlet	16.92	57.20	25.88	15.62	56.05	28.33	14.97	56.35	28.68	14.25	57.75	28.00
Canal	17.57	57.92	24.51	15.97	57.42	26.61	15.12	57.37	27.51	15.47	59.15	25.38
Mixing zone	17.32	58.97	23.71	16.70	58.92	24.38	15.52	57.52	26.76	16.37	60.47	23.16
Upstream	19.42	61.45	19.13	18.70	61.57	19.73	16.70	59.70	23.33	18.20	62.92	18.88
Downstream	19.35	61.30	19.35	17.95	60.42	21.60	16.62	58.77	24.61	17.32	62.07	20.61

more stable inorganic complexes than the ferrous form, Fe (II). As the ferric forms get precipitated out, it is not available for the complex formation so readily. In turn, the stable complex ions have low percentage and free plus labile complex ions show a high percentage. From Tables 1 and 3 it is found that with the increase in pH and Eh values from inlet to downstream, the free plus labile complex ions also increase in the same pattern for all metals. Thus it can be concluded that the proportion of free plus labile complex ions is greatly influenced by the factors pH and Eh. The stable complex ions, on the other hand, are influenced by the presence of inorganic and organic ligands. It is evident that the con-

Table 4. Percentage of metal forms associated with different particle sizes (Leachates)

	>12 μm			>1–12 μm			>0.45–1 μm			>0.25–0.45 μm		
	C+F	CS1	CS	C+F	CS1	CS	C+F	CS1	CS	C+F	CS1	CS
Iron												
Depth 2.5 ft												
Inlet	23.57	55.32	23.11	26.80	58.72	14.48	26.67	58.52	14.81	26.20	56.97	16.83
Outlet	24.25	54.77	20.98	27.05	59.32	13.63	27.45	58.72	13.83	26.47	57.52	16.01
Depth 1.5 ft												
Inlet	24.62	55.55	19.83	28.82	59.37	11.81	27.85	58.87	13.28	26.65	57.82	15.53
Outlet	25.20	55.57	19.23	28.47	59.50	12.03	28.52	58.62	12.86	26.95	58.60	14.45
Manganee												
Depth 2.5 ft												
Inlet	20.42	54.52	25.06	21.60	56.42	21.97	20.82	55.12	24.86	22.45	56.27	21.28
Outlet	21.47	55.00	23.53	22.40	56.75	20.85	22.77	55.47	21.76	23.47	57.10	19.43
Depth 1.5 ft												
Inlet	22.42	55.65	21.93	22.90	57.00	20.10	22.77	55.97	21.25	23.57	57.15	19.28
Outlet	22.25	55.62	22.13	23.52	57.57	18.90	23.57	56.07	20.36	24.42	58.80	16.78
Cadmium												
Depth 2.5 ft												
Inlet	11.17	55.87	32.96	8.82	55.05	36.13	8.15	52.87	38.98	10.02	55.52	34.45
Outlet	11.92	56.10	31.98	9.65	55.60	34.75	8.97	53.85	37.18	10.57	55.87	33.55
Depth 1.5 ft												
Inlet	12.15	57.12	30.73	10.15	55.87	33.98	9.92	54.10	35.98	11.47	56.92	31.60
Outlet	13.10	57.62	29.28	10.80	56.45	32.75	10.6	54.60	34.80	12.40	58.07	29.52
Copper												
Depth 2.5 ft												
Inlet	12.10	55.55	32.35	11.45	55.20	33.35	9.95	53.72	36.33	12.75	56.50	30.70
Outlet	13.85	56.20	29.95	12.57	55.77	31.66	10.37	54.30	35.33	13.05	57.07	29.88
Depth 1.5 ft												
Inlet	14.50	57.57	27.93	12.75	56.62	30.63	11.00	54.97	34.03	14.42	58.40	27.18
Outlet	14.87	58.05	27.08	13.02	57.57	29.41	11.60	55.12	33.28	15.15	59.40	25.45
Zinc												
Depth 2.5 ft												
Inlet	9.67	52.40	37.93	9.80	52.47	37.73	9.05	51.87	39.08	11.72	54.65	33.63
Outlet	10.77	53.05	36.18	10.72	52.92	36.36	10.00	52.12	37.88	11.90	54.82	33.28
Depth 1.5 ft												
Inlet	11.95	54.75	33.30	10.60	52.97	36.43	10.37	53.17	36.46	12.49	55.30	32.30
Outlet	14.05	56.55	29.40	13.05	55.40	31.55	11.55	54.40	34.05	13.90	56.57	29.53
Nickel												
Depth 2.5 ft												
Inlet	9.60	53.82	36.58	12.50	54.42	33.08	11.47	54.50	34.03	12.55	55.47	31.98
Outlet	10.72	54.80	34.48	12.57	54.52	32.91	12.10	54.95	32.95	13.45	56.60	29.95
Depth 1.5 ft												
Inlet	11.85	55.62	32.53	12.82	54.77	32.41	12.97	55.85	31.18	13.85	57.10	29.05
Outlet	12.40	56.15	31.45	14.40	56.12	29.48	14.45	57.15	28.40	14.25	57.50	28.25

centration of these ligands decreases from inlet to downstream and therefore, the proportion of sable complex ions increases. Manganese follows the same pattern as iron but the difference between "C+F" and "CS" forms is less than that of iron. That the behaviour of manganese is different, has been reported earlier also (6,7). Its solubility varies due to its variable valencies +1 to +7. And, manganese unlike iron, does not readily form stable complexes with ligands.

Table 5. Leaching of metals through soil column in different time intervals (Days)

Season/	Depth 2.5 ft (Conc. in ppb)						Depth 1.5 ft. (Conc. in ppb)					
days	Fe	Mn	Cd	Cu	Zn	Ni	Fe	Mn	Cd	Cu	Zn	Ni
PRE												
1	363.24	173.24	24.85	28.54	72.94	13.32	324.10	151.80	20.40	24.30	63.40	12.50
2	344.15	166.73	20.24	25.32	65.65	11.78	313.40	148.90	16.40	19.30	60.10	12.60
3	328.64	164.54	22.46	26.98	67.86	12.46	301.20	147.40	19.20	22.40	56.80	11.10
4	334.68	157.26	17.51	24.46	63.53	11.28	307.10	150.50	17.30	21.30	53.30	10.50
5	325.43	161.85	19.82	21.59	62.73	9.84	303.40	142.80	15.10	20.40	54.70	9.10
6	325.57	158.36	15.65	23.71	56.87	10.91	394.90	136.60	15.40	19.90	59.60	10.40
7	315.81	151.57	16.12	20.87	59.21	11.21	297.50	138.30	13.50	20.80	51.50	9.60
MON												
1	356.74	168.46	22.63	27.25	70.89	11.76	314.50	147.70	19.10	23.50	61.50	11.60
2	336.45	159.34	21.24	24.64	62.25	12.21	302.70	142.60	17.50	22.20	58.70	10.10
3	338.63	150.84	18.96	25.13	56.49	11.51	297.40	137.60	18.70	20.70	51.30	10.80
4	327.86	156.56	17.52	22.35	58.34	10.12	305.10	139.20	18.40	19.40	49.40	9.80
5	321.62	152.73	19.16	21.36	57.61	11.21	301.60	140.80	15.90	17.30	55.10	10.40
6	319.38	149.21	16.96	20.64	52.95	9.41	294.90	134.10	13.40	18.50	47.20	8.80
7	314.78	145.62	14.41	18.98	53.79	9.12	291.20	128.30	12.20	18.90	47.50	9.90
POS												
1	355.35	166.75	23.26	27.67	68.56	13.18	311.90	144.50	18.60	22.90	59.90	11.90
2	331.54	159.92	21.48	24.16	60.87	11.97	308.50	139.90	16.80	21.30	53.40	9.30
3	338.83	150.34	18.43	22.36	67.21	12.35	294.90	143.40	15.40	19.60	54.50	10.10
4	327.96	153.48	19.92	25.42	63.78	11.36	299.70	135.70	14.90	18.40	49.90	9.40
5	334.74	149.36	16.57	20.62	57.97	10.98	288.50	139.20	15.70	17.10	46.20	8.90
6	319.73	144.79	16.91	21.41	53.32	9.31	282.30	132.10	14.10	16.20	51.40	8.20
7	313.49	146.59	14.64	18.43	49.42	9.26	284.80	127.90	12.20	16.60	47.40	7.70
WIN												
1	353.45	171.16	22.15	26.74	65.71	13.48	316.30	142.40	17.90	22.40	58.20	12.10
2	327.68	149.75	19.36	24.81	63.95	12.27	304.90	136.10	15.40	19.20	51.20	11.30
3	305.56	157.63	21.65	21.73	63.53	11.56	301.70	137.80	16.60	21.60	49.50	10.10
4	314.78	154.34	18.83	22.52	55.84	11.52	295.60	137.40	14.30	18.40	46.20	9.50
5	296.73	144.35	17.95	20.25	58.26	10.84	299.30	132.60	12.40	17.40	50.90	10.40
6	297.41	139.97	17.76	19.56	53.84	11.51	291.80	125.60	12.30	19.10	45.50	9.10
7	288.19	133.45	15.17	17.69	51.18	9.97	286.10	123.70	11.70	16.90	40.80	8.40

In case of cadmium, copper, zinc and nickel, a similar pattern is followed, i.e., the stable complex ions are in higher proportion than the free plus labile complex ions. Slowly labile complex ions are always more. In the aquatic system in which oxidation occurs, hydrous iron and manganese oxide constitute a highly effective sink for heavy metals (8). Due to the high adsorptive power of the sink metals the other trace metals get adsorbed on it which in turn may influence the physico-chemical properties like flocculation, precipitation, and sorption etc. Therefore, the sink metals have influence over the transport and fate of the adsorbed trace metals.

Table 3 also shows that iron, manganese and nickel are more concentrated in the particulate size (> 12um) than in the colloidal size (> 0.25—0.45um). This may be due to the association of these metals with the larger particles as a result of flocculation or adsorption, as the affinity of these metals are more than cadmium, copper and zinc. The metals may be bound to the particulates physically or chemically. This is responsible for the transport and fate of these metals in the aquatic system. Metals are often found in natural waters associated with suspended particulate matter. Iron (III) and manganese (IV) are found in many systems as colloidal hydrous oxides whereas many other trace metals are found adsorbed on these oxides (9). Iron and manganese act as sink metals for other heavy metals. Cd, Cu, and Zn get adsorbed on these metals and get precipitated. Since the different chemical forms of Fe and Mn are more in the >12um size, they would be able to adsorb much of the other meals. This, in turn, will result in comparatively lower concentrations of "C+F" form of other metals in this size range, which is experimentally found (Table 3).

It is also found that copper has less stable complex ions (CS) than Cd and Zn. It may be due to greater affinity of copper towards the organic ligands resulting in precipitation as organometallic complexes. Copper exhibits greater adsorption tendency towards humic acids than does cadmium. (10,11). On the other hand, cadmium forms moderately stable complexes with a variety of organic compounds.

In case of leachates (Table 4) the free plus labile complex ions are more in the colloidal sizes and less in the particulate sizes. In soil the residential time of meals is more. From Table 2 it is found that the leachates are more acidic than overflow so me are more acidic than overflow so metals are more soluble in leachates. It is also found that chloride ions concentration is more in leachates. Ferric form has a tendency to form metal-chloro complex ions which are very mobile in the aquatic systems. Due to this complex formation and mobility it does not get adsorbed on the particulate matters so easily. As mentioned earlier, iron (III) and manganese (IV) are found in many systems as colloidal hydrous oxides whereas many other trace metals are found adsorbed on these oxides. From Table 4 it is found that as the concentration of different chemical forms of Fe and Mn increases, the adsorbed metal concentration also increases. So there must be some factors other than adsorption on the sink metals responsible for this deviation. It may be due to the competition of cations Fe and Mn with the other metals (Cd, Cu, Zn and Ni) for the ligands. In leachates containing high concentration of dissolved iron or where there is a substantial contribution to complexion by fairly small concentration of a more powerful ligand, competition effects are more likely to be observed, If the concentration of uncompelexed iron were maintained high relative to other metals it is possible that very little of the complexing capacity would be available to those other metals (12).

From the lysimeter experiments, it is also found that at different pH levels which are prevalent in different seasons, premonsoon (Pre), monsoon (Mon), post monsoon (Pos) and winter (Win), the leachability of different heavy metals Fe, Mn, Cd, Cu, Zn and Ni varies (Table 5). At four pH levels of leachates corresponding to these four seasons, viz., 5.40, 6.01, 6.32 and 6.45, the leachability follows the order:

pH 5.40;	Mn	>	Zn	>	Cu	≥	Fe	≥	Cd	>	Ni
pH 6.01;	Zn	≥	Cu	>	Mn	>	Fe	≥	Cd	>	Ni
pH 6.32;	Cu	≥	Cd	>	Zn	≥	Mn	>	Fe	>	Ni, and
pH 6.45;	Cu	>	Cd	≥	Zn	>	Mn	≥	Fe	>	Ni.

From this it can be concluded that under acidic conditions Cu, Cd, and Zn are sorbed in the soil more rapidly than Fe, Mn and Ni. Moreover Fe and Mn are strongly adsorbed in the soil. So Fe and Mn are not readily released to the bulk of waters at the higher pH range. At higher pH range the solubility of heavy metals Cu, Cd and Zn in water is more than that of Fe and Mn. It is also known that the metals Fe and Mn have a stronger affinity to the ligands than Cu, Cd and Zn. So Fe and Mn are strongly adsorbed in the soil and strongly bonded with the legends available in the soil with high binding energy. Since Fe/Mn oxide are also sinks for Cu, Cd and Zn, these metal are not readily available for the ligands. So these metals are released easily to the bulk of waters.

4. CONCLUSIONS

From the aforementioned results and discussions, we can conclude the following:

In suspended particulates in leachates the percentage of free ions plus labile complex ions of Mn, Cu, Zn and Ni are more in the colloidal sizes i.e. > 0.25—0.45um. Fe is more prominent in the size > 0.45 — 1um and > 0.25 — 0,45um than size > 12um. Cd is distributed equally both in particulate and colloidal sizes. However in case of the overflow and river water, the percentage of free plus labile complex ions of the above said metals are found more in the particulate size of > 12um and > 1–12um. In terms of biological significance and toxicity of heavy metal species, all the studies metals are significantly present in the smallest size fraction, i.e., in the colloidal size. Since they are prominently present in colloidal size, they are readily available for the aquatic biota by different processes like osmosis, diffusion etc. The size distribution of heavy metals is mainly controlled by the water solubility factors, like pH, Eh etc. These parameters are largely dependent on the quality and quantity of the effluents coming out of the thermal power plant, and are mainly responsible for causing toxicity to the aquatic biota. From these observations, it can be said that the leachates have more toxic effects on the aquatic biota than the overflow of the effluents.

From the lysimeter leachate experiment it can be said that the concentration of heavy metals is more at the lower depths. Simultaneously it is seen that over a period of time the leachability decreases. So, over a longer period, the leachability ca be substantially reduced. The leachability is mainly dependent on the pH of water and soil. Since the effluents are found to be acidic the leachabilty is more. This may lead to the metal species leaching out to the lower depths which can contaminate ground waters.

The concentration of different species of metals viz., free ions plus labile complex, slowly labile and stable complexes in the inlet and outlet of ash pond is more than that in the carrying canal. In the river, these values are more in the mixing zone than in the downstream and upstream. This change in speciation can be attributed to the adsorption, change of ambient physico-chemical characteristics and microbial activities in the water course. Since it is found that downstream has more metal concentration than the upstream, it can be concluded that the effluents of the power plant are responsible for this. In addition, the physical factors like dilution, velocity and volume of the waters in the river, temperature etc. also moderate these values.

It is known that the lower the pH value, the higher is the proportion of free ions in the solution. pH of the water controls the toxicity value of the heavy metals. So the toxicity is greater in the acidic waters than in the basic waters. The toxicity of heavy metals in the waters is highly variable. In the present study it is found that the free metal ions plus their labile complexes are more than the stable complexes in the water body. This may result in toxic effects in the waters. The pH of receiving waters decreases due to the effluents coming out of the power plant which are acidic. These effluents may impart higher toxicity to the surface run off and the leachates, which in turn may affect the ground waters. In the present study, it is found that the concentrations of heavy metals Fe, Mn, Cu, Cd, Zn and Ni have a positive correlation. Since it is known (13) that acute toxicity for the aquatic biota arises from competitive interaction between major essential elements, Ca, Co, Cu, Fe and Zn, it is possible that the above mentioned six metals can cause acute toxic effects in the receiving waters, due to their different forms and interaction between themselves.

It can be suggested that the effluents before being discharged to the ash pond can be treated with Ca and Mg hydroxides which can reduce the acidic value of the waters, so that the leachability and toxicity of heavy metals can be reduced. It is known that Ca and Mg are the competitive cations for the inorganic and organic ligands. So by increasing the Ca and Mg concentration in the waters the availability of the ligands for the heavy metals will be reduced which in turn, will reduce their leachability and toxicity. By doing this, we can achieve the desired soil attenuation value. Although the hardness of waters will increase to a certain extent, this is not as harmful as the heavy metal species.

REFERENCES

1. APHA 1980, *Standard methods for the examination of water and waste water*, 18th edition, American Public Health Association, New York.
2. Laxen, D.P.H. and Harrison, R.M., 1981 *A scheme for the physico-chemical speciation of trace metals in fresh water samples*, Science of the Total Environment, 19, 59.
3. Campanella, E., Cardarelli, F., Ferri, T., Petronico, B.M. and Pupella, A., 1987, *Evaluation of heavy metal speciation in an urban sludge*. Science of the Total Environment, 61, 217.
4. Christensen, T.H. and Lun, X.Z., 1989, *A method for determination of cadmium species in solid waste leachates*. Water Research, 23, 73.
5. Allen, S.E. 1989 *Chemical Analysis of Ecological Materials*, 2nd edition, Blackwell Scientific Publications. Rome and London.
6. ASCE and AWWA, 1990, *Water treatment plant design*, 2nd edition. McGraw-Hill, New York.
7. Montgomery, J.M. 1985, *Water treatment principles and design*, John Wiley, New York.
8. Lee. G.F., 1975, *Role of hydroxous metal oxides in the transport of heavy metals in the aquatic environment*. Pergamon Press, Oxford.
9. Jenne, E.A., 1968, *Controls on Mn, Fe, Co, Ni, Cu and concentration in soil and water: The significant role of hydrous Mn and Fe oxides, and trace organic matter*, American Chemical Society Advanced Chemistry Series, 73, 337.
10. Guy, R.D., Chakrabarti, C.L. and Schramm, L.L., 1975. *The application of simple chemical model of natural waters to metal fixation in particulate matter*. Canadian Journal of Chemistry, 53, 661.
11. Nriagu, J.O. and Coker, R.D., 1980 *Trace metals in humic and fulvic acids from lake Ontario sediments*. Environmental Science and Technology, 14, 443.
12. Griffin, R.A. and Shimp, N.F., 1976, *Effect of pH on exchange adsorption or precipitation of lead from landfill leachates by minerals*. Environmental Science and Technology, 10, 1256.
13. Moore, J.W. and Ramamoorthy. S. 1984, *Heavy Metals in Natural Waters*. Springer-Verlag, New York.

4

A POSSIBILITY OF APPLICATION OF CLINOPTILOLITE FOR WATER POLLUTION CONTROL

Eva Horváthová-Chmielewská

Comenius University
Faculty of Natural Sciences
Mlynská dolina, 842 15 Bratislava, Slovak Republic

ABSTRACT

Ammonia removal by means of Slovakian clinoptilolite from ammonium enhanced tap waters was examined in laboratory as well as in field experiments. Field installation consisted of two steel pressure columns operated in serie, each one filled with 56 kg of (0,3–1,0) mm grain -sized clinoptilolite. Hydraulic loading rate of columns was 900 liter per hour. Two storage tanks and a stripping tower constructed from HDPP was used for regenerant recovery. The performed studies verified a high efficiency of drinking water ammonia removal by inland clinoptilolite.

Key words: ion exchange, clinoptilolite, drinking water purification, regeneration, ammonia recovery, air stripping.

1. INTRODUCTION

The use of conventional ion exchange resin for removal of nitrogenous material from waste water has not been found attractive because of lack of suitable selectivity, high costs, difficulties with handling of regenerant effluents, etc.

Above limitations may be overcome by using a clinoptilolite, suitable for selective ammonium removal and enhanced nutrients—water pollution control. This cation exchanger occurs naturally in several deposits in the western United States, Japan, China and the south-eastern Europe[1].

The largest known as well as the highest quality deposit of clinoptilolite was found in southern California within a deposit of bentonite, called Hectorite because of its proximity to Hector, California[2].

Chemistry for the Protection of the Environment 3, edited by Pawłowski *et al.* Plenum Press, New York, 1998

After exploration of huge natural zeolite deposits in the Eastern Europe the interest for industrial exploitation of these minerals has extensively arised also in this region[3,4].

Ion exchange technology using naturally occurring clinoptilolite was originally developed in USA. The Tahoe Truckee Sanitation Agency (TTSA) in California is only one facility in the world which utilizes inland clinoptilolite for municipal wastewater treatment for about last 20 years.

Research on application of the Slovakian clinoptilolite for ammonia removal from drinking water and wastewater started in former Czechoslovakia many years ago at Water Research Institute[5]. A considerable research has been conducted on the basic physico-chemical properties of zeolites including adsorption and ion exchange of metal cations and some radionuclides[5]. This paper summarizes the pilot plant research on removal of ammonia from drinking water.

2. EXPERIMENTAL

Field experiments were performed in autumn 1986. A laboratory model (see Photograph 1) was operated before the field experiments started. Each of two vertical glass columns (high 150 cm, diameter 3,2 cm) filled with 700 g of clinoptilolite having grain size (0,2–0,7) mm were operated down flow mode with loading rate 9 liter per hour (11 BV/hour). Tap water with addition of NH_4Cl as source of ammonium was pumped to the

Photograph 1. Laboratory model for inland clinoptilolite tests.

top of the column and regularly sampled using sampling ports at the bottom of the columns. Initial ammonium concentration in model solution was 3,2 mg per liter. To increase a number of bed volumes throughput while maintaining a low ammonium effluent concentrations semi-countercurrent operation with two beds in series was used. Various laboratory stripping vessels constructed from a plexiglass rectangular tubes and filled with a large specific surface packing material of modified volumes were tested for ammonia recovery of exhausted regenerants[6].

The zeolite ion exchange pilot installation with a hydraulic loading rate of 900 l per hour (12,8 BV/hour) was situated at the field experimental facility near Bratislava, where the tap water enriched with NH_4Cl to initial ammonium concentration of 1 mg per liter was treated using two pressure steel columns operating in serie. Each column was filled with 70 l (56 kg) of clinoptilolite having a grain size as follows:

0,75–1,0 mm	6,95 %
0,5–0,75 mm	52,8 %
0,3–0,5 mm	34,5 %
0,1–0,3 mm	4,2 %
< 0,1 mm	0,95 %

Active mineral content in the clinoptilolite tuff was estimated to about 60%.[5]

Bottom of each column was filled with 14 l of a sand. Columns were operated in down flow mode during the common service cycle and upflow mode in regeneration cycle. For regeneration a solution of 2% NaCl and pH ~ 9 was used. In pilot-scale, an HDPP tower of total high 6,5 m, assembled from 6 modules of ground-plane measures 980 X 650 mm, was used for ammonia stripping. To strip ammonia out of the regenerant brines a large quantities of air through the tower are necessary, therefore the tower was improved by installation of two lateral air blowers at the bottom. Countercurrent tower configuration, by which the entire air flow enters at the bottom, while the treated solution enters the top and falls across the vertical waves shaped laminated glass slats to the bottom of the tower, resulted in above tower design improvement (Fig. 1). All water analyses have been done according to standard methods. For ammonia measurements the Nessler method was chosen and a content of metals was analysed by atomic absorption spectrophotometry.

3. RESULTS AND DISCUSSION

While laboratory model of described structure treated during one service cycle to acceptable effluent concentration 0,5 mg ammonium per liter a volume of 675 liter, the pilot clinoptilolite column treated 85 cubic meter of water, i.e. 4—days operated clinoptilolite bed removed from water 81 g ammonia. To compare zeolite capacity data in both synthetically prepared ammonium enhanced drinking waters, a higher capacity value was achieved at laboratory conditions, where the initial ammonium concentration was higher (a_{lab} = 3 mg/g; a_{pilot} = 1,5 mg/g).

The data on analysis of influent and effluent are depicted in Tab. 1. There is some pronounced exchange between NH_4^+, Na^+ and K^+ ions and slight increasing of NO_3^--component in effluent, probably due to nitrification as well as psychrophil and mesophil bacteria increase in water samples, by reaching of steady-state conditions. However, after alkaline regeneration and backwashing of clinoptilolite bed this phenomenon disappeared. Moreover, drinking water treated on zeolite is assumed to be afterwards hygienically secured.

Tab. 2 shows content of metal ions in effluent. Several Al-enhanced concentrations observed especially in high pH indicates on Al-release, probably due to an attack of accompany-

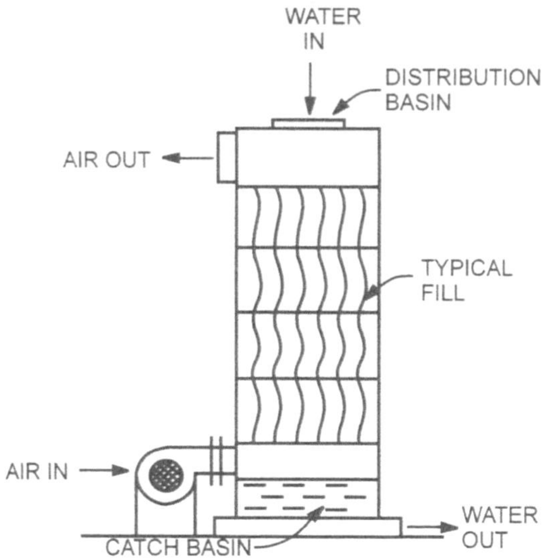

Figure 1. Principle sketch of pilot air stripping tower.

Table 1. Average water quality analyses of influent and after ion

Influent					
component in mg/l					
NH_4^-	3,1	3,5	3,4	3,3	3,3
NO_2^-	0	0	0	0	0
NO_3^-	2,8	18,8	21,6	23,5	15,5
O_2	10,1	14,1	8,6	8,6	9,0
Ca^{2+}	58,1	59,1	65,1	53,1	54,1
Mg^{2+}	17,6	1,2	0	13,9	5,1
Na^+	15,5	15,5	17,0	17,0	16,9
K^+	5,3	4,8	5,3	5,3	5,2
Fe^{3+}	0	0	0	0	0
Mn (total)	0	0	0	0	0
COD	1,5	1,3	1,5	1,5	1,4
Alkalinity, mmol/l	3,8	3,8	3,8	4,0	3,9
Effluent					
component in mg/l					
NH_4^-	0	0	1,2	2,6	3,3
NO_2^-	0	0	0	0	0
NO_3^-	7,8	21,6	27,0	23,5	22,0
O_2	9,5	10,6	8,7	10,9	9,0
Ca^{2+}	52,1	51,1	40,1	64,1	65,1
Mg^{2+}	12,1	5,5	20,6	0	0
Na^+	44,0	25,5	25,5	20,0	17.5
K^+	1,0	1,0	1,0	2,2	2,6
Fe^{3+}	0	0	0	0	0
Mn (total)	0	0	0	0	0
COD	1,5	1,8	1,5	1,5	1,4
Alkalinity, mmol/l	4,1	4,4	3,8	4,1	4,2

Table 2. Metal concentrations in mg/l of water samples at various levels of ammonia removal by zeolite ion exchange technology

Solution type	Al	Fe	Cu	Ag	Pb	Zn	Cd	Ni	Co	Cr	Sr	Cs
Recycled regenerant	3,120	0,360	0,020	<0,001	0,010	0,290	<0,001	0,650	0,100	0,032	0,002	0,004
Backwash pH 10,2	1,120	0,150	0,010	<0,001	0,005	0,070	<0,001	<0,005	0,004	0,002	0,008	<0,002
Backwash pH 7,3	0,150	0,070	0,009	<0,001	0,005	0,060	<0,001	0,015	<0,002	0,002	0,050	<0,002
Influent	0,070	0,280	0,009	0,001	0,005	0,390	<0,001	0,020	0,020	<0,002	0,320	<0,002
Clino bed effluent	0,160	0,060	0,005	0,002	0,005	0,040	<0,001	0,010	0,010	<0,002	0,020	<0,002
Clino bed effluent steady-state	0,080	0,100	0,005	<0,001	0,006	0,090	<0,001	0,005	<0,002	0,054	0,260	<0,002
Tap water extract after 150 minutes mixing with pulverized zeolite	0,420	0,090	0,009	0,002	0,004	0,060	0,012	0,010	0,008	0,002	0,320	0,002

ing minerals and inpurities (cristobalite, mica). Therefore use of the regenerant having lower pH is recommended. The principle of two pilot beds operation was following:

Single, one clinoptilolite bed was in service till the ammonium concentration in effluent reached the limit value 0,5 mg/l, then the second column was connected in series. After exhaustion of the first column, it was disconnected and regenerated. During a regeneration of the first column, the second one was in service. The more exhausted column was always at the influent end backed up by one column having lower loading. By this way a number of bed volumes throughput in service cycle was increased by 50%.

Similar operation was used for ammonia removal of tannery waste water at Shoe Manufactury Water Reclamation (Zlín, former Czechoslovakia).[7] A higher ammonium concentration in waste water (in average 50 mg/l) leads to higher ion exchange capacity.

Three columns operation (two beds on stream while the third bed being regenerated) as well as more complicated valving and piping was required at this facility. Against the sufficient long–4 days operating drinking water service, waste water service cycle lasted only 15 hours. Consequently, to shorten regeneration part of waste water treatment, a fractional-3 portion regeneration and ammonia recovery according to Scheme 1 was proposed.

In drinking water process, exhausted bed was regenerated with 2% NaCl solution having pH ~ 9, at flow rate 900 liter per hour, in upflow mode. During the regeneration a 85%-ammonia elution from clinoptilolite bed capacity was sufficient. In 2 hours passed 25 BV of regenerant were satisfactory for maintaining a solid time schedule and required water quality in drinking water purification. A half hour lasted backwashing with tap water by about 30% bed expansion followed the regeneration cycle, additionaly.

Regenerant was air stripped, without mass closed loop operation. Stripped gases were discharged to the atmosphere, because the pilot study did not affect air pollution significantly. Total volume of a regenerant (1800 l) of pilot facility was recycled through the tower during 150 minutes to decrease the initial ammonia concentration of 50 mg/l to less than 10 mg/l. In the first stripping cycle 47% of ammonia was removed, therefore second cycle was performed additionaly. The stripping tower was operating countercurrent wise. Regenerant was pumped to the top of the tower by the same loading rate as in clinoptilolite bed service (900 liter per hour) and falled against blowed air from the bottom of the tower (by capacity 800 liter per second). The major factors which affected design and process performance of tower were tower configuration, air flow and pH. For renovation of 1 liter pilot elutriant (regeneration effluent) a consumption of 3,8 cubic meter air was estimated. To maintain pH = 11 of elutriants during the stripping process as well as to compensate a loss of natrium ions in regeneration effluents, regularly addition of NaOH was provided.

4. CONCLUSION

The study on application of clinoptilolite for ammonia removal from drinking water indicated a progress in the whole ion exchange-regenerant recovery operation.

Following aspects of proven technology can be summarized finally:

Drinking water with enhanced ammonium concentration was efficiently purified on naturally occuring Slovakian clinoptilolite.

The ion exchange capacity of used clinoptilolite tuff in pilot-scale facility reached 1,5 mg / g (tuff having about 60% clinoptilolite). The low capacity of zeolite is a function of the low ammonium concentration in purified water.

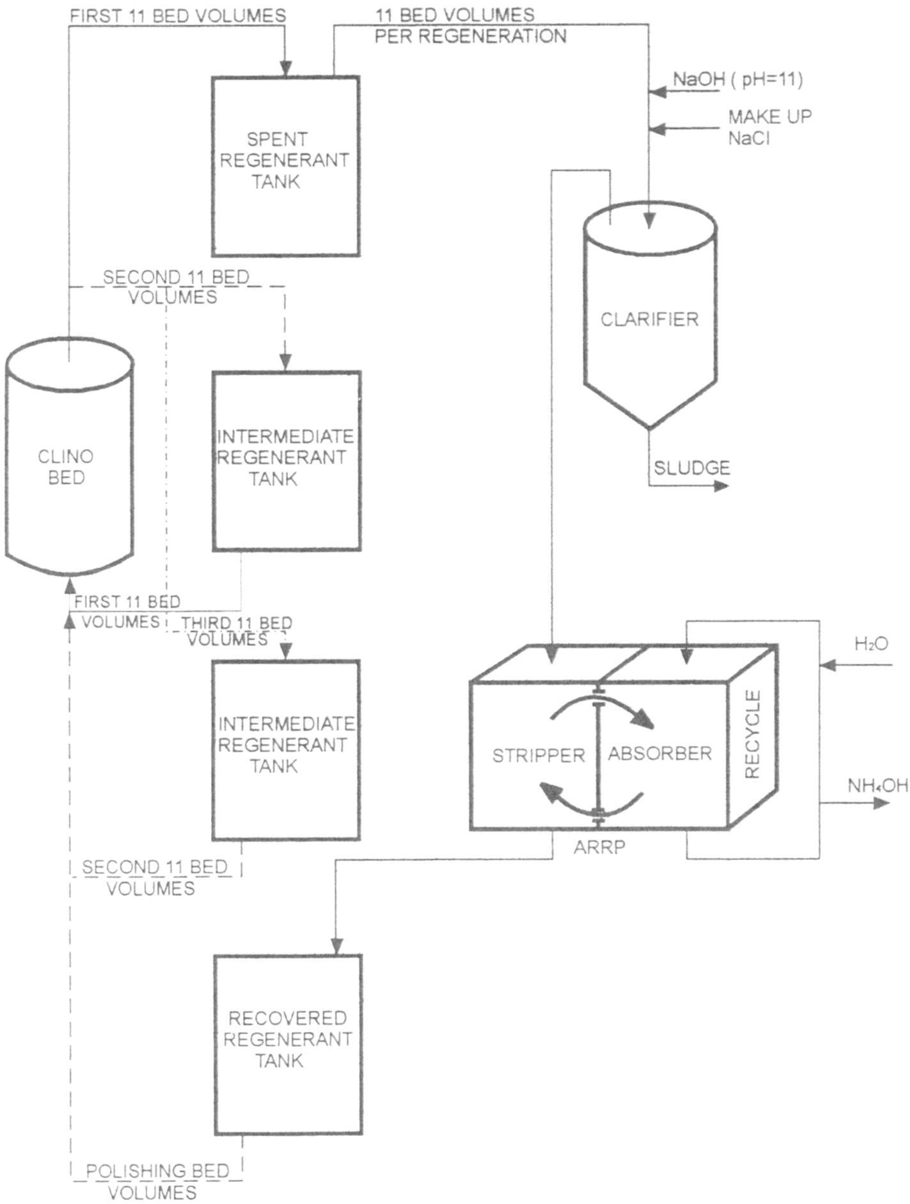

Scheme 1. Principle diagram of 3 portions regeneration and ammonia recovery system applied for tannery wastewater treatment in Zlin (former Czechoslovakia).

The exhausted clinoptilolite beds were regenerated with 2% NaCl solution (having pH ~ 9, adjusted by NaOH) countercurrent wise and finally backwashed with tap water to keep the neutral pH.

Spent regenerants have been recovered by air stripping method. The pilot-scale system was operated without mass clossed loop, stripped gases were discharged into atmosphere.

To maintain pH ~ 11 of stripped solution as well as to compensate a loss of natrium ions after regeneration, regularly addition of NaOH to stripped solution was provided.

However, the cost for treatment of 1 cubic meter of water has not overpassed at that time 1 crown, any full-scale clinoptilolite facility has appeared since the year 1986, when the study was performed.

ACKNOWLEDGMENT

This studies, currently supported by National Science Council VEGA under grant EFA 1/3067/96 were performed at author's previous employer, Water Research Institute.

REFERENCES

1. Gottardi, G.; Galli, E.: Natural Zeolites, Springer Verlag, Berlin 1985.
2. Mumpton, F.A.; Sand, L.B.: Natural Zeolites, Occurrence, Properties and Use, Pergamon Press Oxford 1978.
3. Rudenko, G.; Tarasievitch, J.; Kravtchenko, B.: Utilization of clinoptilolite as filtration material at industrial water treatment facility, Chimia Technol. vody 1, pp. 54, 1983.
4. Deltchev, B.: One-layer Bulgarian clinoptilolite filters, NNTI pri Vodokanalprojekt, BPPB Sofia 1988.
5. Chmielewská-Horváthová, E.: Reverse Equilibrium Study for Metal Uptake on Conditioned Zeolite, Environment Protection Eng.; Vol. 22, No. 3–4, pp. 37–44, 1996.
6. Chmielewská, E.: Modelling an air stripping process for ammonia removal, Environment Protection Eng.; Vol. 21, No. 1–4, pp. 41–49, 1995.
7. Horváthová, E. et al.: Technology for wastewater treatment in tannery processing plants, CS-Patent 02 604–89, No. 274 068.

5

EFFECT OF LAND MANAGEMENT IN WINTER CROP SEASON ON METHANE EMISSION FROM THE FOLLOWING RICE GROWTH PERIOD

Z. C. Cai and H. Xu

Institute of Soil Science, Chinese Academy of Sciences
Nanjing 210008, China, email: zccai@ns.issas.ac.cn

ABSTRACT

A pot experiment was carried out to investigate the effect of land management in winter crop season on CH_4 emission from the following rice growth period. The results showed that the mean CH_4 flux from the soil flooded and fallowed in the winter crop season (FF) was the highest (25.93 mg CH_4 m^{-2} h^{-1}), followed by the soil planted with alfalfa and incorporated with 3.4% green manure in the top layer (AL) before it was flooded and prepared for rice transplanting (21.44 mg CH_4 m^{-2} h^{-1}). The CH_4 emissions from the soil planted with winter wheat (WW) and the soil drained and fallowed (DF) were only 19.6 and 17.4% of that from FF, respectively. Drainage in the winter crop season led to an increase in the oxidant content in the soil and slowed down the decrease rate of soil Eh after the soil was flooded. There was a significant relationship between the simultaneously measured CH_4 flux and soil Eh for DF and WW treatments. The soil redox potential was low enough for CH_4 production when the rice was transplanted and was not a critical factor influencing CH_4 emission in FF treatment during the rice growth period. Incorporation of green manure provided extra electron donors and carbon sources, which accelerated soil reduction rate after the soil was flooded and stimulated CH_4 emission. It is concluded that the water management not only in the rice growth periods but also in the non-rice growth period plays a very important role in governing CH_4 emission from rice fields.

Key words: methane emission, water management, winter crop season.

1. INTRODUCTION

China is an important rice production country with 22.6% of the world rice harvested area and 36.3% of the world rice grain production[1]. Early field measurements showed that the CH_4 emissions from Chinese rice fields were very high; the highest mean CH_4 flux recorded throughout rice growth period in the world. Based on these early measurements, the annual CH_4 emission from Chinese rice fields was estimated to be as high as around 30 Tg CH_4[2]. The majority of the mean CH_4 fluxes from Chinese rice fields measured thereafter, however, were much lower than these early measurements, mostly ranged from <1 to 15 mg $CH_4\,m^{-2}\,h^{-1}$[3] and did not significantly differ from those measured in the rice fields elsewhere in the world. Up till now no studies have been done to explain why the CH_4 emissions from the rice fields in the early measurements were unusually high. It is important to understand the underlying mechanisms that result in unusual large CH_4 emission for figuring out strategies for reducing CH_4 emission from rice fields.

Rice is widely produced in China, from South to North, from East to West and from plain to mountainous areas. The crop rotation and water management in the rice production areas in China vary considerably according to local climate, soil properties and irrigation and drainage facilities. The annual water regime of the rice fields in China can be roughly divided into i) intermittent irrigation and ii) continuous flooding during the rice growth period but drainage in the winter crop season and iii) flooding not only during the rice growth period, but also during the non-rice growth period. Cai summarized the data available and found that CH_4 emissions from the rice fields drained in winter crop season were much lower than those from the fields flooded annually[3]. The highest mean CH_4 emission was also recorded in an annually flooded rice field[2]. This clearly indicates that water management in winter crop season is a very important factor influencing CH_4 emission from the following rice growth period. The previous studies on mitigation options for CH_4 emission from rice fields focus on the rice growth period[4]. Research on the effect of land management, including water management, in winter crop season on CH_4 emission from the following rice growth period may induce the strategy for mitigating CH_4 emission from rice fields. For investigating the effect, a greenhouse experiment was conducted and the results are presented in this paper.

2. MATERIALS AND METHODS

The pot experiment was conducted in a greenhouse. The soil with organic matter content of 17.02 g kg^{-1}, total N 1.18 g kg^{-1}, and pH(H_2O) 6.3, was collected from a rice field in the Experimental Farm of Jurong Agricultural College, Jiangsu Province, China, just after rice was harvested in 1995, air-dried and passed through a 5 mm sieve. Four kilogram prepared soil was filled into experimental pots and 2 kg prepared soil mixed with 30 g rice straw containing 413 g kg^{-1} organic carbon was filled on the top. The soils were then treated as follows: drained and fallowed (DF), flooded and fallowed (FF) with *ca* 20 mm floodwater layer, planted with winter wheat (WW) or alfalfa (AL). All treatments were triplicated. The winter wheat and alfalfa were harvested on 26 May, 1996. The wheat straw was removed in the treatment WW and 68 g green manure was incorporated with the surface soil in the treatment AL (approximately 3.4% of top soil). All experimental pots were flooded with a floodwater layer of *ca* 20 mm on 1 June and rice seedling was transplanted on 14 June. A floodwater layer of *ca* 20 mm was maintained during the rice growth period in all treatments. Methane fluxes were measured during the rice growth period using the method of static chambers with 50x50x100 cm in the interval of 3–10 days[5]. Methane concentration in the gas samples was determined with GC/FID. Soil Eh was simultaneously determined by three Pt electrodes inserted at 5 cm depth when CH_4 flux was measured and averaged soil Eh was shown in Fig.2.

Before the soils were flooded, amorphous Mn (Mn-o) and Fe (Fe-o), organic matter contents, and total N in the soils were analyzed. Amorphous Mn and Fe were extracted with 0.2 mol/L ammonium oxalate (pH 3.2)[6] and determined with AAS.

3. RESULTS AND DISCUSSION

The variations of CH_4 fluxes among the triplicates of a treatment were not very large except when the fluxes were very low (less than 1 mg CH_4 m^{-2} h^{-1}). On the average over the all measurements, the standard deviation coefficients were 58.8%, 75.4%, 31.1% and 39.8% in the treatments DF, WW, FF and AL, respectively, with the maximum of 166.7% occurred in the treatment WW. For avoiding making the figures too messy, the averaged fluxes were shown in Fig. 1. Methane fluxes were significantly different among the treatments. The seasonal variations of CH_4 fluxes from the experimental pots could be clearly divided into two groups. The CH_4 fluxes were substantial on the day of rice transplanting and there were three peak fluxes observed in the treatments FF and AL during the rice growth period. The first one occurred 7 days after rice transplanting, the second in the tillering stage and the third in the early ripening stage. In the treatments WW and DF, the fluxes were very low at the beginning and slowly increased with time, but were always lower than those from the treatments FF and AL during the rice growth period. No distinct peak fluxes occurred in these two treatments throughout the rice growth period. Therefore, the mean CH_4 fluxes were significantly lower for the treatments WW and DF than those for the treatments FF and AL, only 19.6 and 17.4% of that from FF, respectively. There

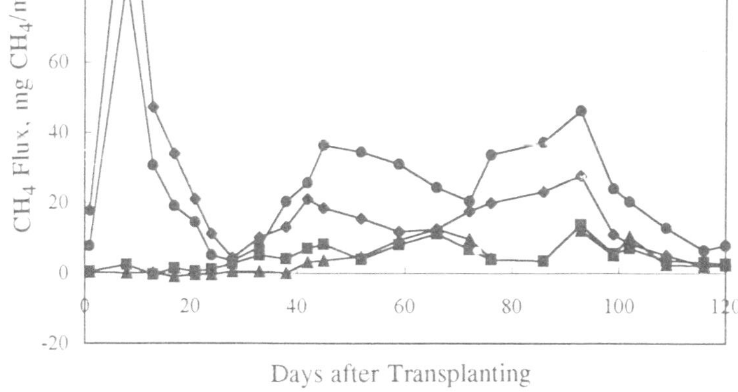

Figure 1. Seasonal variations of CH_4 fluxes from rice pots, which were treated in the winter crop season as drained and fallowed (DF), flooded and fallowed (FF) with *ca* 20 mm floodwater layer, planted with winter wheat (WW) or with alfalfa (AL).

Table 1. The mean CH_4 fluxes from the rice pots, which were treated in the winter crop season as drained and fallowed (DF), flooded and fallowed (FF) with *ca* 20 mm floodwater layer, planted with winter wheat (WW) or with alfalfa (AL)

Treatment	Mean CH_4 fluxes from the individual rice pots ($mg\ CH_4\ m^{-2}\ h^{-1}$)			Average
DF	4.53	5.04	3.98	4.52B
FF	23.85	27.55	26.38	25.93A
WW	4.84	3.34	7.09	5.09B
AL	25.00	23.54	15.79	21.44A

Different letter (A and B) shows that the difference is significant at the 0.01 probability level.

were no significant differences of the mean CH_4 flux between the treatment FF and AL and between the treatment WW and DF (Table 1).

Methane production is caused by methanogenic bacteria, which are active only under very strict anaerobic conditions[7]. Although Asakawa and Hayano found that methanogenic populations able to use H_2-CO_2, methanol, and acetate were almost constant in the soils with a summer rice under flooded conditions and a winter wheat under unflooded conditions during 2-years irrespective of soil moisture regime[8], it is possible that the methanogenic populations in soils may be reduced if the soils are drained and exposed to atmospheric oxygen in winter because methanogenic bacteria are very sensitive to oxygen[7]. Effects of land management in winter crop season on methanogenic bacteria might be one of the reasons why the soil flooded in winter crop season emitted the largest amount of CH_4, but this was not investigated in these experiments. Methane emission from flooded rice fields is the balance of CH_4 production and oxidation. Drainage in winter crop season probably benefits methanotrophic bacteria and stimulates the oxidation of produced CH_4, and in turn further reduces CH_4 emission. But the effect of water management in winter crop season on CH_4 oxidation during the following rice growth period was not investigated in these experiments either.

However, the differences in soil properties related to the soil oxidation-reduction status were noticeable among the treatments due to different management in the winter crop season. Drainage in the winter increased the oxidation potential in the soil. Fig. 2 showed the seasonal variations of soil Eh at 5 cm depth throughout the rice growth period. The soil Eh was below −200 mv 7 days after rice transplanting in the treatments FF and AL. It is well known that CH_4 production occurs when soil suspension Eh is below -150 mv[9,10]. The soil redox potential might be low enough for CH_4 production on the day of rice transplanting (not measured). Afterwards the soil Eh fluctuated between -80 mv and -250 mv during the rice growth period. Wang *et al.* found that CH_4 production increased exponentially with the decrease in soil Eh from −150 to −250 mv[10]. However, for the treatments FF and AL regression analysis showed that there was no significant relationship between simultaneously measured CH_4 flux and soil Eh ($R^2 = 0.0337$ for AL and $R^2 = 0.0878$ for FF using the equation: $y = ax^2 + ax + c$ to simulate experimental data), indicating that soil Eh was not a critical factor controlling CH_4 flux. In the treatments WW and DF, the soil Eh was much higher than those in the treatments FF and AL in the early stage of rice growth. It took approximately 40 day (DF) and 60 days (WW) after the rice transplanting before the soil Eh decreased to below 0 mv. In striking contrast with the treatments FF and AL, there were significant relationships between simultaneously measured CH_4 flux and soil Eh (Figs. 3 and 4), suggesting that high soil Eh was a critical factor constraining CH_4 fluxes from the treatments WW and DF.

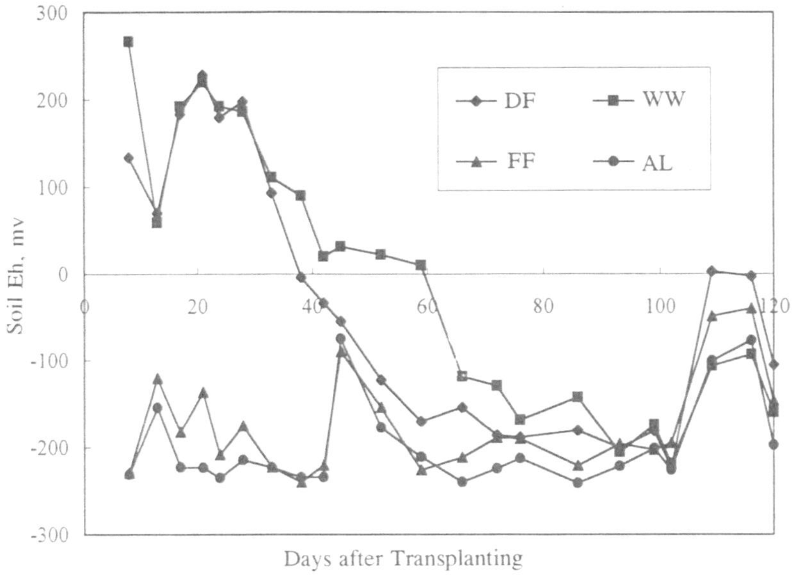

Figure 2. Seasonal variations of soil Eh at the depth of 5 cm in rice pots, which were treated in winter crop season as flooded and fallowed (FF), drained and fallowed (DF), planted with wheat (WW) or with alfalfa (AL).

Soil Eh status depends on the relative concentration of electron donors and receptors. Easily decomposable organic carbon is a main electron donor and NO_3^-, reducible Mn^{4+} and Fe^{3+}, and SO_4^{2-} are electron receptors. Laboratory anaerobic incubation demonstrated that there was a good relationship between CH_4 production and soil organic matter or easily decomposable organic carbon[11]. Added green manure plays a similar role as soil organic matter in flooded soil. Although the soil was unflooded, as in the treatments WW and DF, in the winter crop season, the addition of green manure before flooding provided the soil of the treatment AL with extra electron donors and carbon sources, which made the soil Eh decrease faster after flooding and the seasonal variation of soil Eh was very

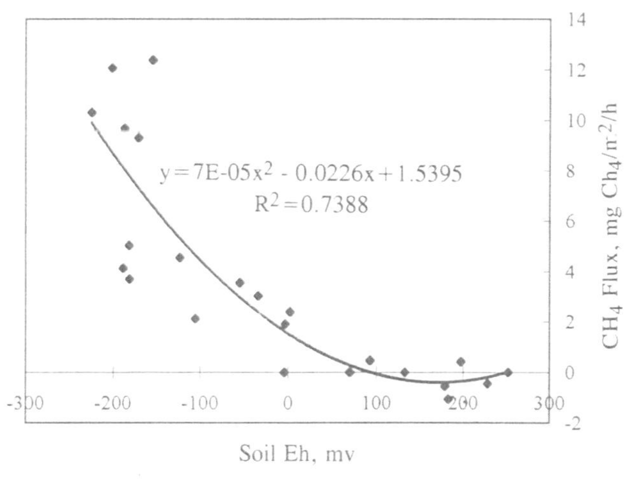

Figure 3. The relationship between simultaneously measured CH_4 flux and soil Eh in the treatment DF.

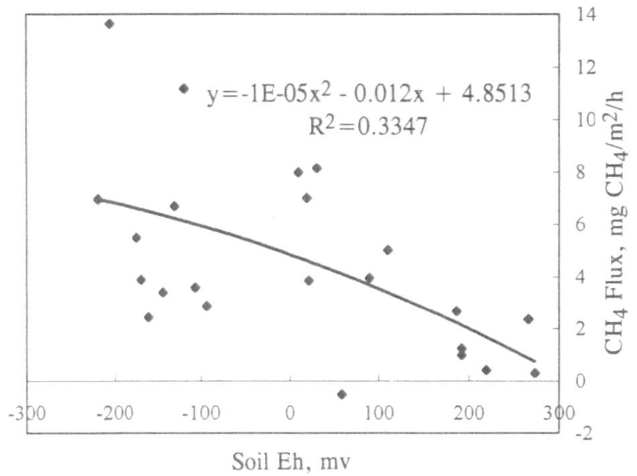

Figure 4. The relationship between simultaneously measured CH_4 flux and soil Eh in the treatment WW.

close to that of the treatment FF (Fig. 2) during the rice growth period. This is probably the main reason that the mean CH_4 flux from the treatment AL was close to that from the treatment FF (Table 1).

In contrast, the increases in the concentration of the electron receptors slow down the decrease in soil Eh after flooding. It has been demonstrated that CH_4 production is inhibited by addition of NO_3^-, Fe^{3+} and SO_4^{2-}[12]. In this experiment, soil organic matter contents, Mn-o and Fe-o were analyzed before flooding. These soil variables were not significantly different among the treatments, but on the average, the content of Mn-o was lowest in FF and highest in DF, while the organic matter content was highest in FF and lowest in DF. There was a very significant negative relationship between the content of Mn-o and the mean CH_4 flux over the rice growth period (R^2 = 0.8150, significant at the 0.01 probability level). The mean CH_4 flux was also very significantly correlated with soil organic matter content (R^2 = 0.8825, significant at the 0.01 probability level). But there was no significant relationship between the content of Fe-o and the mean CH_4 flux.

It has been demonstrated that CH_4 production will not recover immediately if a flooded soil is dried once in a while[11,13]. Based on declines in CH_4 production by artificially drained soils it has been suggested that climatic change may lead to decreased CH_4 production if global warming causes lowered water tables in northern peatland[13,14]. Our study showed that the effect of flooding in winter crop season on CH_4 emission during the following rice growth period was the same as that of green manure incorporation at approximately 3.4% of soil (Table 1). In China, the rice fields flooded in winter account for 8–12% of total rice harvested area, mainly distributed in southwest China, and are the dominant contributor of total CH_4 emission from Chinese rice fields[3]. If the irrigation and drainage facilities in the winter flooding rice fields could be improved substantially and the flooding water could be drained completely in winter, total CH_4 emission from Chinese rice fields would be significantly reduced.

ACKNOWLEDGMENT

The project was supported by Chinese Academy of Sciences and the National Natural Science Foundation of China (grant No. 49771073).

REFERENCES

1. IRRI-International Rice Research Institute, World Rice Statistics 1990. International Institute, Manila, The Philippines. 1991.
2. Khalil, M. A. K., R. A. Rasmussen, M. X. Wang, and L. Ren, Methane emissions from rice fields in China. Environ. Sci. Technol, 1991, *25*: 979–981.
3. Cai, Z. C., A category for estimate of CH_4 emission from rice paddy fields in China. Nutrient Cycling in Agroecosystems, 1997, *49*: 171–179.
4. Wassmann, R., H. Papen, and H. Rennenberg, Methane emission from rice paddies and possible mitigation strategies. Chemosphere, 1993, *26*: 201–217.
5. Yagi, K. and K. Minami, Spatial and temporal variations of methane flux from a rice paddy field. In: Biogeochemistry of Global Change, ed. R. S. Oremland, Chapman & Hall, New York, 1993, pp. 353–368.
6. Chen Jia-fang and He Qun, Chemical distinction of oxide minerals. In: Methods for Soil Colloid Research (in Chinese), ed. Hseung Yi, Science Press, Beijing, 1985, pp. 251–269.
7. Min Hang (ed), Anaerobic Microbiology (in Chinese), Zhejiang University Press, Hangzhou, 1993, p. 274.
8. Asakawa, S. and K. Hayano, Populations of methanogenic bacteria in paddy field soil under double cropping conditions (rice-wheat). Biol. Fertil. Soils, 1995, *20*: 113–117.
9. Masschelegn, P. H., R. D. DeLaune, and W. H. Patrick Jr, Methane and nitrous oxide emission from laboratory measurements of rice soil suspension: effect of soil oxidation-reduction status. Chemosphere, 1993, *26*: 251–260.
10. Wang, Z. P., R. D. DeLaune, P. H. Masschelegn, and W. H. Patrick Jr, Soil redox and pH effects on methane production in a flooded rice soil. Soil Sci. Soc. Am. J., 1993, *57*: 382–385.
11. Crozier, C. R., I. Devai, and R. D. DeLaune, Methane and reduced sulfur gas production by fresh and dried wetland soils. Soil Sci. Soc. Am. J., 1995, *59*: 277–284.
12. Achtnich, C., F. Bak, and R. Conrad, Competition for electron donors among nitrate reducers, ferric reducers, sulfate reducers, and methanogens in anoxic paddy soil Biol. Fertil. Soils, 1995, *19*: 65–72.
13. Freeman, C., M. A. Lock, and B. Reynolds, Fluxes of CO_2, CH_4, and N_2O from a Welsh peatland following simulation of water table drawn-down: potential feedback to climatic change. Biogeochemistry, 1993, *19*: 51–60.
14. Moore, T. R., and R. Knowles, The influence of water table levels on methane and carbon dioxide emissions from peatland soils. Can. J. Soil Sci., 1989, *69*: 33–38.

6

STUDIES ON N_2O EMISSIONS FROM AGRICULTURAL LAND OF RICE-WHEAT ROTATION SYSTEM IN THE TAI-LAKE REGION OF CHINA

Xu Hua and Xing Guangxi

Laboratory of Material Cycling in Pedosphere
Institute of Soil Sciences,
Academia Sinica, Nanjing, 210008, China

ABSTRACT

This paper presents the results obtained from studies on N_2O emissions from rice and wheat fields of Tai-Lake region in 1992 and 1993. Effects of chemical N and P fertilizers, organic manure, water management and a nitrification inhibitor(thiourea) on N_2O fluxes were studied. Total annual N_2O emission from agricultural land of rice-wheat rotation in Tai-Lake region was also estimated. The results showed that application of chemical N, P fertilizers and organic manure greatly increased N_2O emissions from rice or wheat fields. The N_2O emission induced by chemical fertilizer N accounted for 0.19–0.48% and 0.56–0.63% of the total N applied in the rice-growing and wheat-growing season respectively. N in degradable organic compounds induced N_2O-N loss rates were in the range of 0.81–1.90%. Applying a nitrification inhibitor and retaining a layer of water on the field surface reduced N_2O emission significantly from rice fields. Based on our two years' observation data in Wu county of Tai-Lake region, it is calculated that the total annual N_2O emission from agricultural land of rice-wheat rotation in Tai-Lake region was about 8.2×10^9g N_2O-N.

Key words: Rice Field, Wheat Field, Nitrous Oxide, Flux.

INTRODUCTION

The greenhouse effect due to atmospheric nitrous oxide[1] and the destruction of the ozone layer by nitrous oxide[2] have attracted worldwide attention. There are many sources of N_2O, both natural and anthropogenic, which are difficult to quantify. A best estimate of the current anthropogenic emission of N_2O is 3–8 Tg(N)/yr. The atmospheric concentration of N_2O in 1994 was about 312 ppbv, and it was increasing annually by 0.25% in the late 80s and early 90s[3]. Soils have been identified as the major source of N_2O, accounting for 65% of total global emissions[4]. So far the studies on N_2O emission from soil have been focused on upland soils rather than rice paddy soils, and it has been generally thought that rice paddy field emits only small amounts of N_2O while flooded[5–8], however, rice paddy field may have a high potential to emit considerable N_2O to the atmosphere when it is under irrigation and drainage alternative condition[9,10].

China is one of the major agricultural countries in the world. The most important crop rotation in our countries is rice-upland crop rotation in which rice-wheat rotation is the most popular one. So, studies on N_2O emissions from agricultural land of rice-wheat rotation are undoubtedly of great significance for estimating accurately total N_2O emissions from Chinese farmland and finding mitigation options. But unfortunately little has been done in this respect, particularly on N_2O emission from rice field.

Tai-Lake region with several thousands years' history of agricultural cultivation is one of the high-yielding agricultural regions in China, and it is also one of the areas where the largest amounts of chemical fertilizer have been used in recent years. The most popular crop rotation in Tai-Lake region is rice-wheat rotation, with a cultivated area of 1.88 million hectares, about 80% of the region's total cultivated land area. For this reason, we selected two localities in Wu county, located in the center of Tai-Lake region, as our experimental sites where observation and determination of N_2O emission fluxes from rice and wheat fields were conducted in 1992 and 1993.

MATERIALS AND METHODS

1. Soil Properties and Treatments for Rice and Wheat Field Experiments

In 1992 and 1993, measurements of N_2O emission fluxes from rice and wheat fields were conducted in the farm of the agricultural extension station of Wu county or in the farmland of Changqiao village near the agricultural extension station. The soils tested are well permeable paddy soils. Some selected properties of studied soils and experimental treatments of rice and wheat fields are listed in Table 1 and Table 2 respectively. For the rice field experiments in 1992 and 1993, all the pig manure and 40% of urea or ammonium sulphate were used as basal application on June 21, 60% of urea or ammonium sulphate (20% each on June 24, July 14 and August 12 for urea and 20% each on June 28,

Table 1. Selected soil properties of rice and wheat experimental fields

Year	Site	Crop	Total N mg/g	OM mg/g	pH	Texture
1992	A*	Rice	1.97	32.0	6.80	Light clayey
1993	B**	Rice	2.22	38.0	6.30	Light clayey
1992–1993	A	Wheat	1.97	32.0	6.80	Light clayey

*A=The farm of the agricultural extension station of Wu county;

**B=The farmer's field of Changqiao village near the agricultural extension station of Wu county.

Table 2. Experimental treatments of rice and wheat fields

Rice 1992	Rice 1993	Wheat 1992–1993
Control	Control	Control
450 kg/ha urea (N)	1050 kg/ha AM* (N)	900 kg/ha AM (N)
675 kg/ha urea (HN)	N + 15 t/ha pig manure (O)	1350 kg/ha AM (HN)
N + 15t/ha pig manure (O)	N + 4.05 kg/ha thiourea** (H)	N + 2250 kg/ha rice straw (S)
	N + continuously flooded (W)	N + 225kg/ha SP*** (P)

*AM=Ammonium sulphate;
**Thiourea is a kind of nitrification inhibitor, and was mixed with the basal;
***SP=Superphosphate.

July 15 and August 15 for ammonium sulphate) was added as topdressing. Rice seedlings were transplanted on June 21 and harvested on October 28.

As to wheat field experiment in 1992–1993, all the rice straw, superphosphate and 40% of ammonium sulphate were applicated as the basal on Nov. 4 1992, and 60% of ammonium sulphate (17% each on Nov.5, 1992, Dec.20, 1992, and Feb. 15, 1993, 9% on March 20, 1993) was added as topdressing. Wheat seed was sowed on Nov. 5, 1992 and wheat was harvested on May 28, 1993.

2. Water Management of the Experimental Rice Fields

For all the treatments, except treatment W, the water management was the same as that practiced by the local farmers. That is, the soil was covered with a water layer, or at least remained saturated with water, until the tillering of rice plants, then the soil was alternately wetted and dried, and in the mature period the rice field was drained so as to facilitate reaping the rice.

3. Gas Sample Collection and N_2O Measurement

The close chamber method[11] was used to collect the gas sample at an interval of 3–7 days for rice fields and 7–30 days for wheat field. To minimize the effect of diurnal variation, sampling was always carried out between 9.00 a.m. and 11.00 a.m. A square plastic water trough which had been designed to be inserted by the open side of chamber was installed in every plot of the experimental wheat field to make sure that the sample-collecting system was completely airtight. N_2O concentration in sample was analyzed by Gas chromatography (SQ-207) equipped with a ^{63}Ni electron capture detector(ECD). Before the sample entered the GC/ECD, it passed through a tube filled with caustic soda and asbestos to remove the CO_2 and H_2O, then the purified sample entered into a cold trap(liquid nitrogen and pentane being stirred into a glutinous mass), The N_2O became well enriched in the cold trap, and finally it was carried into the chromatographic column through a 6-port valve, being separated and detected. The precision of the N_2O analysis was ±1.40% for repeated analysis of standard gas having N_2O concentration of 300.3 ppbv.

4. Equation for Calculating the Emission flux of N_2O

The equation for calculating the emission flux of N_2O is as follows:

$$F = \rho H \cdot dc/dt \cdot 273/T$$

where F is the emission flux of N_2O, whose unit is $\mu gN_2O\text{-}N/m^2/h$. ρ represents the density of N_2O under the standard condition, the numerical value being 1.25 kg/m^3. H stands for height of the close chamber, whose unit is m. dc/dt is the change in N_2O concentration in the close chamber per unit time, ppbv/h. T denotes the temperature in the close chamber, whose unit is K.

RESULTS AND DISCUSSION

1. Effects of Different Treatments on N_2O Emission fluxes from Rice and Wheat Fields

Three field experiments were conducted in 1992 and 1993 to investigate the effect of different treatments on N_2O emission fluxes from rice and wheat fields. The results obtained are listed in Tables 3, 4, and 5.

As can be seen from the tables, N_2O emissions increased with the application rate of chemical N fertilizer. For example, application of 450 kg/ha and 675 kg/ha urea increased N_2O fluxes by 48.3 and 65.0 percent respectively compared with the control (Table 3). Though applying chemical N fertilizer did increase the N_2O emission, there was no linear correlation between N_2O fluxes and the fertilizer application rate (Table 3 and Table 5). This may be due to high spatial variability of N_2O emissions from agricultural soils and changing plant uptake rates, soil residual N and N loss rates through other processes at different chemical N fertilizer application rates. Spatial variability of N_2O emissions from soils has been recognized by several investigators[10,12,13]. It was reported by Xu et al.[10] that spatial variabilities of N_2O fluxes from two Chinese rice paddy fields located in Nanjing and Fengqiu were in the range of 38.3–96.3%. Duxbery et al. reported that the spatial variability is reduced when fluxes were summed and averaged over important flux periods[13]. It seems that spatial variability of N_2O flux can only partly influence the effect of experimental trentments.

N_2O emissions from rice and wheat fields also increased with organic manure application, but N_2O emission increasing rates were associated with the kind of organic manure used (Tables 3, 4 and 5). 15 t/ha pig manure application greatly increased N_2O emissions from rice fields by 34.5 and 63.6 percent in 1992 and 1993 respectively (Table 3, 4), and applying 2250 kg/ha rice straw only slightly increased N_2O emission from wheat field by 7.7 percent(Table 5), which indicated that the C/N ratio in the organic matter is very important for N_2O emission from agricultural soils. Bremner and Blackmer reported that the

Table 3. Effect of different treatments on N_2O emission fluxes from rice fields in 1992

Treatment	Mean flux ($\mu gN_2O\text{-}N/m^2/h$)	% of the N added
Control	34.6	
450 kg/ha urea (N)	51.3	0.22*
675 kg/ha urea	56.8	0.19*
N + 15 t/ha pig manure	69.0	0.81**

*% of the inorganic N added.
**Organic N induced N_2O-N loss rate.

Table 4. Effect of different treatments on N_2O emission fluxes from rice fields in 1993

Treatment	Mean flux ($\mu gN_2O\text{-}N/m^2/h$)	% of the N added
Control	30.5	
1050 kg/ha AM*(N)	65.6	0.48*
N + 15 t/ha pig manure	107.3	1.90**
N + 4.05 kg/ha thiourea	35.1	
N + continuously flooded	17.0	

*% of the inorganic N added.
**Organic N induced N_2O-N loss rate.

nitrification rate, N_2O/NO_3^- ratio and the N_2O emission rate all increased with decreasing C/N ratio in organic amendments to soil[14].

15 ton pig manure is only equivalent to 67.5 kg/ha organic N at most, assuming 0.45% nitrogen in pig manure (on a wet basis). The great increase in N_2O emissions by application of pig manure with relatively low amount of organic nitrogen indicates that a given amount of N applied in organic fertilizer may generate more N_2O than the same amount of N applied as mineral fertilizer. As shown in Table 3, 4, organic N induced N_2O-N loss rates were 0.81–1.90%, and inorganic N induced N_2O-N loss rates were only 0.19–0.48%. Apparently, the increased N_2O by organic manure application cannot be accounted for only by the increase in organic nitrogen. Duxbury et al.[13] noted that the range of annual emission fluxes of N_2O from a tilled mineral soil and from a cultivated organic soil was 0.49–0.90 kgN_2O-N/ha and 7–165 kgN_2O-N/ha, respectively. Thus, the N_2O emission flux from the organic soil is far higher than that from the mineral soil. Organic soil and soils heavily fertilized with organic manure contain more soil organic matter, thereby increasing the N source for nitrification and denitrification in the soil and supplying adequate nutrition and energy to soil microorganisms involved in nitrification and denitrification. This may be one of the major reasons for the greatly increased N_2O flux by applying organic manure. Organic manure application can lead to production of more N_2O than application of the same amount mineral fertilizer N. Therefore, a good deal of attention should be given to what contribution the nitrogen of different sources will make to total N_2O emission from agricultural soils for estimating accurately global N_2O emission and finding mitigation options.

Table 5 shows that phosphate application can also increase N_2O emission from wheat field. N_2O emission from the plot receiving normal chemical N fertilizer plus superphosphate was 25.1 percent higher than that from the plot just applied with routine chemical N fertilizer,

Table 5. Effect of different treatments on N_2O emission fluxes from wheat fields in 1992–1993

Treatment	Mean flux ($\mu gN_2O\text{-}N/m^2/h$)	% of the N added
Control	29.8	
900 kg/ha AM(N)	51.8	0.56
1350 kg/ha AM	67.0	0.63
N + 2250 kg/ha rice straw	64.8	
N + 225 kg/ha SP	55.8	

which is consistent with the results reported by Minami and Fukushi[15]. Phosphate application can enhance nitrite accumulation in soil. Nitrite ions react chemically with organic molecules forming nitroso-groups (-N=O) which are unstable. Gaseous products (N_2, N_2O) can be formed from this group. Furthermore, phosphate creates favorable conditions for both enzyme activity responsible for N_2O production and the nitrifying population[15].

Application of a nitrification inhibitor can effectively reduce the N_2O emission flux from rice fields. In the treatment of 1050 kg/ha ammonium sulphate plus nitrification inhibitor, the N_2O emission flux was 35.1 $\mu gN_2O\text{-}N/m^2/h$, which is only slightly more than half compared with the treatment applying only the same amount of nitrogen fertilizer and is close to the result of control (Table 4). This is a further verification that application of nitrification inhibitor can effectively reduce the N_2O emission from soils[16]. Results in Table 4 also show that continuously flooding the rice field during the rice-growing season strongly reduced the N_2O emission flux. When the soil is submerged continuously with a water layer, nitrification proceeds slowly, while denitrification proceeds increasingly towards N_2, and N_2O diffusion is severely hindered by the water layer. Therefore, N_2O flux is very low.

Total N_2O emission during a rice-growing season accounted for 0.19–0.48% of the total amount of chemical N applied in the season (Table 3, 4). The figure is consistent with the results (0.33–0.55%) obtained by Minami[9], but higher than Smith's[5]results, 0.01–0.05%, indicating that the factors controlling N_2O emission fluxes from rice fields are rather complicated. Ammonium sulphate induced N_2O-N loss rate from wheat field were 0.56–0.63% which is only slightly higher than that obtained from the rice fields (Tables 4 and 5). So, not only upland soil but also rice paddy soil can contribute a great deal to atmospheric N_2O.

2. Temporal Variations of N_2O Emission fluxes from Rice and Wheat Fields

2.1. Temporal Variations of N_2O Emission fluxes from Rice Fields. Fig. 1 shows temporal variations of N_2O fluxes from differently treated plots during the rice-growing season in 1993. Factors controlling the N_2O emission flux are numerous and varied, but the occurrence of N_2O emission peaks was closely associated with fertilization and soil moisture condition. Altogether, gas samples were collected 26 times during the period of 129 days from rice transplanting to its harvest. Collection of gas samples was regularly done between 9.00 a.m. and 11.00 a.m. of every sampling date, so as to avoid the influence of diurnal variations on temporal variations of N_2O emission. As can be seen from Fig. 1, there appeared 4, 5 or 6 N_2O emission peaks of different heights in all the treatments during the period of rice growth. The first three peaks, corresponding mainly to fertilizer application of 4 times in the earlier stage of rice growth, occured about one week after fertilizer application. However, the last 1–3 N_2O emission peaks were influenced dominantly by soil water management and had little to do with fertilizer application. During the period when the last 1–3 peaks occured, the soils were drained or alternately irrigated and drained. The soil water content was close to field capacity. At such a soil water content, both nitrification and denitrification can proceed, and the proportion of N_2O in reaction products is high. Smith et al.[5] observed an N_2O emission peak in the late rice-growing season when the rice field was drained, which is in line with our results.

2.2. Temporal Variations of N_2O Emission fluxes from Wheat Fields. Fig. 2 shows temporal variations of N_2O fluxes from differently treated plots during the wheat-growing

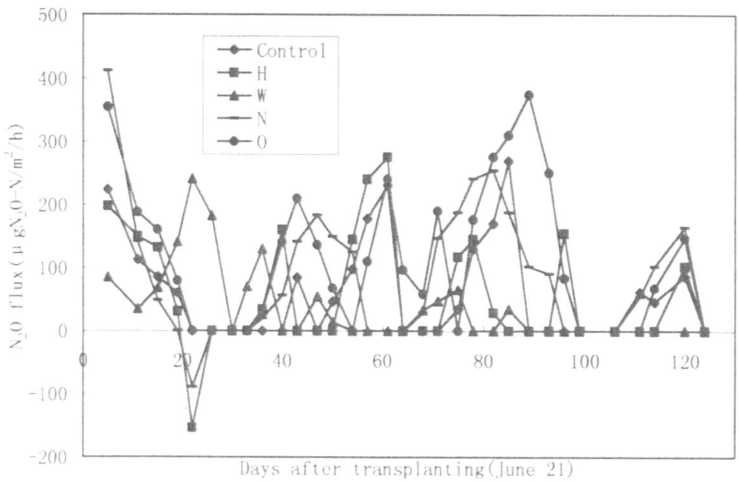

Figure 1. Temporal variations of N_2O emission fluxes from rice fields of different treatments in 1993.

season in 1992–1993. Gas samples were collected 19 times during the period of 204 days from wheat seed sowing to wheat harvest. We also collected gas sample one time after wheat harvest. Sample-collecting time was fixed at 9.00 a.m.-11.00 a.m.of every sampling date to minimize the influence of diurnal variations on temporal variations of N_2O fluxes. There are 5 N_2O emission peaks which occured during wheat-growing season for all the experimental plots. Fertilizer N application may be the major reason for the first 4 N_2O emission peaks, although the third N_2O emission peak may also be due to physical release of N_2O trapped in frozen soil when the topsoil thawed in early spring. Bremner et al.[17] hypothesized that N_2O was being produced in the winter in the unfrozen subsoil and released as the soil thaws. The last N_2O emission peak occured in early May when soil temperature and rainfall were relatively high. N_2O production in soil is favored by high temperature

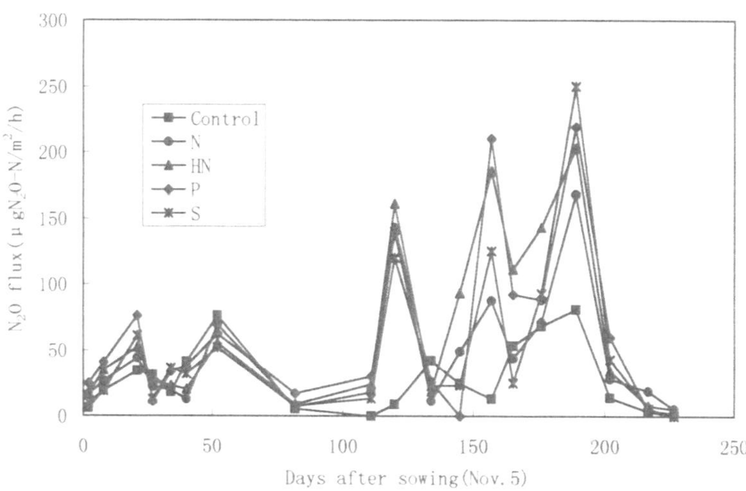

Figure 2. Temporal variations of N_2O emission fluxes from wheat fields of different treatments in 1992–1993.

and water availability to some extent if there are not other limiting factors for N_2O formation. N_2O emission decreased with decreasing soil nitrogen availability in the very late stage of wheat-growing season, even though soil temperature at that time was also relatively high. There were almost no N_2O emissions after wheat harvest, indicating that soil nitrogen availability is one of the most important factors influencing N_2O production and emission from agricultural soil.

3. Preliminary Estimation of Annual N_2O Emission from Agricultural Land of Rice-Wheat Rotation in Tai-Lake Region

As described above, this experiment was conducted in Wu County which is located in the center of Tai-Lake region. Our experiment included the common fertilizer rate (treatments N), and tillage, irrigation and other field management were comparable to that practiced by the local farmers. Based on data obtained through the observations in two rice-growing seasons and one wheat-growing season, we have made a preliminary estimation of annual N_2O emission from agricultural land of rice-wheat rotation in Tai-Lake region. The total area of agricultural land with rice-wheat rotation in Tai-Lake region is about 1.88 million hectares. The mean N_2O emission fluxes from rice and wheat fields with the normal treatment which local farmers practice were 58.5 and 51.8 $\mu gN_2O\text{-}N/m^2/h$ respectively. The rice-growing and wheat-growing periods were 129 and 204 days respectively. So, annual N_2O emission from agricultural soil of rice-wheat rotation in Tai-Lake region was $8.2\times10^9 gN_2O\text{-}N$.

It is obvious that there were uncertainties in mean N_2O fluxes of rice-growing or wheat-growing season and annual N_2O emissions due to diurnal, spatial and temporal variations of N_2O fluxes from agricultural soils. The diurnal variation in N_2O flux poses the problem of establishing a representative sampling time.The peak N_2O flux usually occurred in the early afternoon, and the lowest N_2O flux appeared in the early morning. The most reliable estimate of the N loss as N_2O is obtained using samples collected during the mid-morning or mid to late afternoon[12,18]. Sampling in our experiment was always carried out in mid-morning(between 9.00 a.m. and 11.00 a.m.). So, uncertainties in mean N_2O fluxes of rice-growing or wheat-growing season and annual N_2O emissions due to diurnal variation have been minimized. However, further experiment with replication and higher sampling frequency should be done to estimate uncertainties resulting from spatial variability and reduce uncertainties due to temporal variation.

REFERENCES

1. Yung, Y.L., Wang, W.C. and Lass, A. A., Greenhouse effect due to atmospheric nitrous oxide, Geophys. *Res. Lett.* 1976, *3*: 619–621.
2. Crutzen, P. J., Atmospheric chemical processes of the oxide of nitrogen, including nitrous oxide in C. C. Delwidhe(ed.) *Denitrification, Nitrification and Atmospheric Nitrous Oxide*, Wiley and Sons, New York, 1981, P: 17–44.
3. Houghton, J.T., L.G. Meira Filho, B.A. Callander, N. Harris, A. Kattenberg and K. Maskell. *Climate Change 1995. The Science of Climate Change*, Cambridge University Press, 1996, p. 8.
4. Prather M. Derwent R. Ehhalt D, Fraser P. Sanhueza E, Zhou X., Other trace gases and atmospheric chemistry, In: Houghton JT. et al.(eds) *Climate change 1994; radiative forcing of climate change and an evaluation of the IPCC IS92 Emission Scenarios*, Cambridge University Press, Cambridge, 1995.pp 77–126.
5. Smith, C. J., M. Brandon, and W. H. Patrick, Jr., Nitrous oxide emission following urea-N fertilization of wetland rice, *Soil Sci. Plant Nutr.*, 1982, *28*: 161–171.

6. De Datta, S.K., R.J. Buresh, M.I. Samson, W.N. Obcemea and J.M. Real, Direat measurement of ammonia and denitrification fluxes from urea applied to rice, *Soil Sci. Soc. A.J.*, 1991, *55*:543–548.
7. Buresh, R.J., S.K. De Datta, M.I. Samson, S. Phongpan, P. Snitwongse, A.M. Fagi and R. Tejasarwana, Dinitrogen and nitrous oxide flux from urea basally applied to puddled rice soils, *Soil Sci. Soc. A.J.*, 1991, *55*:268–273.
8. Freney, J.R. and O.T. Denmead, Factors controlling ammonia and nitrous oxide emissions from flooded rice fields, *Ecol. Bull.*, 1992, *42*:188–194.
9. Minami, K., Emission of nitrous oxide from agro-ecosystem. *JARQ, Japan Agricultural Research Quarterly*, 1987, *21*: 21–27.
10. Xu Hua, Xing Guangxi, Zu-Cong Cai and Haruo Tsuruta, Nitrous oxide emissions from three rice paddy fields in China, *Nutrient Cycling in Agroecosystems*, 1997, *49*:23–28.
11. Yagi, K. and Minami, K., Emission and production of methane in the paddy fields of Japan. *JARQ*, 1991, *25*: 165–171.
12. Ryden, J.C., L.J. Lund and D.D. Focht, Direct in-field measurement of nitrous oxide flux from soils, *Soil Sci. Soc. A.J.*, 1978, *42*:731–737.
13. Duxbury, J. M., D. R. Bouldin, R. E. Terry, and R. L. Tate, Emission of nitrous oxide from soil. *Nature*, 1982, *298*: 462–464.
14. Bremner, J. M. and A. M. Blackmer, Terrestrial nitrification as a source of atmospheric nitrous oxide, in C. C. Delwiche(ed.), *Denitrification, nitrification and atmospheric N_2O*. John Wiley & Sons Ltd., Chichester, 1981, pp. 151–170
15. Minami, K. and S. Fukushi, Effect of phosphate and calcium carbonate appication on emission of N_2O from soil under aerobic conditions, *Soil Science and Plant Nutrition*, 1983, *29*: 517–524.
16. Bremner, J. M., G. A. Breitenbeck, and A. M. Blackmer, Effect of nitrapyrin on emission of nitrous oxide from soil fertilized with anhydrous ammonia, *Geophys. Res. Lett.*, 1981, *8*: 353–356.
17. Bremner, J. M., Robbins, S. G. and A. M. Blackmer, Seasonal variability in emission of nitrous oxide from soil. *Geophys. Res. Lett.*, 1980, *7*: 641–644.
18. Denmead, O.T., J.R. Freney and J.R. Simpson, Nitrous oxide emission from a grass sward, *Soil Sci. Soc. A.J.*, 1979, *43*:726–728.

7

ATMOSPHERIC DEPOSITION MEASUREMENTS IN NORTHERN POLAND

K. B. Mędrzycka[1], O. Westling,[2] and S. Strzałkowska[1]

[1]Technical University of Gdańsk
Chemical Faculty
Gdańsk, Poland
[2]IVL, Swedish Environmental Research Institute
Aneboda, Sweden

INTRODUCTION

Acid depositions have caused severe damage to the forest ecosystems of Poland, due to the direct effects on the vegetation and changes to the forest soil. The development of these forested areas depends on the emission of air pollutants, their transport in the atmosphere, and their deposition on vegetation and soil.

After transport, the pollutants are deposited by wet, dry or fog deposition.

Wet deposition is the process by which the atmospheric pollutants (earlier attached to and dissolved in cloud and precipitation droplets or particles) are delivered to the earth's surface as rain, hail or snow. Dry deposition is the process where gases and particles are deposited directly from the atmosphere onto soil, vegetation and other surfaces. When cloud or fog water droplets are directly intercepted by the receptor surface, then we have cloud or fog deposition.

Together with wet deposition, particle dry deposition is responsible for the atmospheric load to ecosystems of such pollutants as sulphate, nitrate, chloride and ammonium. However, it has been found, that the dry deposition contribution to the potential acidification of the ecosystem is relatively small[1].

Measurements of pollutant deposition may be made using different methods. Wet deposition is measured by collecting the precipitation in the samplers placed in open fields. There are two types of samplers: bulk samplers and wet only samplers. A bulk sampler has no provision to exclude dry deposition during dry periods, and this may influence the chemical composition of precipitation passing through the funnel contaminated during a dry period. Wet-only samplers are closed during the dry periods and the funnels are opened only during atmospheric precipitation events.

Chemistry for the Protection of the Environment 3, edited by Pawłowski *et al.* Plenum Press, New York, 1998

Dry deposition can be measured directly (measuring the deposition at the surface itself) or indirectly (surface accumulation methods). Direct methods comprise the measurement of deposition to natural surfaces using the throughfall method or to surrogate surfaces e.g. dustfall buckets, petri dishes etc. Throughfall is the most widely used method.

Throughfall refers to anything from the water dripping from canopies (leaves, needles, branches) and falling through gaps in the canopy. Stemflow is the part of precipitation running down tree trunks. The concentration of the ions is usually higher in stemflow than in throughfall; however in the case of a mature spruce forest the stemflow contribution is small. Thus, in the case when the monitoring station is located in a mature spruce forest, the stemflow is not measured[2].

However, the throughfall method has several weak points, among which the most important are canopy exchange processes (diffusion and ion exchange between the surface water and the internal parts of plants)[3]. Schaefer and Reiners have found that diffusion is the major cause of elevated anions concentration in throughfall[4]. The rate of canopy exchange depends on tree species. For example, deciduous trees tend to lose more nutrients from crown foliage through leaching than coniferous tree species. However, the last ones lose nutrients throughout the year as they stay green all year[5]. The presence of certain pollutants may enhance the canopy leaching. For example ozone increases the permeability of cell membranes thus increasing ion leakage from foliage[6].

As it has been stated[1,2], the throughfall is generally a good method for the estimation of sulphur, sodium and chloride deposition and is less suitable for the estimation of nitrogen compounds of atmospheric input. The reason is that the deposited inorganic nitrogen retained in the canopy may be converted into organic compounds due to biological interaction.

Another weak point in using the throughfall method is the large spatial variability of throughfall fluxes within forest stands. Besides, the deposition of contaminants such as insects, bird excrements, etc. onto the measurement devices may obscure the measured concentrations.

AIM OF THE STUDY

The investigations described in this paper were started in the autumn of 1994 as a result of the cooperation between the Blekinge Air Quality Association (Blekinge Luftvardsforbund) in Sweden and the Chemical Faculty of the Technical University of Gdańsk. The idea was to investigate the concentration of atmospheric pollutants in the southern part of the Baltic coast. It was found earlier[2] that in case of certain ions, their deposition in Sweden exhibited a pattern of increase from north to south. So, one may expect that the highest concentration in southern Sweden resulted partly from pollutant transfer from the south. Thus, the northern part of Poland has been included in the monitoring net covering all Scandinavian countries.

The aim of this study was to evaluate the content of the ions responsible for strong acid rains to a forest site, far from inhabitated and industrialized areas.

The pH and the concentration of the following ions were determined in throughfall and bulk precipitations: SO_4^{-2}, NO_3^-, Cl^-, NH_4^+.

The results from the station in northern Poland were compared to the results from precipitation and throughfall in Sweden.

SAMPLING PROCEDURE

The monitoring of the depositions was carried out in the field station PL1, which was sited inside a mature spruce (Picea abis) forest, 15 km from Gdynia (Fig.1). The sampling of rain and snow took place under the canopy of the spruce stand (throughfall, PL11) and in an open field (bulk wet, PL10) nearby. Sampling was carried out according to the procedure applied in the EC monitoring system in permanent forest plots.

For throughfall sampling the summer equipment consisted of a funnel (15.5 cm diameter) placed on a 2 litre polyethylene bottle (10 collectors). The funnels were equipped with a net to prevent litter fall entering the sample bottle. A polyethylene plastic bag was inserted into the bottle to collect the sample. The bottles were placed on a wooden pole with the bottle openings 0.8 m above the ground.

The equipment for open field (bulk wet) consisted of a funnel (20.3 cm diameter) mounted on 5 litre polyethylene bottle, placed on a wooden pole with the opening approximately 1.5 m above the ground. The winter equipment was designed to collect snow, and was used from January to March. The equipment consisted of a 5 litre bucket mounted on the same pole as the summer equipment.

In order to compare the results of atmospheric pollution from an afforested area with the effects of an inhabitated area, the samples were collected also on the roof of the Technical University building, which is located in an urbanized area of Gdańsk (PG station). The third sampling station was located in a Gdańsk suburb (green gardens, allotment area, GS station). These two last stations were in operation from October 1994 until January 1996. The PL1 station is still operating.

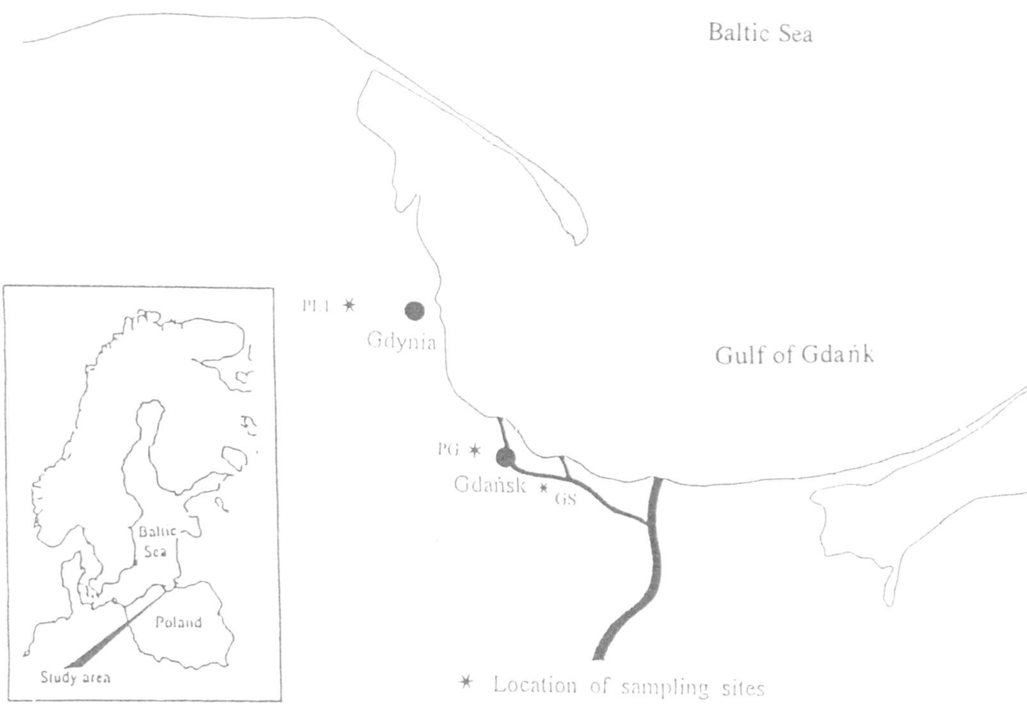

Figure 1. Location of the study area with the sampling station.

RESULTS

Spectrophotometric methods were used for the determination of particular ion concentrations. The preconcentration of a sample was needed for analysis of sulphate ion. The particular analysis was performed using the equipment offered by Merck, Palintest and Hach. The analytical procedures were suited to our needs[7].

The deposition of particular ions was calculated by multiplying their concentration by the amount of water precipitation (PL10) or by throughfall water (PL11).

The results are presented in the diagrams (Figs 2–9). Some of the diagrams cover the whole reported period, some only the example from one year.

Fig. 2 presents the diagram for 94/95 precipitation results. A large variation may be observed for different months and it is difficult to find any regularity. However, the maximum appeared during the spring and autumn seasons.

Fig.3 presents pH values for bulk wet (PL10) and throughfall (PL11) in samples collected from 20.10.94 till 01.10.95. Generally, it has been found that the pH in throughfall varies from 3.4 to 6.4 and usually is lower than pH in bulk wet (4.4–7.1). The highest values were observed during the hot season, due to the less emission and due to possible neutralization by atmospheric dust. Similar results were received during the following two years.

Considering acidifying pollutants, the most important results are for sulfur and nitrogen deposition (Figs 4–9).

It has been found in our investigations that the concentration of strong acids as well as of the other components is dependent on the amount of precipitation. There has not been observed a seasonal regularity in the precipitation amount, thus the deposition of pollutants does not show any dependence related to the season of the year. For example, in the 94/95 period the highest sulfur deposition (~2–2.5 kgS/ha) was observed during the winter months (Fig. 4), while in 96/97 the highest deposition values were reached during the spring months (2.5–3.8 kgS/ha, Fig. 6).

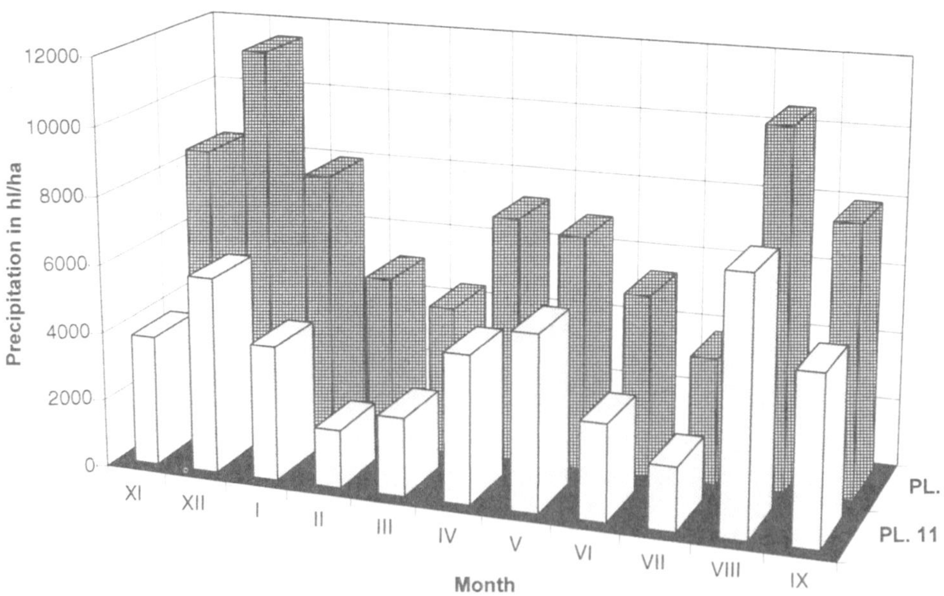

Figure 2. Precipitation results from 1994/95.

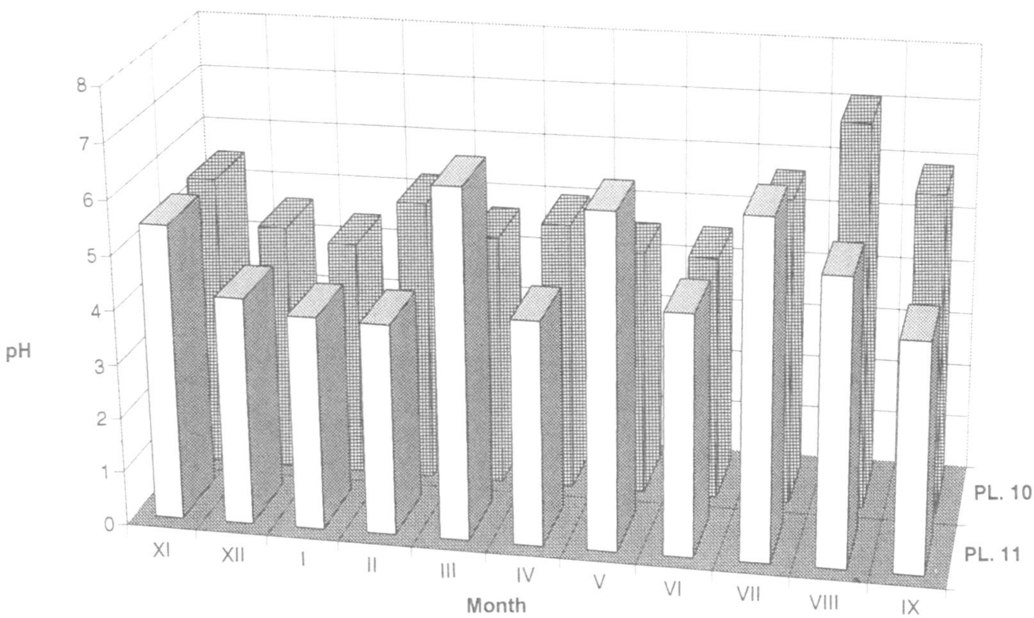

Figure 3. pH values for the samples collected during 1994/95.

Nitrates monthly deposition was usually below 1 kgN/ha, however, in some cases it was higher, especially during the summer of 96 (2.5–4 kgN/ha). We did not observe, in our investigations, the constant ratio sulphate/nitrate, as it was found by Barbolani[8]. In the case of our station, it is possible that the origin of sulphate and chloride is partly marine; however, it is difficult to evaluate its share on the basis of our investigations.

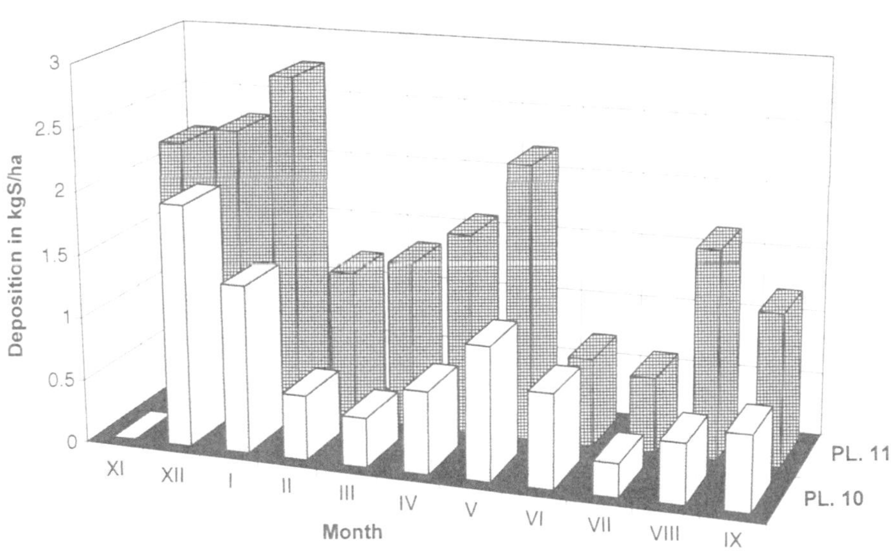

Figure 4. Sulfur deposition from October 94 to September 95.

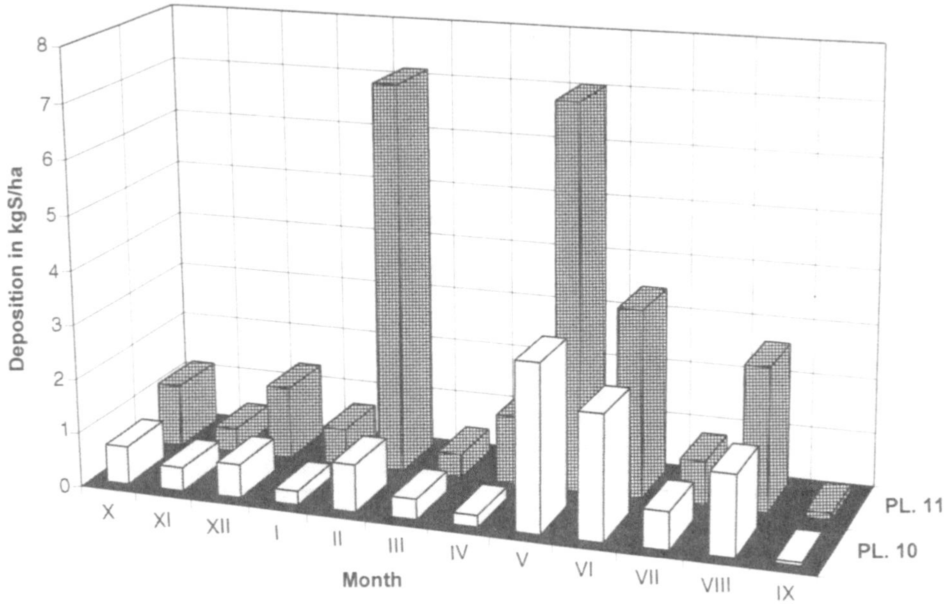

Figure 5. Sulfur deposition from October 95 to September 96.

In Table 1 are presented the annual depositions for all stations. The results relate to the period from 20.10.94 till 01.10. 95.

One may observe that the deposition of all ions is smaller in the open field (PL10) than the deposition in Gdańsk–city, an inhabitated and industrialized area (PG station). Simultaneously, it was found that at PL11 station, the results are the highest, even higher than in the urban area (except the NH_4-N deposition, which is highest in the city). This re-

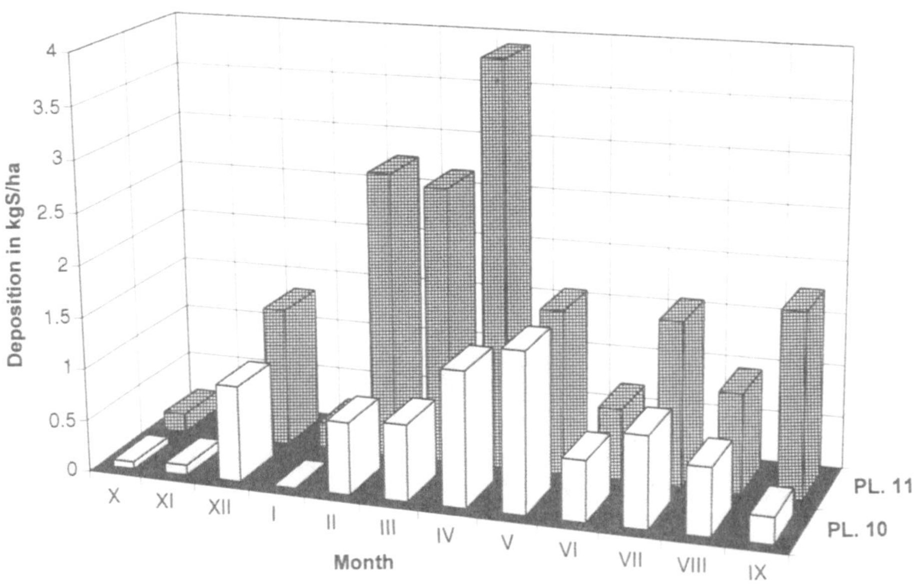

Figure 6. Sulfur deposition from October 96 to September 97.

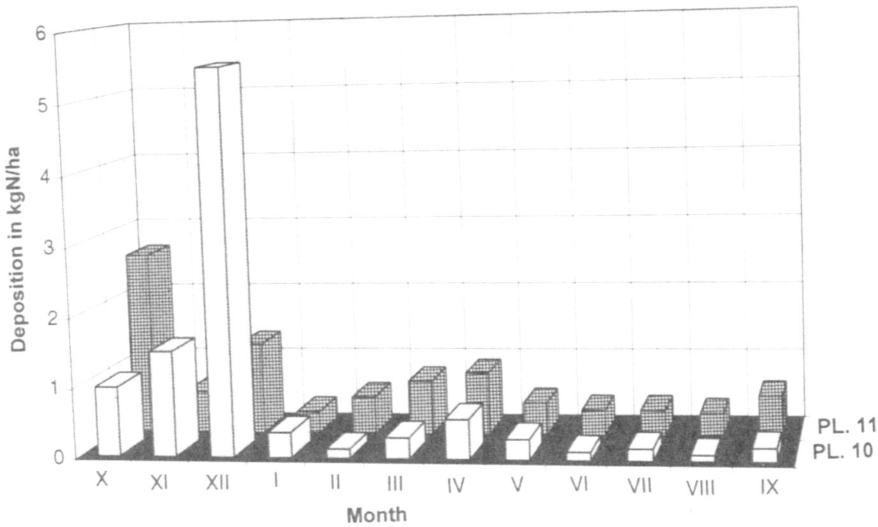

Figure 7. Nitrates deposition from October 96 to September 97.

sults from the leaching from needles and branches and from biological interaction. The least polluted area appeared to be the "small gardens" (allotment district), located to the East from Gdańsk (GS station). The concentration of Cl^- ions higher that are in this area than in PL10 station, may result from the closer distance to the Baltic Sea coast, which is shorter than for PL10 station.

In Table 2, the results of annual deposition from 20.10.94. till 04.10.97 are presented for PL10 and PL11 stations.

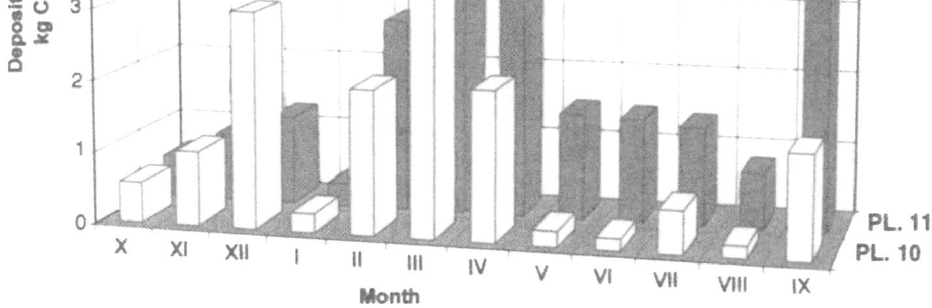

Figure 8. Chlorides deposition from October 96 to September 97.

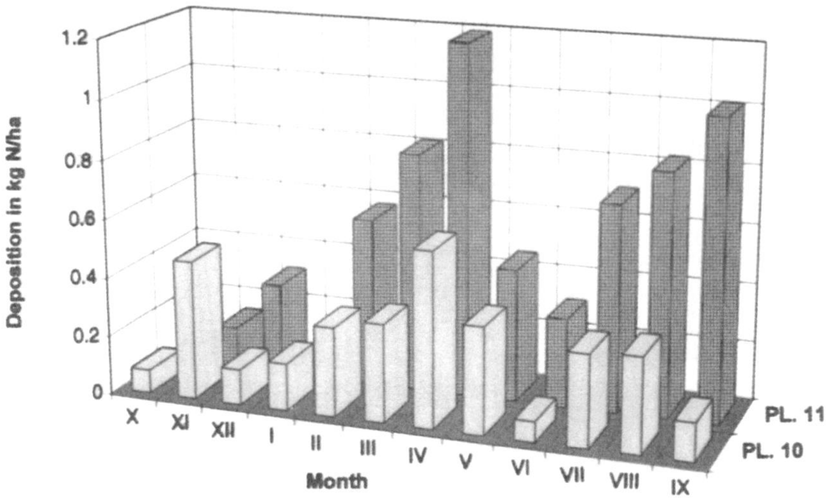

Figure 9. Ammonia nitrogen deposition from October 96 to September 97.

DISCUSSION

Poland belongs to a region dominated by large old industrial complexes, which are the main source of emission to the atmosphere. It has been presented[1] that the wet deposition of sulphate is large in eastern and central Europe and for Poland, the annual fluxes of non-marine sulphate ions range between 200 and 400 mol/ha (eastern Poland), or between 400 and 700 mol/ha (western Poland). These relate to 6–13 kg of sulfur per hectare or

Table 1. Annual deposition at different sampling stations

Parameter	Sampling station PL10	PL11	PG	GS
SO_4^{-2} [kgS/ha]	7.91	17.72	12.25	6.68
Cl^- [kgCl/ha]	8.91	25.45	13.19	13.15
NO_3^- [kgN/ha]	2.46	6.11	2.91	2.32
NH_4^+ [kgN/ha]	3.65	4.57	6.07	2.08

Table 2. Deposition results for PL10 and PL11 sampling stations

	20.10.94–01.10.95 346 days		02.10.95–05.10.96 370 days		06.10.96–04.10.97 364 days	
Parameter	PL10	PL11	PL10	PL11	PL10	PL11
Precipitation [hl/ha]	78973	43992	71080	48629	66684	35302
SO_4^{-2} [kgS/ha]	7.91	17.72	10.95	26.48	7.37	16.28
Cl^- [kgCl/ha]	8.91	25.47	8.55	14.82	17.12	22.98
NO_3^- [kgN/ha]	2.46	6.11	13.88	13.42	10.18	9.13
NH_4^+ [kgN/ha]	3.65	4.57	6.59	7.30	3.24	6.71

13–21 kgS/ha, respectively. The annual average dry deposition of sulphates is evaluated at 200–1000 mol/ha (6–32 kgS/ha) and as it was shown the dry deposition in areas with forested terrain increases relative to the values calculated according to the long-range transport model concentration (EMEP model)[1]. The dry deposition variation is determined by the characteristic of the terrain, land use and meteorological parameters (wind speed, radiation, temperature, humidity, precipitation).

Considering nitrates deposition, the fluxes are evaluated at 200–400 mol/ha (2.8–5.6 kgN/ha) in central and eastern Poland and 400–700 mol/ha (5.6–9.8 kgN/ha) in western Poland[1]. In Poland these emissions arise mainly from industry, power stations and motor vehicle exhausts. In southern Scandinavia the emission of nitrates is evaluated at 300–700 mol/ha (4–9 kgN/ha), and the emission of sulphates is about 200–400 mol/ha (6–21 kgS/ha)[1]. Thus, similar levels of both emissions have been presented for southern Scandinavia and for northern Poland.

Ammonia emissions, which arise mainly from livestock waste, may be to a large extent in areas with intensive agricultural land use. For Poland it is evaluated at 200–400 mol/ha (eastern Poland) and 400–700 mol/ha (western Poland)[1]. These relate to 3.6–7.2 kgN/ha or 7.2–12.5 kgN/ha.

Our results are generally similar to those published by Erisman[1]. The annual wet deposition of sulphur in Polish station PL10 is 7.4–11 kgS/ha (Table 2), while in southern Sweden it is about 6–8 kgS/ha[3, 9]. Sulphur deposition in throughfall in the Gdańsk region is about 16–26 kgS/ha (table 2), and in Sweden it is about12–20 kgS/ha[9, 10].

Nitrates deposition in PL10 station varies from 2.5 to 14 kgN/ha and in PL11 station (throughfall) it varies from 6.1 to 13.5 kgN/ha (Table 2). Simultaneously, the deposition of ammonia nitrogen was 2.8–6.6 kgN/ha in the open field and 4.5–7.3 kgN/ha in throughfall. Thus, the cumulative nitrogen deposition in the Polish field station (nitrate + ammonia nitrogen), varies from 6 to 20 kgN/ha in the open field and from 11 to 21 kgN/ha in throughfall (Table 2). In southern Sweden the total nitrogen deposition in 1993 was about 10 kgN/ha in the open field and about 20 kgN/ha in throughfall[9].

Thus, one may conclude that in the northern part of Poland, the atmospheric deposition of nitrogen is similar to that in southern Sweden, while the deposition of sulphur is higher in Poland than in Sweden, even in afforested, non-inhabitated areas.

On the basis of critical loads for different areas[11], it has been stated that in the northern part of Poland, where the PL1 station is located, the critical deposition of sulphur should be 500–1000 eq/ha (16–32 kgS/ha), and the critical deposition of nitrogen is also 500–1000 eq/ha (7–14 kgN/ha). It has been also reported[11] that in reality, in this region the critical load of sulphur is exceeded by 200–500 eq/ha (6.4–16 kgS/ha) and the critical load of nitrogen is exceeded by 0–200 eq/ha (0–1.4 kg/ha). In order to improve the state of the environment it is necessary to minimize emissions born in the industrialized and urbanized areas to the atmosphere, because the pollutants are usually transported for a long distance from the source and may be deposited on the ground as well as in the Baltic, increasing their contaminant loading.

ACKNOWLEDGMENT

The authors wish to thank to Blekinge Air Quality Association for its financial support of the investigations and personally to Bengt Norman and Lars Bengtson for their personal engagement in this cooperation.

REFERENCES

1. J.W.Erisman, G.P.J.Draaijers, Atmospheric Deposition in Relation to Acidification and Eutrophication, Elsevier, Amsterdam, 1995.
2. G.Lovblad, M.Hovmand, A.Reissel, O.Westling, D.Aamlid, A.Hyvarinen, J.Schaug, Throughfall Monitoring in the Nordic Countries, IVL-Report, Gothenburg, 1994.
3. O. Westling, H.Hultberg, G.Malm, Total Deposition and Tree Canopy Internal Circulation of nutrients in a Strong Acid Deposition Gradient in Sweden, as Reflected by Throughfall Fluxes. In:L.O.Nilsson, R.F.Huttl, U.T.Johansson (eds), Nutrient Uptake and Cycling in Forest Ecosystems, Kluwer Academic Publishers, Netherlands, 1995.
4. D.A.Schaefer, W.A.Reiners, Throughfall Chemistry and Canopy processing mechanisms. In:S.E.Lindberg, A.L.Page, S.a.Norton (Ed.), Acid precipitation, Vol.3: sources deposition and canopy interactions, Springer Verlag, 1990.
5. W.H.Smith, Forest Vegetation as a Sink for Gaseous contaminant, Springer Verlag, New York, 1981.
6. L.S.Evans, I.P.Ting, Ozone-induced Membrane Permeability Changes, Am.J.Botany, 60, 115 (1973).
7. K.B.Medrzycka, L.Dworznik, B.Urban, P.Konieczka, O.Westling, V Polish Conference of Analytical Chemistry, Gdańsk, 1995, Proc.p.560.
8. E Barbolani, S.Del Panta, R.Udisti, F.Pantani, Chemical composition of atmospheric precipitation in the Tuscan Appennines, Annali di Chimica, 78, 405(1988).
9. E.H.Larsson, O.Westling, Luftfororeningar i sodra Sverige. IVL - raport, Aneboda, 1994.
10. W.Ivens, G.Lovblad, O.Westling, P Kauppi, Throughfall monitoring as a means of monitoring deposition to forest ecosystem. Evaluation of European data, Nordic Council of Ministers, Nord 1990, Copenhagen.
11. R.J.Downing, J.-P.Hettelingh, P.A.M.deSmet, Calculation and Mapping of Critical Loads in Europe. Status Report 1993. RIVM Report No 259101003, Bilthoven, the Netherlands.

CONTROL OF VOLATILE ORGANICS EMISSION TO THE ATMOSPHERE DURING THE SOLVENT SUBLATION PROCESS

Krystyna Mędrzycka, and Sebastian Pastewski

Chemical Faculty
Technical University of Gdańsk
G. Narutowicza 11/12, 80-952 Gdańsk, Poland

ABSTRACT

The solvent sublation technique has been applied for the removal of mesitylene from aqueous emulsion. The aim of the investigations was to determine the method of mesitylene emission control during air stripping or solvent sublation processes. It has been found that compared to air stripping, solvent sublation reduces to a great extent the emission of mesitylene to the atmosphere (91–95%). The residual emission (about 5 to 9%) could be minimized by the continuous flow of the organic layer, or totally avoided by carbon adsorption on a charcoal trap.

INTRODUCTION

Solvent sublation, the process originated by Sebba[1] for the removal of ionic-surfactant complexes, was later applied for removing a large number of hydrophobic molecular compounds[2–8] and ion-pair complexes[9–11]. The substances in water are transported to the organic solvent placed on the top of the aqueous head. The transport medium is gas bubbles rising through the aqueous phase. The solvents used in solvent sublation should have large affinities for the solutes and should be nontoxic, immiscible with water and nonvolatile.

In solvent sublation, the main solute transport between aqueous and organic phase is a unidirectional transport of solute by gas bubbles[12,13]. This transport depends on the specific surface area of the bubbles, the air flow rate, the mass transfer coefficient for the solute from the aqueous phase to the air bubbles, and the equilibrium relationship for the solute between the gas-water interface provided by gas bubbles and the water phase. The mechanism of transport may be based on evaporation of the solute into the gas bubbles

Chemistry for the Protection of the Environment 3, edited by Pawłowski *et al.* Plenum Press, New York, 1998

(volatile compounds), formation of air bubble-particle aggregates (dispersed substances), surface adsorption of the solute at the air-water interface (surface active compounds), or adsorption of the complexes formed with surfactants (metal ions, dyes).

Solute transport, with the water droplets taken into the organic phase is the second most important mechanism in solvent sublation. It results from the fact that a bubble rising in water is not exactly spherical.

Solute transport can also take place by molecular diffusion through the boundary between aqueous and organic layers, driven by the solute activity gradient between the two phases. The extent of this transport depends also on the mass transfer coefficient for solute from the aqueous to the organic phase and on the contact area between the organic layer and the aqueous layer. In the initial stages, we observe only transport of solute from aqueous phase to organic layer and, as solute concentration in the organic phase increases, the reverse transfer becomes important.

Our previous investigations show[14] that solvent sublation can be a good method for protection of the atmosphere against volatile organic pollutants, because the organic phase can prevent the atmosphere from being polluted by the vapours, which otherwise could be released from the gas streams coming out of the aqueous layer during flotation or other processes where aeration takes place. However, in some cases, depending on the process parameters and properties of the organic solvent, as well as the properties of the pollutant, emission of both to the atmosphere can take place. The emission depends also on the gas flow rate, temperature and ratio of the aqueous phase to the organic solvent volumes.

Ososkov[15–16] investigated the emission of volatile compounds to the atmosphere during processes where aeration took place. He has found that when compared to air-stripping, solvent sublation decrease emissions by 34–82% for toluene, by 22–69% for trichloroethylene, by 45–87% for chlorobenzene and by 57–89% for 1,3-dichlorobenzene. Thus, the emission reduction was more effective for less volatile and more hydrophobic pollutants. The emission depends also on the process parameters e.g. gas flow rate, temperature, the type of the organic phase and on the presence of the other solutes. Decyl alcohol used as an organic solvent gave better results in emission reduction than paraffin oil.

The objective of this research was to determine the efficiency of mesitylene (1,3,5-trimethylbenzene) removal from water, as well as its emission to the air during solvent sublation. Mesitylene is a volatile hydrocarbon with its Henry's law constant equal to 0.2244, vapour pressure equal to 1.95 mm Hg, solubility in water equal to 57 ppm[17] (data for temperature 20°C).

EXPERIMENTAL

Solvent sublation processes were carried out in the apparatus presented in the previous work[14]. The glass column (120 cm length, 4.5 cm in diameter) was filled with mesitylene emulsion (1500 cm^3) and the organic solvent was poured on the top of the aqueous layer (25, 50 or 100 cm^3). Dodecane (99%+, Lancaster Synthesis), tetradecane (99%+, Lancaster Synthesis), 1-decanol (98%+, Lancaster Synthesis) and mineral oils (white light and heavy light both from Aldrich Chemical Co.) were used as the organic layers.

A sintered glass sparger, fitted at the lower end of the column, was used to produce air bubbles, which rose through the column. Runs were conducted in a batch system. The gas flow rate was set at 4 dm^3/h.

All runs were carried out at room temperature.

Emulsions, prepared with mesitylene (98%, Aldrich Chemical Co Ltd.) and distilled water, were stabilized with sodium dodecylobenzenesulfonate (NaDBS), whose concentration was 10^{-5} Mole/dm^3.

The concentrations of mesitylene in the aqueous phase was determined by GLC. From water, mesitylene was previously extracted with hexane.

The analysis of organic vapours in gas leaving the column was done by the method consisting of the collection of organic vapours on the adsorbent, desorption and subsequent analysis by gas chromatography[18,19]. For collection of the organic vapours, glass tubes filled with charcoal (Fisher) were used. Glass tubes of 13 cm length and 4 mm of the inner diameter were subdivided into two sections of 100 mg and 50 mg of charcoal. The sections were separated by glass wool. The front position of 100 mg was used to collect the vapours, while the 50 mg backup section trapped the solvent if breakthrough occurred on the front charcoal portion. The tubes were recommended by Polish Standards[20]. For kinetic investigations, the tubes were changed every 30 min. Desorption of the collected substances from the charcoal was accomplished by CS_2 (10 cm^3) and the chromatographic analysis was carried out directly.

All chromatographic analyses were accomplished on Chrompack 9001 Gas Chromatograph, equipped with a flame ionization detector (FID). A 30 m long capillary column, with an internal diameter 0.32 mm, and coated with XTI-5 (5% diphenyl–95% polysiloxane), was used. The injector temperature was 200°C, and the detector temperature was 200°C. Analyses were carried out employing a temperature programming mode ranging from 60 to 160°C.

RESULTS

The results of mesitylene sublation to different organic layers are presented in Table 1. The final mesitylene concentration (c_f) depends on the initial concentration (c_i) and on the type of the organic solvent used in the sublation process. In the table, the amount of mesitylene emitted to the atmosphere (m_m) is also presented. The emission is also expressed as a percentage of the initial mesitylene concentration (m_m/c_iV * 100%) or as a percentage of the mesitylene removed from water ($m_m/(c_i - c_f)V$ * 100%), where V is the aqueous layer volume, and c_i and c_f are the initial and final concentrations of mesitylene in the aqueous phase, respectively.

Table 1. Mesitylene sublation results. Volumes: aqueous emulsion—1500 cm^3, organic solvent 50 cm^3

Organic layer	Mesitylene concentration in water		Cumulative mesitylene emission		
	c_i [mg/dm^3]	c_f [mg/dm^3]	m_m [mg]	$\frac{m_m}{c_i \cdot V} \cdot 100\%$ [%]	$\frac{m_m}{(c_i - c_f) \cdot V} \cdot 100\%$ [%]
Dodecane	135	34	8	4.0	5.3
Tetradecane	169	49	14	5.5	7.8
Decanol	159	8	17	7.1	7.5
Mineral oil					
light	180	33	31	11.5	14.1
heavy	250	30	67	17.9	20.3

Table 2. The effect of thickness of the organic layer on mesitylene emission to the atmosphere

Organic layer	Organic layer thickness [cm]	Mesitylene concentration in water		Cumulative mesitylene emission		
		c_i [mg/dm^3]	c_f [mg/dm^3]	m_m [mg]	$\frac{m_m}{c_i \cdot V} \cdot 100\%$ [%]	$\frac{m_m}{(c_i - c_f) \cdot V} \cdot 100\%$ [%]
Dodecane						
	6	189	73	11	3.9	6.3
	3	122	24	9	4.9	8.6
	1.5	125	8	9	4.8	5.3
Tetradecane						
	6	180	24	21	7.8	9.0
	3	169	49	14	5.5	7.8
	1.5	190	22	38	13.3	15.1
1-decanol						
	6	169	6	12	4.7	4.9
	3	159	8	17	7.1	7.5
	1.5	156	13	23	9.8	10.7
Mineral oil light						
	6	138	6	37	17.8	18.7
	3	180	33	31	11.5	14.1
	1.5	219	4	92	28.0	28.5

As may be observed, in all cases the emission of mesitylene to the atmosphere takes place. However, the extent of the emission strongly depends on the nature of the organic solvent applied as a top layer. The best results were received when dodecane was used (only 8 mg of mesitylene was emitted), and much worse results were achieved with the mineral oils, especially heavy oil (67 mg of mesitylene was trapped in the carbon trap). In this case, 14–20% of the total mesitylene amount removed from water was released to the atmosphere. Thus, one may conclude that the mineral oils are not suitable organic solvents to be applied in the solvent sublation process for the removal of volatile compounds.

It is obvious that the retention of mesitylene in the organic layer depends on the amount of the organic solvent used. In Table 2 the effect of thickness of the organic layer is presented.

It is evident that solvent sublation reduces emissions to the atmosphere. However, this improvement strongly depends on the thickness of the organic layer. There is no substantial difference between 6 and 3 cm thick layers, but a 1.5 cm thick layer evidently does not provide sufficient organic solvent because the emission of mesitylene increased markedly compared to the results when 3 and 6 cm thick layers were provided. (The volumes of organic solvent relating to the thickness 1.5, 3 and 6 cm are 25 cm^3, 50 cm^3 and 100 cm^3, respectively).

The evaluated removal efficiency as well as the emission to the atmosphere depends on the initial concentration of pollutants in water. This effect is presented in Table 3.

As it can be seen from Table 3, the initial concentration of mesitylene strongly affects emissions during solvent sublation, and the net amount emitted to the atmosphere is greater, the higher the initial mesitylene concentration (c_i) is. However, the relative emissions, calculated as a percentage of the initial concentration, show the opposite relation; the higher the initial concentration, the smaller percentage of mesitylene is emitted to the atmosphere. In the last column, the emission is expressed in relation to the amount of mesitylene removed from water. The data show that 5–8.6% of the removed mesitylene

Table 3. The effect of initial mesitylene concentration on mesitylene emission to the atmosphere after 2 h of the process run. Organic phase: dodecane, volume: 50 cm^3

Mesitylene concentration in water		Cumulative mesitylene emission		
c_i [mg/dm^3]	c_f [mg/dm^3]	m_m [mg]	$\frac{m_m}{c_i \cdot V} \cdot 100\%$ [%]	$\frac{m_m}{(c_i - c_f) \cdot V} \cdot 100\%$ [%]
450	210	18	2.7	5.0
260	138	14	3.6	7.7
135	34	8	4.0	5.3
122	52	9	4.9	8.6

was released to the atmosphere, which means that 91–95% of the mesitylene removed from water was trapped in the organic layer.

In Figures 1 and 2, the kinetics of emitted mesitylene versus the time of aeration are presented.

Analysing the diagram on the kinetics of mesitylene emission, one may observe that at the beginning of the process, when the organic phase did not contain mesitylene, the interception of mesitylene by the organic solvent was much better than at the middle of the process, when the content of mesitylene was the highest. As the mesitylene concentration in the organic phase increased due to its transfer from the aqueous phase, the emission of

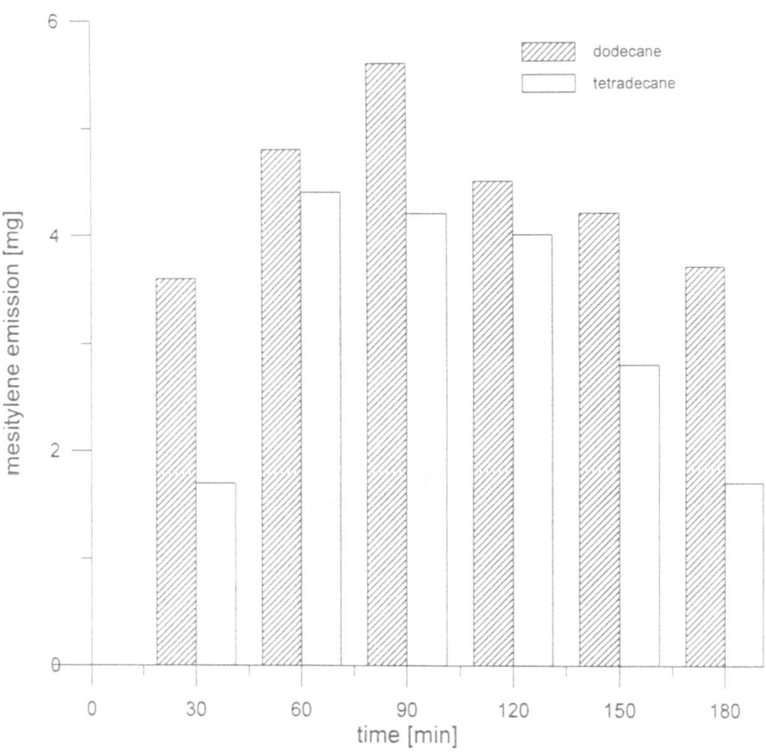

Figure 1. Kinetics of mesitylene emission to the atmosphere. Organic phase volume: 50 cm^3.

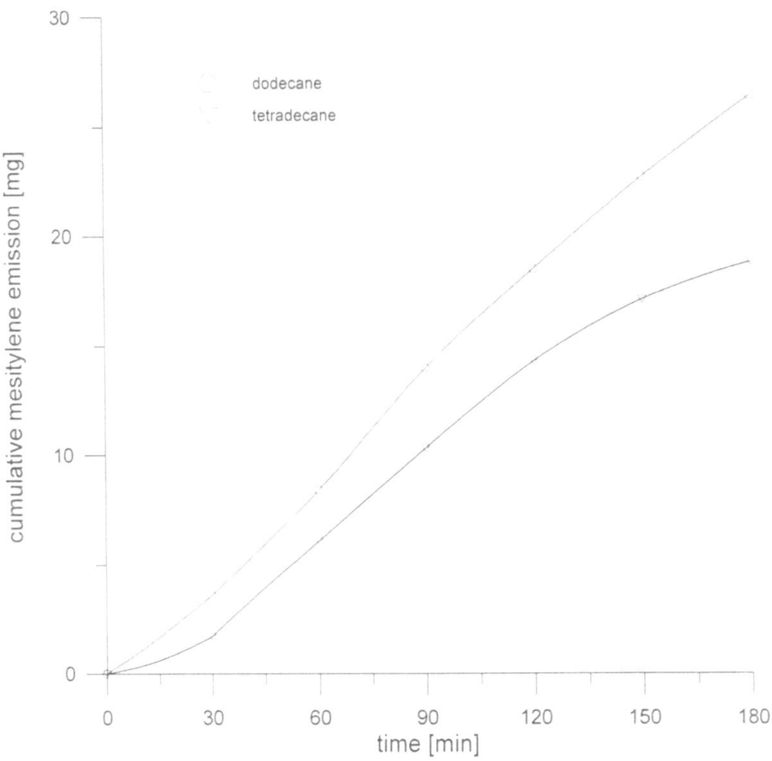

Figure 2. Cumulative mesitylene emission to the atmosphere. Organic phase volume: 50 cm^3. Run 1: c_i= 450 mg/dm^3, organic phase: dodecane. Run 2: c_i= 169 mg/dm^3, organic phase: tetradecane.

its vapour to the atmosphere also increased. However as the process continued, the mesitylene emission to the atmosphere decreased, which results from the decrease of the amount of mesitylene transferred to the organic layer from the aqueous phase (due to the decrease of mesitylene content in the aqueous phase).

CONCLUSIONS

Generally, one may state that, compared to air stripping, solvent sublation reduces to a great extent the emission of volatile organics to the atmosphere. However, the emission is not reduced to zero. The observed emission depends on the nature of solvent used as an organic layer, and was the lowest in the case of dodecane and the highest in the case of mineral oil. Reasonably good results were achieved also for tetradecane and decanol as organic layers, while mineral oils did not give satisfactory results.

The volume of the organic phase affects the emission also. The reduction of the emission increases with the increase of the thickness of the overlying organic layer.

The emission depends on the initial concentration of mesitylene in the water and was greater when this concentration was higher.

In order to avoid or control the emission to the atmosphere, the exchange of the organic layer must be considered. This avoidance of control may also be achieved by the continuous flow of the organic layer (in order to not exceed the saturation concentration of solute in the organic phase).

Finally, another possibility for controlling the emission is using a charcoal trap, through which the air leaving the solvent sublation unit will be passed. The analysis reported in this paper confirms the choice of such a solution.

REFERENCES

1. Sebba F., Ion Flotation, Elsevier Publishing Co., New York, 1962.
2. Valsaraj K. T., Springer C., Sep. Sci. Technol., 21, 789 (1986).
3. Huang S., Valsaraj K. T., Wilson D. J.,Sep. Sci. Technol., 18, 941 (1983).
4. Foltz L. K., Carter K. N., Wilson D. J.,Sep. Sci. Technol., 21, 57 (1986).
5. Valsaraj K. T., Wilson D. J., Colloids Surf., 8, 203 (1983).
6. Tamamushi K., Wilson D. J., Sep. Sci. Technol., 19, 1013 (1985).
7. Wang W. K., Huang S. D., Sep. Sci. Technol., 23, 375 (1988).
8. Shih K. Y., Han W. D., Huang S. D., Sep. Sci. Technol., 25, 477 (1990).
9. Womack J., Lichter J. C., Wilson D. J., Sep. Sci. Technol., 17, 897 (1982).
10. Huiri D., Shouzhen Y., Jialai L., Anal. Sci., 5, 601 (1989).
11. Szeglowski Z., Bittner-Jankowska M., Mikulski J., Nukleonika, 18, 299 (1973).
12. Valsaraj K. T., in Flotation Science and Engineering, Ed. Matis, K.A., Marcell Dekker, New York, 365 (1995).
13. Smith J.S., Valsaraj K.T., Thibodeaux L.J., Ind. Eng. Chem. Res., 35, 1688 (1996).
14. Mędrzycka K., Pastewski S., in Chemistry for the Protection of the Environment II, Ed. L. Pawłowski, W.J. Lacy, C.G. Uchrin, M.R. Dudzińska, p. 67–74, Plenum Publ. Corp., New York, 1996.
15. Ososkov V., Kebbekus B., Chen M., Sep. Sci. Technol., 31(2), 213 (1996).
16. Ososkov V., Kebbekus B., Chou Ch. Ch., Sep. Sci. Technol., 31(10), 1377 (1996).
17. Mędrzycka K.B., Sep. Sci Technol., 25(7&8), 825 (1990).
18. Rothweiler H., Wager P.A., Schatter Ch.,Atmospheric Environment, 25B(2), 231 (1991).
19. Determination of Organic Vapors in the Industrial Atmosphere. Bulletin Supelco Inc., 769C.
20. Polish Standard PN-78/Z-04116/01 (Air purity protection)

A METHOD OF REDUCING THE SO_2 EMISSION FROM POWER BOILERS

Jan Jędrusik,[1] Eugeniusz Kalinowski,[2] and Maria Jędrusik[2]

[1]Power Generation Station of Wrocław, Ltd., Poland
[2]Technical University of Wrocław, Poland

ABSTRACT

It is generally known that reduction of SO_2 emission involves significant expenses. In most of the installations all over the world, the cost of removing 1 kg of SO_2 is of the order of 2,0–7,5 US$. Therefore, the SO_2 emission reduction isn't in practice pushed beyond what is allowed by appropriate standards.

The method of lowering the SO_2 emission, presented in the paper, is a two-stage process which allows any level of emission reduction within the limits of 20% through 80%. The process consists in introducing a lime or dolomite sorbent into the boiler's combustion zone. Chemical reduction occurs at the temperature of 900–1000°C that lets reduce the SO_2 concentration by ca. 35%. After leaving the boiler and passing through a rotational air-heater, the combustion gas enters a reactor, which is a special-purpose variation of the boiler flue's vertical segment.

Further reduction of sulfur dioxide contents occurs in the reactor due to the introduction of an alkaline water solution and the combustion gas being cooled down. Results of the installation tests are presented as plots of SO_2 concentration versus heat capacity of the boiler, and operating parameters of the desulphurization installation. The SO_2-removal technology, described above, ensures not only the required reduction of sulfur dioxide content, but also high efficiency of the electrostatic precipitator, which removes dust from the post-desulphurization combustion gas, as evidenced by the appropriate diagrams and tables.

Keywords: desulphurisation; electrostatic precipitator.

1. INTRODUCTION

Particulate concentrations in flue gases coming from power plant boilers are usually reduced in electrostatic precipitators. The problem of joint action of existing electrostatic precipitators and desulphurisation or denitrification plants becomes very significant.

Chemistry for the Protection of the Environment 3, edited by Pawłowski *et al.* Plenum Press, New York, 1998

The decision concerning the choice of method of desulphurisation is usually a compromise between expected efficiency of the flue gas cleaning process, its cost, and limitations caused by the location of the plant.

In Poland it is often necessary to forgo high efficiency wet methods of desulphurisation for approaches which are simpler in design and can easily be inserted into already existing technologies.

This kind of solution was favored in the FGD (WAWO) installation for the desulphurisation of water boiler flue gases. The principle of the method is based on the injection of calcium into the boiler combustion chamber where the temperature varies within the range 800–1200°C. Here SO_2 and SO_3 bond with calcium compounds. The efficiency of this process does not exceed 35%. An improvement on this value can be achieved by humidification of flue gases in a reactor. Processes such as moistening of ash, reduction in flue temperature, particulate agglomeration and precipitation occur here. In order to prevent condensation inside electrostatic precipitators, flue gases are reheated with hot air [1]. Figure 1 shows a typical electrostatic precipitator connected to an FGD installation.

Full-scale operational tests, carried out in 1991 through 1995 in the desulphurization plants of the WP-120 and WP-70 boilers of the Power Generation Station of Wrocław, Ltd., allowed to collect all the information required to assess the WAWO's method operating and capital costs. Four technological variants of the WAWO method were also worked out with various abilities to adapt their SO_2-emission-limiting devices to the SO_2 emission standards, both currently valid and those expected in the year 2010 [2] [3].

2. DESCRIPTION OF TECHNOLOGY

The process of limiting the SO_2 emission is a two-stage operation which allows any level of emission reduction within the limits of 20% through 80%. Emission level control is only possible if analyzers for continuous measurement of SO_2 concentration and desulphurized flue gas flow intensity are installed.

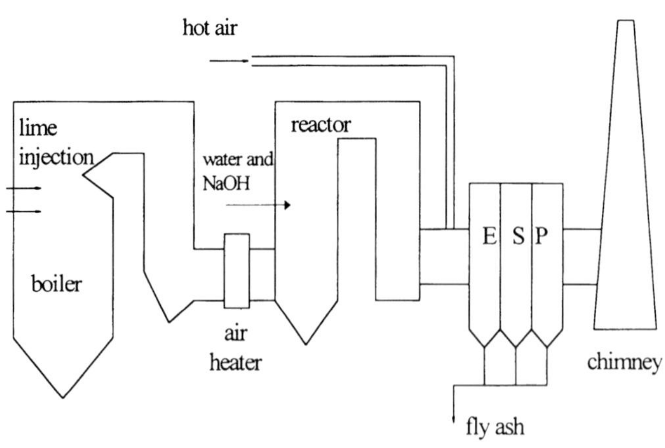

Figure 1. FGD WAWO process schematic.

As can be seen from the schematic in Fig. 1, the process of limiting the SO_2 emission consists in introducing a lime or dolomite sorbent into the boiler's combustion zone. The main chemical reaction

$$CaCO_3 + 1/2\,O_2 + SO_2 \rightarrow CaSO_4 + CO_2,$$

which occurs at the temperature of 900–1000°C, lets reduce the SO_2 concentration by ca. 35%. After passing through a rotational air-heater, the combustion gas from the boiler enters a reactor, which is a specially designed vertical segment of the boiler flue.

The following processes occur in the reactor:

- cooling and evaporation of water, due to a controlled addition of alkalie water solution through pneumatic nozzles,
- agglomeration of dust particles, which settle at the bottom of the reactor,
- reduction of SO_2 emission as a result of the following reactions:

$$NaOH + SO_2 + H_2O \rightarrow NaHSO_3 + H_2O$$

$$Ca(OH)_2 + NaHSO_3 \rightarrow CaSO_3{\bullet}1/2H_2O + NaOH + 1/2H_2O$$

The NaOH solution, by reacting with SO_2 molecules, forms $CaSO_3$, which is sparingly solvable and thermally stable in the reactor's operating conditions (100–140 °C); this product settles, partially at the reactor bottom, partially in the precipitator.

All the measurement, monitor and control circuits are linked to the computerized control system of the Master 3 boiler.

Production wastes in the form of ash/calcium sulfate mixture are stored. Opinions, whether the wastes are suitable for the production of building materials, differ as yet.

3. RESULTS OF INVESTIGATION

The results presented in the work refer to the desulphurization of flue gases resulting from the combustion of coal with the following parameters:

• calorific value	20000 to 24000 kJ/kg
• total sulfur	0,72 to 1,05%
• ash content	18,0 to 22,0%
• total water content	7,0 to 10,0%

Figure 2 shows the relationship between SO_2 concentration in the flue gas leaving the desulphurization plant, and the thermal capacity of the boiler [5] [6]. An increase in the boiler's heat capacity and higher coal consumption result in increased concentration of SO_2 after the FGD installation.

Figure 3 shows the desulphurization effectiveness versus the boiler capacity for optimum operational parameters of the desulphurization plant. The application of alkaline liquid (NaOH + H_2O) injection into the reactor ensured an increased desulphurization efficiency (ca. 80%), only slightly dependent on the boiler's thermal capacity. In the case of only water being injected into the reactor, the efficiency of FGD is lower (ca. 60%) and the dependency on boiler operating parameters more pronounced.

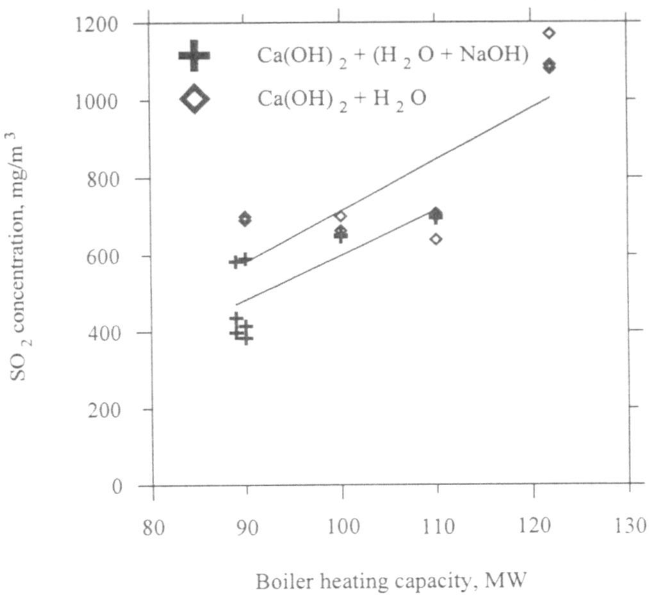

Figure 2. Relationship between SO_2 concentration and the heating capacity of boiler.

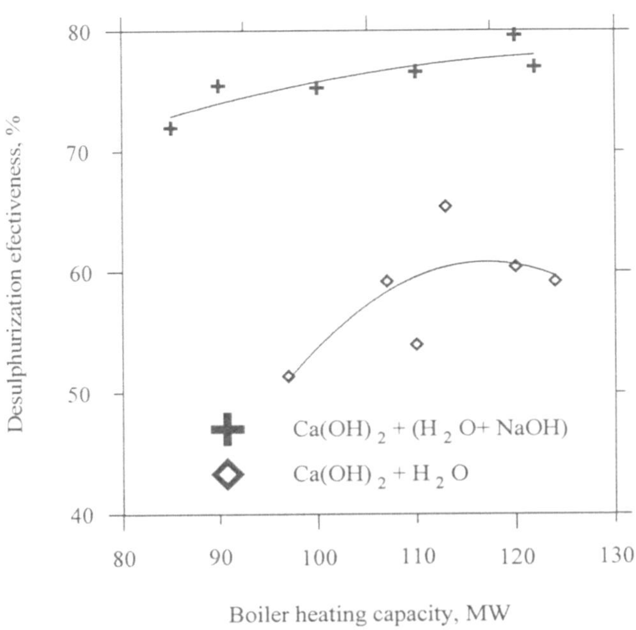

Figure 3. Relationship between the desulphurization effectiveness and the heating capacity of boiler.

4. EFFECT OF FLUE GAS DESULPHURIZATION ON THE EFFECTIVENESS OF DUST REMOVAL BY PRECIPITATORS

Experiments carried out with the FGD WAWO unit (which was situated downstream a 140 MW boiler) included the application of three sorbents: $(CaOH)_2$, CaO and $CaCO_3$, but preliminary results showing an FGD efficiency of 70% revealed that calcium hydroxide would be best suited.

The operating parameters were as follows:

• excess lime	Ca : S = 2 : 1
• temperature in reactor	353 K to 363 K
• temperature in precipitator	363 K
• maximum water demand for flue gas cooling	16 m^3/h
• amount of lime needed	700 to 1600 kg/h.

The next step in the experiment included an attempt at increasing the FGD efficiency by the application of additives to the liquid which was used to disperse the reducing agent. One of these was sodium hydroxide and it was introduced into the reactor in the form of a solution. For the above parameters of the FGD installation, the performance of the electrostatic precipitator was investigated [7], [8]. The plant under scrutiny was a two-zone low efficiency (92 %) precipitator. Without the FGD system the particulate concentration at the outlet of the precipitator varied between 180 mg/Nm^3 and 1200 mg/Nm^3 . Analysis of flue parameters was carried out based on data sets obtained from continuous monitoring of gaseous and particulate composition. The measurement devices included DR-280 photometric dustmeters (made by DURAG), UNOR 6N units for the measurement of gaseous composition (NO, SO_2 , CO) and an OXOR 6N oxygen analyzer. Dust meters were calibrated and tested gravimetrically in the course of the FGD process [9], [10].

The investigation allowed to assess the effect of FGD WAWO installation on the degree, in which the dust concentration in the flue gas dropped after the precipitator. This is shown in Figure 4.

During operation of the FGD (WAWO) installation the sulfur content and calcium content in fly ash increased threefold (up to 1.2 %) and tenfold (up to approximately 7 %) respectively. Figure 4 shows the fly ash particles produced when the flue gas is desulphurized by the WAWO method.

5. SUMMARY

- Full-scale tests of the "WAWO" desulphurization installation confirmed the ability to achieve operational flue gas desulphurization effectiveness of 20% through 80%, for a constant capital cost.
- The operational cost of the process depends on the standards of admissible SO_2 emission. Total operational cost of the plant run 6000 hours per year amounts to 2.8US$/kg SO_2 for 50% efficiency, and 5.2US$/kg SO_2 for 90% efficiency [5].
- Desulphurisation by means of the WAWO technology changed the properties of flue gases carrying particles in channels before the precipitator. Cooling and humidification of flue as well as particulate removal in the reactor played an important part in the process.

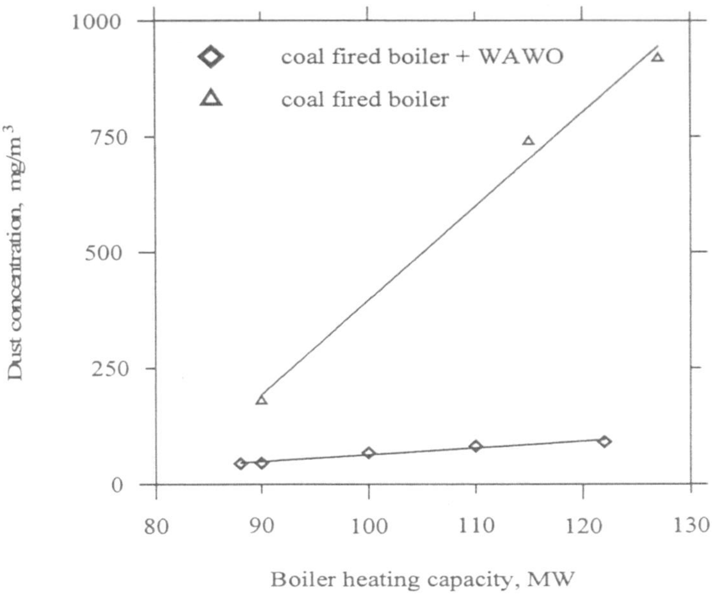

Figure 4. Relationship between dust concentration and the heating capacity of boiler.

- Application of the WAWO desulphurisation installation downstream power plant boilers increased the precipitation efficiency. This finding is of prime importance to electrostatic precipitators with small collection areas, because it eliminates the need of redesigning the precipitator, which involves additional expenditure.
- Investigation of the precipitator after the WAWO FGD unit showed that, for a boiler power of 88 to 122 MW, lime injections of 860 to 1200 kg/h and water of 4 to 10 m^3/h brought about a drop in particulate concentration after passage through the electrostatic precipitator from approximately 970 mg/m^3 to 100 mg/m^3.

REFERENCES

1. Patent of Polish Republic, No 168505, "A method of removing gaseous sulfur compounds from flue gases."
2. Gostomczyk M.A., Sieczkowski J., "Treatment of dust-boiler flue gases with the WAWO method," Proceedings of the Conference "Limitation of pollution from power generating equipment," PZITS–Poznań, April 1995 (in Polish).
3. Gostomczyk M.A. et al., "Dry methods of SO_2 emission limitation," Proceedings of the Conference "Limitation of pollution emission from grate and dust boilers," ODiTK, Sopot, February 14–16 1996 (in Polish).
4. Gostomczyk M.A., Ciasnocha Cz., Jędrusik J., "Operational experiments on implementing the WAWO technology in the Power Generation Station of Wrocław, Ltd.," 3rd Symposium on the Reduction of Pollution Emission to the Environment POL-EMIS '96, Szklarska Poręba, June 1996 (in Polish).
5. Ciasnocha Cz., Jędrusik J., Starnawski W., Operation of the WAWO-type desulphurization installations, Scientific-Technical Conference "Low Emission Combustion Techniques" Ustroń-Zawodzie 28–30.03.1996 (in Polish).
6. Gostomczyk M.A. *et all*, Optimization tests of the SO_2 removal process from boiler flue gas by means of the WAWO method in the Wrocław Power Generation Station, Ltd, Wrocław Technical University Report, Series SPR I-15 No 56, 1995.

7. Jędrusik M., Jędrusik J. The Effect of the Desulphurisation System on the Electrostatic Precipitator: A Case Study. Fourth CSIRO Conference Gas Cleaning, Jamberoo NSW Nov. 10 - 15, 1991.
8. Jędrusik M., Ulatowski G., Investigation of the Effect of WAWO Desulphurization Technology on the Precipitator Operational Effectiveness in a KW - 3 boiler, part I, Report of the Technical University of Wrocław, Series SPR J-20 No 22, 1992 (in Polish).
9. ISO 9096: Air quality - Stationary source emissions—Determination of concentration and mass flow rate of particulate in gas carrying ducts—Manual gravimetric method
10. Jędrusik M. Teisseyre M., Świerczok A., Effect of the WAWO Desulphurization Installation on the Operational Efficiency of a KW-3 Boiler in the Power Generation Station of Wrocław, Ltd. Report Series SPR I-20 No 2 1996 Technical University of Wrocław

10

ATMOSPHERE PROTECTION THROUGH ENERGY LOSS MINIMIZATION

Eugeniusz Kalinowski, Anna Krawczyk, and Maria Jędrusik

Technical University of Wrocław
Wrocław, Poland

ABSTRACT

Most of the electrical energy produced all over the world (ca. 80%) is derived from the fuel combustion process. The process involves an immense atmosphere pollution by waste products from power generation stations, i.e. solid wastes (ashes), but predominantly gaseous ones (combustion gases). Combustion gas constituents, i.e. CO_2, SO_2, and NO_x, are the main energy conversion wastes that pollute the atmosphere. Reduction of that sort of pollution is done based on the two fundamental provisions:

1. increase in the efficiency of the power station, this being limited by the second law of thermodynamics,
2. decrease of electrical energy consumption by a particular industrial process or a household to an allowable minimum.

The paper emphasizes the fact that non-reversible phenomena result in a loss of the maximal work, especially of electrical energy, in the process of both generation and application.

To evaluate the loss of work, the concept of exergy is employed in thermodynamics. Exergy is defined as the highest possible work obtainable from a particular system of bodies in their natural environment.

Thus the loss of exergy due to non-reversibility of phenomena is equivalent to the loss of work (and especially to the loss of electrical energy as the main energy medium). Means of reducing the losses, and consequently, the pollution level as a method of atmosphere protection, constitute the subject of the paper.

Keywords: atmosphere protection; energy protection; Gouy-Stodola law.

Chemistry for the Protection of the Environment 3, edited by Pawłowski *et al.* Plenum Press, New York, 1998

1. THE LOSS OF MAXIMUM WORK—GOUY–STODOLA LAW

1.1. The Second Law of Thermodynamics

Of the many formulations of the second law of thermodynamic, the most known is that given by Ostwald, which states that no 2nd order perpetual motion machine is possible. If, however, mathematical presentation is required, the proper formulation of that law of thermodynamics is the principle of entropy increase.

$$\pi = S_{syst2} - S_{syst\,1} = \sum_{i=1}^{i=n} \Delta S_i \tag{1}$$

We shall employ this formulation to determine analytically the loss of the maximum work. This, however, requires that the concept of reversible (ideal) and non-reversible phenomena be discussed.

Reversible phenomena are defined as those, in which the return of the system of bodies to its initial state is possible. This means that the increase of entropy in such phenomena is zero ($\pi = 0$).

Allowing for the fact that technology is dominated by fluid-flow phenomena, the loss of maximum work will be determined based on a fluid-flow machine presented in Fig. 1, in which exchange of heat Q_o with the environment having temperature T_o occurs.

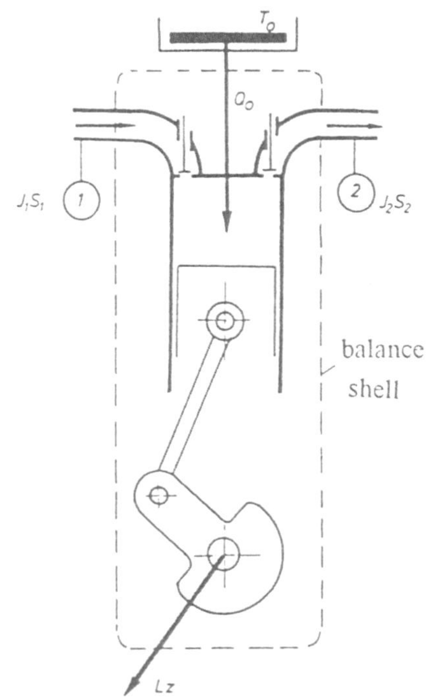

Figure 1. Model of a fluid-flow machine operating in contact with environment.

By denoting the entropy of the inflow fluid flux by I_1, and that of the outflow fluid flux by I_2, the external work L_Z can be expressed according to the first law of thermodynamic [1]:

$$L_Z = I_1 - I_2 + Q_0, \qquad (2)$$

while, according to the 2nd law of thermodynamics, the increase π of the system's entropy is expressed as

$$\pi = S_2 - S_1 - Q_0/T_0. \qquad (3)$$

By eliminating Q_0 from the set of both equations and rewritting it appropriately, the following is obtained:

$$L_Z = I_1 - I_2 - (S_1 - S_2)\,T_0 - \pi\,T_0. \qquad (4)$$

If all processes that occur in the machine are reversible, i.e. $\pi = 0$, then external work L_Z is the greatest possible. If such maximum work is denoted by $L_{Z\,max}$, then

$$L_{Z\,max} = I_1 - I_2 - (S_1 - S_2)\,T_0. \qquad (5)$$

The loss of work due to non-reversibility of phenomena is then determined by

$$\Delta L = L_{Z\,max} - L_Z \qquad (6)$$

or

$$\Delta L = \pi\,T_0, \qquad (7)$$

which amounts to the formal expression of the Gouy–Stodola law.

In order to minimize the loss of work in the process of work conversion, the value of π, i.e. the entropy increase in the system of bodies, should be minimized. This generally requires that the process be performed with minimum friction, transfer of heat occur at economical (not maximized) temperature differences, and the so called pseudothermal phenomena (not resulting from temperature differences but producing heat) be kept to minimum.

Energy itself and its losses do not constitute a sufficient criterion to evaluate the loss of maximum work. To this aim, the concept of exergy has been introduced into thermodynamics. Exergy is defined as the maximum capacity of a body to perform work.

1.2. Exergy

The natural environment is the place where both unusable materials and unusable heat come, unusable meaning 'with no ability to perform work'. These are substances with temperatures equal to the ambient temperature and partial pressures equal to the partial pressures of the same substances in the environment. This is expressed by saying that the substances remain in thermodynamical equilibrium with the environment.

Since fluid-flow processes prevail in technological equipment, exergy B of the fluid flux is assumed as fundamental.

$$B = B_k + B_p + B_t, \tag{8}$$

where:
B—kinetic exergy,
B_p—potential exergy,
B_t—thermal exergy.

Also other sorts of exergy are allowed for, where appropriate, e.g. nuclear exergy. With kinetic, potential and other exergies neglected, the exergy of flux is given as

$$B = B_t = I - T_0\, S, \tag{9}$$

this having been obtained from the equation (5).

The decrease in exergy ΔB_t can be expressed as

$$\Delta B = \Delta B_t = I_t - I_2 - T_0\,(S_1 - S_2) = L_{Z\,max}, \tag{10}$$

so it is equivalent to the maximum work.

1.3. Exergy Balance

Although exergy is not conserved like energy, the exergy balance equation can be written by closing the balance with the exergy loss δB_u, the latter resulting from the irreversibility of processes in the system being discussed. Exergy is conserved, if all processes are reversible.

The equation of exergy balance is as follows:

$$\delta B_u = B_l - B_2 - \Delta B_u - \Delta B_{zr} - L_Z. \tag{11}$$

where:
B_l—exergy of substance coming in
B_2—exergy of substance going out
ΔB_U—increase of exergy in system
L_2—external work
The graphical form of the equation (11) is presented in Fig. 2.

A more detailed discussion of the concept of exergy and its balance can be found in [2] and [3]. Nevertheless, it should be noted here that exergy is generally not conserved, as is the case with energy for example.

2. CONSUMPTION OF ENERGY

Another means of environment protection is economy in the consumption of energy, mainly electrical energy. Over 80% of electrical energy produced world-wide comes from both thermal power-generation stations and thermal power/heat-generating stations, which are intensive atmosphere polluters due to CO_2, SO_2, and NO_x emission.

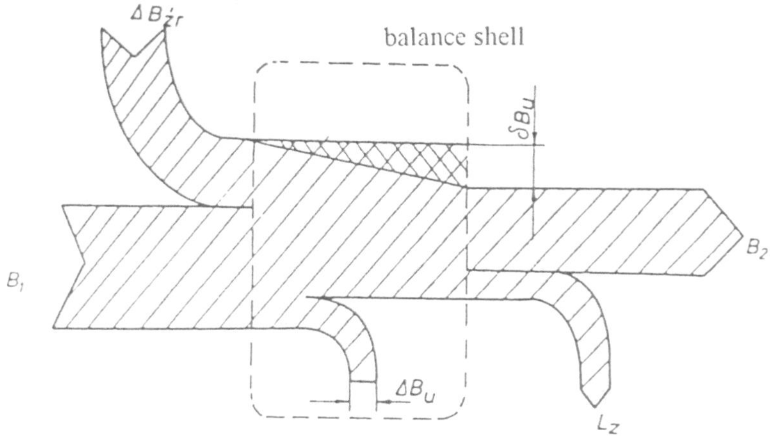

Figure 2. Exergy balance (graphically): B_I—exergy of substance coming in, ΔB_{zr}—increase of exergy in external sources, ΔB_U—increase of exergy in system substance, B_2—exergy of substance coming out, L_Z—external work, δB_u—loss of exergy due to the irreversibility of phenomena in a system.

This type of pollution can be limited not only by increasing the efficiency of power stations (e.g. by constructing power/heat generating plants rather than pure electric power plants), but also by designing technological processes so as to achieve minimum heat consumption per product unit, which is essentially to minimize heat losses and dispense with the necessity to utilize the so called waste heat. Until this is arrived at, waste heat utilization is economically justified.

Another provision of the atmosphere protection is the elimination of excessive energy consumption in households by using less energy-demanding domestic appliances like refrigerators, washing machines, dryers, mechanical devices, lighting, televisions, etc.

All these provisions are evidently beneficial to the atmosphere protection.

REFERENCES

1. Msieh Jui Sheng, Principles of Thermodynamics, New York Mc Graw-Hill Book Comp., 1975
2. E.Kalinowski, Thermodynamics, Editors of the Wrocław Technical University, Wrocław 1994 (in Polish).
3. J.Szargut, R.Petela, Exergy, WNT, Warszawa 1965 (in Polish).

11

PROBLEMS OF THE IMPLEMENTATION OF ENVIRONMENTAL MANAGEMENT SYSTEM ACCORDING TO ISO14001 IN POLAND

Robert Pochyluk

Center for Environmental Studies
Technical University of Gdańsk

ABSTRACT

Environmental management systems become an significant issue among Polish enterprises. Experiences from Polish companies indicate that background of great part of problems related to their implementation are of human origin—misunderstandings, resistance to change, attitudes and inability of long-term planning. The most common problems to be met in our practice are:

1. lack of thoroughly performed environmental review,
2. improper identification of significant environmental aspects,
3. difficulties in delegating authority to the right people,
4. improper training of employees,
5. difficulties to communicate within an organisation,
6. lack of personal involvement of top management,
7. improper use of external consultants.

This paper aims at presenting our ideas how those problems could be solved. make the content of this paper.

Key words: environment, management, ISO 14001.

1.INTRODUCTION

Poland, as one of those countries which have serious chance to join European Union at the very beginning of the next century, is currently trying to comply with European standards in many fields. Environmental protection is one of them. At the same time factors like consumer pressure, stronger environmental legislation and ethical issues start to force companies to consider environmental issues as strategic issues. It is getting more and more evident that the proper management is what helps most to prevent environment from day-to-day deterioration. The integrated environmental management system may help in improving environmental performance and avoid accidents which could have significant influence on environment. Therefore standards have been issued in order to give guidelines how to manage industrial activities to reduce influencing environment. Nowadays, apart from EMAS[1] (Eco-Management and Audit Scheme) Regulation, ISO14001[2] is the most important and well known of them. Experiences gained in implementing Environmental Management System (EMS) in Polish companies indicate that background of great part of problems is of human origin—misunderstandings, resistance to change, attitudes and inability of long-term planning. "Command and control" type of environmental regulations enforcement is still commonly seen and companies' reaction is adequate to that. While according to our survey environmental awareness in Polish enterprises is really high and everyone admits that something needs to be done in order to protect the environment it is clear that only limited number of companies try to do something more than meeting basic legal requirements if doing anything at all. As R. Welford notes "there is always an incentive (...) for profit-maximising firms seeking short-term rewards to opt out and become free riders (assuming that everyone else will be environmentally conscious such that their own pollution will become negligible)"[3]. One must remember that companies trying to implement EMS in Poland belong to those most active and aware and besides they have difficulties in changing their attitudes both to environment and management.

2. PROBLEMS

Great part of problems related to implementation of EMS in Polish enterprises are caused by lack of understanding of what is EMS and why it is implemented. The main objective and indirect result of implementing EMS should be improved environmental performance of a company. The goal of EMS is to manage environmental problems and not to hide them, as it often took place in the old days. Companies having EMS in place should include environmental issues in any decision making process. It does not mean that all the problems should be necessarily solved at once nor become most significant decision factor. It is sufficient if environmental issues are taken as seriously as other problems. The experience indicates that this is much easier for companies having serious environmental problems than those having only little problems. They simply understand that solving environmental problems may be question of surviving. It is much easier to motivate staff and convince them that changes are necessary and some additional work will benefit in future.

It should be clear that EMS outlined in ISO14001 is not the only or best way to manage environmental problems. Certainly the system specially designed for the specific enterprise could be more effective as some of Japanese quality management systems but as ISO 14001 is recognised world wide it gives also the opportunity to demonstrate the environmental efforts to the public. In most of the cases it would not be of any advantage to design own system. ISO14001 gives a lot of freedom to make EMS fit to a company's specifics. If it is well tailored it can easily meet the needs. On the other hand it has to be understood that ISO14001 is just a set of tools which improperly used may bring minor or even adverse results. However demonstration aspect is of great value it may sometimes be not worthy to certify the system. There are advantages of using tools presented in ISO14001 also from business point of view. These are finding potential for savings, anticipation of legal breaches, staff motivation and better image.

The problems described below are examples of improper EMS functioning. They do not act separately. In most of the cases they overlap each other and together make EMS ineffective.

2.1. Initial Environmental Review

Initial environmental review, though not required by ISO14001, is regarded as necessary step in order to make EMS effective. The standard requires that "organisation shall establish and maintain (a) procedure(s) to identify the environmental aspects of its activities, products or services..." (clause 4.3.1). The best way to start with is performing environmental review which will help to assess the current environmental position and will be foundation for all planning phase steps. The review shall be made very thoroughly as all the mistakes will affect further steps—using results of the review the significant environmental aspects/effects will be identified. What has been observed is that the impact of every-day activities and indirect effects are often neglected. "Cradle to grave" approach is hardly used even if a company claims it takes it into consideration. Environmental impacts taking place out of production phase of product life cycle are rarely considered relevant. This is caused probably by old style regulations focusing on the "site impacts." Good example is electricity use, quite often not regarded as influencing environment. As stated before, everyday impacts are sometimes underestimated in comparison with emergencies. Well designed procedure of environmental aspects evaluation should include risk factor which represents probability of impact occurrence. This would bring the right proportion between everyday impacts (100% probability) and emergencies (usually very low probability).

The environmental review should cover not only environmental aspects/effects of activities but also existing environmental procedures which already are in place. They are usually not written or formally recognised. Sometimes they are not even realised but they exist and often work effectively. While identified they should be used as a basis for well structured procedures to be introduced in EMS. They often need to be modified, improved and approved but still they will work better than newly developed procedures.

2.2. Significant Environmental Impacts

Difficulties with identification of significant environmental impacts influences proper objectives and targets formulation. ISO 14001 (clause 4.3.3) say that "when establishing and reviewing its objectives, an organisation shall consider the legal and other requirements, its significant environmental aspects, its technological options and its financial, operational, and business requirements, and view of interested parties." The standard does not mention that targets should be achievable and challenging (it was mentioned in BS7750). Perhaps the authors assumed that any company taking burden of implementing EMS wants to make environmental improvements. This is not always the case. Some companies set very low targets in order to easily achieve them. It happens that targets are already met before they have been set! This problems happens especially when investments are needed. This is not really the way to make EMS implementation easier. First of all, certifiers, who do not wish to loose respect, will find very quickly that continual improvement is not accomplished. Secondly with no challenging targets staff motivation will drop soon. After that it will probably become obvious that maintaining this kind of fake system is more burden than advantage to a company. It has to taken into account that loosing certificate will bring much more bad publicity that certification has brought.

The other issue to be taken into account while setting environmental targets is use of BATNEEC (Best Available Technology Not Entailing Excessive Costs). It is not directly addressed in ISO14001 but it seems obvious that companies having certified environmental management should use it. In EU member states where EMAS Regulation requires to comply to BATNEEC rule, there are objective measures to indicate what best available and not entailing excessive costs really is. In Poland it is not easy but quite often that using just common sense it is clear that technology used is far from being "best."

2.3. Delegating Authority

It is not only environmental matters where Polish managers have difficulties to delegate authorities. In comparison to Scandinavian countries the responsibility and authority given to the lower management and line management is really low.

In terms of implementing EMS it is very important that the relevant people have not only responsibility but also authority given. The ISO14001 standard (clause 4.4.1) requires to appoint a person responsible for environmental management (EMS manager). This person should have enough authorities given to enforce the planned action. This is rarely the case, though often on organisational charts those persons are subordinate to top management. Usually it is quality manager or environmental manager (person responsible for reporting to regional authorities) appointed to position of EMS manager. None of them is the perfect choice. Quality managers have a lot of experience in maintaining quality management systems, they are skilled to prepare the right documents and they tend to build the perfectly functioning management system but they rarely understand the aim of EMS implementation—actual environmental performance improvement.

The environmental managers lack those managerial skills. They are used to react to regulations rather than be proactive. Furthermore they are not regarded managers related to the main stream of company's activities but rather as clerks coping with legal requirements. Therefore they are usually not in the right position to influence employees. Implementing EMS is about changing the company culture, re-thinking methods and procedures used for years. It requires a lot of inter-personal skills and self-confidence to convince people that what they have been doing for long time is wrong and need to be changed.

The easy way to find out if the EMS manager has enough authority is to check to what extent he/she are able to take any financial decisions, for instance in emergency in order to avoid or mitigate potential risks. In most cases it is evident that environmental managers are not in charge for this decisions and that is a significant limitation to his/her authority.

2.4. Training of Employees

Training of all employees according to their needs is a basic requirement of the standard (clause 4.4.2). In Polish enterprises it is often forgotten especially as far as lower level employees are concerned. In most of the cases there is no proper training needs identification procedures. Therefore resources are not used effectively. External training is often part of rewarding system which is only partly right. Participation in external training should be more related to actual company's needs than personal performance or even worse—relations. Some of the employees are thus trained more than necessary but training needs are not met. The above is due to top management view that EMS may be implemented by few highly trained persons. It may be implemented but it will not be possible to run it effectively if all relevant employees are not sufficient competent and aware of their duties and responsibilities.

2.5. Communication

Internal and external communication is crucial to effective EMS implementation (clause 4.4.1). All the employees should be involved in environmental activities at their workplace. The company should find a way to motivate employees and inform them what are the consequences of not conforming to EMS requirements. Very often employees in Polish companies are not aware of reasons of implementing EMS in their company and do not know more general objectives. All the employees, including floor workers, could bring a lot to effective maintaining EMS. They usually know best what is happening on the production and are the best source of environmental findings there. This is the case only if the employees know that it is for in the best company benefit to get this information. Especially the middle management has difficulties to change from caused by "command and control" way of thinking. They prefer to hide some problems rather than discuss their reasons openly. The information flow concerning environmental matters upwards the organisational structure is usually rather weak. Employees are afraid of making any comments or criticism to middle management or top management. The solution to this problem is to train the employees, increase their environmental awareness and provide them with the safe system of upwards information flow.

With external communication Polish companies deal much better. The only difference after EMS is implemented is that all correspondence, telephone inquiries etc. are directed to EMS manager or other authorised person. In good working system this person is to decide and is authorised to pass the information higher, including top management when it is necessary.

2.6. Involvement of Top Management

Top management of Polish companies, though declaring their involvement in environmental management, do not take environmental matters as their top priority. It happens that there is no visible interest shown by managers which could in turn motivate the rest of the staff. Without clear commitment from the top the system may not be efficiently implemented. All the efforts done by EMS manager are not seen as a corporate priority but as idea coming from "environmental freak" who is not taken too seriously. This is crucial that top management authorise the relevant staff publicly so that everybody knows that the "push" come from the top. That helps to convince persuade the reluctant people. Unfortunately, in the most cases if the managers have a clear goal in implementation EMS it is certification of their system rather than improving their environmental performance. This problem will probably solved when there are more environmental management systems certified in Poland and there will be closer public attention to real improvements not only to certificates. Top management, board of directors should be constantly informed on environmental matters. Environmental performance as well as financial performance of an enterprise should be discussed and assessed during top management meetings. If it is not accepted, that simply means a company is not ready for EMS. And there is nothing here to be criticised but not any company may hold EMS certificate.

2.7. Use of External Consultants

The role consultants in EMS implementation is very interesting from many points of view. They are either overestimated or underestimated, rarely used in the effective way. Large enterprises which can afford consultants prefer those names which have direct connection to certifiers and in this way they buy "a package"—system and certificate. This is the fastest way to reach the primary goal. At the same time some companies find this solution unacceptable for two main reasons. Firstly high costs and secondly they prefer to be more independent and build their "own" system.

Systems built entirely by consultants are often too complex for the needs of a company and they do not fit to a company specifics as well as those self designed. Requirements regarding EMS documentation (clause 4.4.4 and 4.4.5) are not as excessive as some of the consultants try to say in order to highlight their input in the process. On the contrary Annex A (clause A.4.5) indicates that "the primary focus of organisations should be on effective implementation of the environmental management system and on environmental performance and not on complex documentation control system." It is very common fact that documentation is much more complicated than necessary while actual system performance is not as good as it could be. That is the main source of negative opinions regarding effectiveness of both EMS and quality management systems.

The above is not to say that using consultants is not the right way. They should be used whenever applicable. We only have to keep in mind that they will leave after the job is finished and the company's' staff has to be able to continue and maintain what has been implemented by consultants. The company has to think what methods and solutions proposed by consultants is acceptable taking into account personnel capabilities. It has to be noted that staff members are the best experts in their field and consultants may add value by looking from a distance, experience and introducing some order to the knowledge and resources that company possesses. And this, as it was said before, should be exploited.

3. SUMMARY

The above described problems can be probably met in well-developed countries as well but 50 years of non-market economy in Poland may be of significance here. Self-regulation is hardly understood term for Polish managers. Though they do not find governmental environmental regulations properly working they often do not want to solve problems without push from the top. This will hopefully change while green image of a company will play more and more important role in attracting customers and certified EMS will be bottom line requirement to have access to attractive markets.

REFERENCES

1. Council Regulation No. 1836/93 allowing voluntary participation by companies in the industrial sector in a Community eco-management and audit scheme (EMAS)
2. ISO14001 Environmental management systems—Specification with guidance for use.
3. R. Welford, Corporate Environmental Management, Earthscan Publishing 1996

12

INNOVATIVE TECHNOLOGY FOR MUNICIPAL WASTE UTILIZATION FOR RZESZÓW CITY

B. Jamróz[1] and J. A. Tomaszek[2]

[1]"Maczki-Bór" Sand Mine, L. L. Company
Katowice, Poland.
[2]Rzeszów University of Technology
Rzeszów, Poland

ABSTRACT

This paper presents a modern system of simultaneous utilization of municipal waste and sewage sludge, based on the complex "MB-PREKO" system carried out by the "Maczki-Bór" Sand Mine, L. L. Company and a new Finnish technology the "Wabio" process which is being introduced for the first time in Poland for Rzeszów City. The advantages of the technology compared with conventional methods of waste utilization, especially composting and burning are shown.

Key words: waste, utilization, sewage sludge, biodegradation, mesophilic fermentation, refuse derived fuel, humus, biogas.

INTRODUCTION

The constantly increasing amount of waste is forcing both researchers and authorities to seek effective methods for its utilization. Organic waste is a special problem because of the appropriate technology required for its mineralization[1]. In Poland, the most common way to deal with municipal waste is to deposit it in dumping grounds. Composting is less popular. Burning in appropriate furnaces seems to be an emerging method. To date, in Poland there has not been a modern system of treating municipal waste in which the mechanical separation of the organic fraction biologically decomposable is followed by biological processing. The systems used currently are based on depositing and composting the waste without preliminary segregation. On the one hand, this adds to the mineralization of organic matter and improves the sanitary properties of the compost. On the other hand, some hazardous, inorganic elements may be transferred to the compost and increased leaching of heavy metals would be observed.

The analysis of the above mentioned ways of waste treatment i.e. depositing, composting and burning indicates that it is biological methods that should be used, in addition they have social approval. Methane fermentation can be used to neutralize the organic waste instead of composting. Its main advantage over composting is that it is the source of a reasonable amount of energy. It is commonly used to stabilize sewage sludge after biological treatment, so it could also be applied to the treatment of municipal waste. Simultaneous utilization of sewage sludge and the organic biofraction of municipal waste is especially recommended[2, 3].

This paper describes a modern system of treatment of municipal waste and sewage sludge currently being introduced in Rzeszów, based on a Finnish technology called "Wabio."

DESCRIPTION OF "WABIO" TECHNOLOGY

"Wabio" is a mesophilic, one-step process of fermenting the organic fraction of waste with sewage sludge through the use of a "wet method" (10–15% s.m. in the load). The diagrams illustrating the technology are shown in Fig. 1. There are three main stages in the process: (1) preparation of the feed, (2) fermentation of the biomass and (3) final treatment. After separating the large and hazardous components, the waste is taken to a hammer mill (grinder). Afterward it is transported, by a conveyor belt, to a trommel, where the biological and inert fractions are separated from the fraction of refuse derived

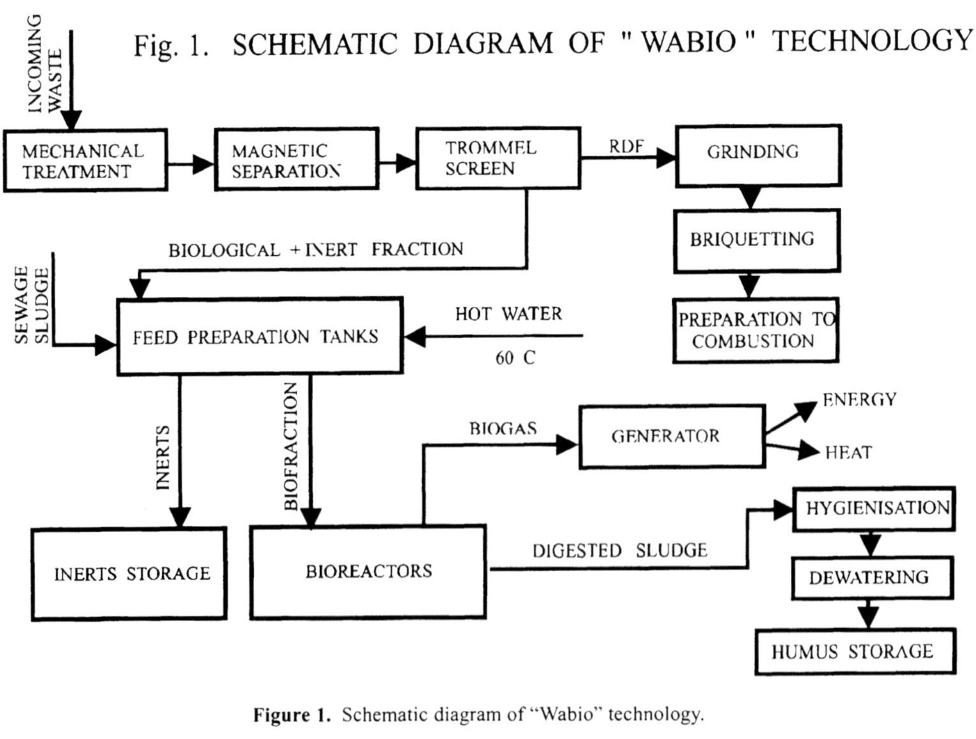

Figure 1. Schematic diagram of "Wabio" technology.

Table 1. Characterisation of refuse derived fuel (RDF)

	Chemical composition: % of wet weight
Moisture content	24,2
Carbon, C	35,9
Oxygen, O	22,3
Hydrogen, H	4,8
Nitrogen, N	0,4
Sulphur, S	0,2
Chlorine, Cl	1,0
Ash	11,2
Lump size: 90 w—% less than 80 mm	
Lower heat value: 13,8 MJ/kg	

The main elements in ash are silica, aluminium, and calcium.

fuel (RDF). RDF is then ground, briquetted and adjusted for burning. The biological and inert fractions are transferred by a conveyor belt to the feed preparation containers. In these containers the biofraction (i.e. the organic fraction of municipal waste containing unsortable mineral contaminants) is mixed with hot water (60 °C) and sewage sludge. During mixing the floating and sedimentary contaminants (inerts) are separated. The homogenous feed prepared in this way is then pumped to bioreactors where it undergoes a process of fermentation for 18–23 days. In the mesophylic fermentation process the mineralization of organic compounds takes place and results in the release of methane and carbon dioxide, called "biogas." This gas is used to generate electricity and heat. Part of the biogas is used to continually mix the content of the reactor. The fermented slurry from the reactors is pasteurized at 70 °C for 30 minutes to rid it of feacal bacteria. Then it is dehydrated in centrifuges or in screw presses. The final product, humus, is transported on conveyors to designated places and prepared for sale. The by product of the process, water is re-used for technological purposes and the excess is redirected to the sewage treatment. The key important parameters of Wabio process are presented in Table 1, 2, 3 and 4.

The Municipal Waste Utilization Works for Rzeszów City has been designed for simultaneous treatment of 51500 Mg/year of municipal waste and 17000 Mg/year of unfermented sewage sludge dehydrated to 82–85%. The Wabio technology is particularly recommended for Rzeszów City because of the considerable amount of organic matter in the municipal refuse. It is also possible to use the existing power boilers of the thermal-electric power station to burn the combustible fraction of waste materials.

Table 2. Quality characterisation of biogas

Moisture after water separation	%	< 2
Methane (CH_4, dry gas)	vol–%	55–65
Carbon dioxide (CO_2, dry gas)	vol–%	45–35
Hydrogen sulphite (H_2S)	mg/nm^3	< 12
Chloride (Cl^-)	mg/nm^3	< 0.3
Fluoride (F^-)	mg/nm^3	< 0.4
Ammonia (NH_3)	mg/nm^3	< 0.05
Dust		0
Higher caloric value	kWh/m^3	5.4..6.4

Table 3. Quality of biohumus produced from organic fraction

pH		7..8
Moisture after dewatering	%	< 70
Volume weight	kg/m^3	890
Max. size of inert particles	mm	10
Organic matter content of dry matter	%	40..50
Biodegradable matter		0
Maturity (German classification)		IV..V
C/N ratio		10..15
Total nitrogen (N)	g/kgDS	10..25
Total phosphorus (P)	g/kgDS	15..20
Potassium (K)	g/kgDS	4...7
Calcium (Ca)	g/kgDS	15..25
Magnesium (Mg)	g/kgDS	2...4
Iron (Fe)	mg/kgDS	70..100
Manganese (Mn)	mg/kgDS	700..900
Chromium (Cr)	mg/kgDS	100...130
Nickel (Ni)	mg/kgDS	30....40
Copper (Cu)	mg/kgDS	230..250
Cobalt (Co)	mg/kgDS	15...20
Zinc (Zn)	mg/kgDS	1000..1200
Lead (Pb)	mg/kgDS	100..400
Cadmium (Cd)	mg/kgDS	2...3
Mercury (Hg)	mg/kgDS	0,8...2

CONCLUSIONS

Wabio technology has many advantages compared with conventional methods of waste utilization, especially composting and burning. The most important among these are:

1. Profitable energy balance:

- It is possible to separate RDF which constitutes 20% of the raw waste mass and produces as much as 80% energy of the flammable substances of the raw waste mass.
- In Wabio technology the main gaseous product is methane, which can be utilized as source of energy whereas the main product of composting is carbon dioxide which is not energy gas.

2. The lack of negative influence on the environment:

  - The process is carried out in a closed reactor which prevents the emission of odour to the atmosphere and contact with birds, rats and other pests.
  - Contrary to composting the Wabio technology eliminates the greenhouse effect (there is no uncontrolled emission of CO_2 and CH_4 to the atmosphere).
  - Pathogenic microorganisms are eliminated.
  - Fixing of heavy metals in the humus as stable chelate compounds.

3. High quality product (humus) which can be used in farming.

Table 4. Parameters of the "Wabio" process

	Stable waste	Sludge
Parameters of the process		
total solids, (TS)	15	9–13
temperature:		
bioreactor, °C	35	35
hygienisation tank, °C	70	70
processing water, °C	40–70	60–70
retention time, d	15–20	20–25
biogas efficiency, m^3/t of waste	150–200	40–70
fermented sludge, TS-%	10 ± 1	8 ± 1.5
dewatered humus, TS-%	35–45	30–35
dehydrated humus, TS-%	55–60	50–55
Product balance per 1000 kg of waste		
biogas, kg	145	60
humus, kg	430 (TS 40%)	400 (TS30%)
sewage, kg	425	540
Energy balance per 1000 kg of dry waste		
biogas, kWh	1960	1760
production of heat in gas turbines, kWh	960	850
heat used, % of generated heat	30	50
energy generated in gas turbines, kWh	700	620
energy used, % of the generated energy	13	13

4. Small area: the compost heap occupies about 20% of the total area, which is several times smaller than the area of a waste dump.
5. Short processing time: 18 - 23 days for the mesophilic process.
6. Low C/N ratio: contrary to the aerobic process this does not require additional substances increasing humidity and C/N ratio.
7. Flexibility: the installation can be constructed stage by stage, depending on the capacity of the bioreactor.
8. Simultaneous treatment: both the solid waste biofraction and the nondigested sludge can be treated in the same technological process.

REFERENCES

1. Baran S. Composting: A Natural Method of Waste Utilization. Conference : *Hazards Caused by Inappropriate Waste Management; Technologies of Waste Utilization*. Czudec-Kaczarnica, May 28–29, 1996. Edited by "Proeco" Foundation. Rzeszów p. 72.
2. Jamróz B. Waste Utilization Factory as an example of Biological Method Used in The Modular System of Complex Waste Utilization. Conference: *Hazards Caused by Inappropriate Waste Management; Technologies of Waste Utilization*. Czudec-Kaczarnica, May 28–29, 1996. Edited by "Proeco" Foundation. Rzeszów pp. 73–92.
3. Prechtl, S., Chwistek, M., Jung, R. and F. Bischof, Examinations for Reducing Organic Polutants in Sewage Sludge by Aerobic Sludge Treatment with Thermophilic Conditions. Conference: *Wastewater Sludge Waste or Resource?* The Publishing Office of Technical University of Częstochowa, 1997, pp 187–194.

13

BIOFILM REACTORS

A New Form of Wastewater Treatment

J. A. Tomaszek and M. Grabas

Rzeszów University of Technology
2 W. Pola Street, 35-959 Rzeszów, Poland

ABSTRACT

The paper presents a new generation of Mobile Bed Biofilm Reactors: Fluidized Bed Biofilm Reactors (FBBRs), Airlift Biofilm Reactors (ALBRs) and Moving Bed Biofilm Reactors (MBBRs). The technical and technological characteristics as well as the efficiency of nitrification and denitrification processes are the main features which have been taken into consideration. The major advantages of the advanced biofilm systems are high specific biofilm surface, high reaction rates, high shear stress that controls biofilm thickness, improved mass transfer and mixing, operation stability, compact technology, application to existing and new plants. The disadvantage of biofilm systems is the problem of phosphate removal.

Key words: biofilm, biofilm reactors, mobile bed, fluidized bed, moving bed, air-lift reactors, nitrification, denitrification, innovative technologies.

INTRODUCTION

Conventional biological technologies of wastewater treatment have several negative aspects. Because of low sludge concentration and low settling rate of biomass, large reactor volume and large settler area are required. Low sludge age causes low treatment efficiency and high surplus sludge production. In aerobic processes, high energy consumption occurs. Large, open reactors emit noise, odour and aerosols. In these technologies, it is increasingly difficult to achieve the new European Union (EU) standards, in the effluent of wastewater treatment plants. Because of increased quality standards, the application of sewage sludge to agricultured crops is restricted. Also the standards of acceptable emission of noise, odour and aerosols from treatment plants are being tightened. This leads to

Chemistry for the Protection of the Environment 3, edited by Pawłowski *et al.* Plenum Press, New York, 1998

the upgrading of some of the existing sewage treatment plant and to the development of new efficient technologies which fulfil the requirements mentioned above.

In moving bed biofilm reactors the biomass is attached to a mobile bed. This results in a close contact of the biomass with the substrate and leads to higher capacity and efficiency. These processes are characterized by high biomass concentration, improved phase mixing, better oxygen transfer and phase separation in the reactors. The realization of high treatment efficiency, a low sludge production and nitrification, all point to the need of a long sludge residence time. This can most efficiently be obtained by immobilization of biomass[1].

Advanced biological processes used in wastewater treatment can be divided into three main groups according to the state of biomass: (1) suspended cultures, (2) mixed cultures and (3) fixed cultures[2]. A large number of studies have shown that the specific activity of fixed microorganisms and higher treatment capacity, compared to the suspended or mixed culture processes[2,3,4,5,6]. Fixed bed processes can be divided into two group depending on the state of medium: fixed bed with immersed granular medium and mobile bed.

The main purpose of this paper is to give the current state-of-the-art and information on mobile bed biofilm reactors (MbBBRs), to present an overview of the principal characteristics of these reactors (performance, scale-up, etc.) completed by a review of their advantages and disadvantages.

CLASSIFICATION OF MOBILE BED BIOFILM REACTORS (MbBBRs)

There are three main groups of MbBBRs whose media are kept moving continuously[2]. In the first group, the movement of the medium is maintained (fluidized) by high water velocity. They are called liquid-lift reactors or fluidized bed reactors. In the second group of MbBBRs the driving force maintaining media in continuous movement is gas (air). They are called gas-lift (air-lift) reactors. The third group of MbBBRs consist of the reactors whose biofilm carriers have the density slightly less than the density of water. The movement of the medium is caused by aeration (turbulent beds) in the aerobic processes and by a mechanical stirrer in the anoxic/anaerobic processes (stirred beds). In these groups of reactors there are reactors with constant feeding and portion feeding (SBRs).

FLUIDIZED BED BIOFILM REACTORS (FBBRs)

Fluidized bed biofilm reactors are the earliest fully developed group among the technologies using mobile beds. The first fluidized bed process developed for wastewater treatment in 1990 was patented in the USA[7,8]. At present there are many two-phase and three-phase FBBRs built in the USA and Europe. The most often used medium is sand[7], granulate active carbon[8], anthracite[9] and plastic[10]. Usually they have diameters from 0.5 mm to 1.0 mm, occasionally larger with a max. 3.0 mm. In the two-phase FBBRs, the homogenous expansion of the medium is caused by high liquid velocity. They are recommended for the following treatments: industrial waster water with low level of organic pollutants[8], tertiary treatment with nitrification[11,12] and methanization of food industry wastes[13]. Recently FBBRs were used for the denitrification of municipal wastewater[14,15,16]. Semon[14] tested upflow fluidized denitrification reactor in the following operating conditions: sand effective size 0.46 mm, methanol feed 3.3g per g nitrate in influent, temperature range 11°C–25°C, average 15,2°C, contact time less than 5 minutes. For nitrate loading as high as 6.41 kg $N\text{-}NO_3^-/dm^3$ of expanded reactor bed, the nitrate removal was

95%. Despite the major advantages of two-phase FBBRs: high efficiency, high biomass concentration and compact installation, their application to industry has been limited by the necessity to control bed expansion and biofilm thickness, influent distribution device and oxygen saturation system. Investment costs have been reduced (up to 50% of the cost of activated sludge reactor) but operating costs are higher[2].

Three-phase FBBRs partly solve these problems by using simultaneous gas and liquid injection. Addition of the gas phase significantly alters the hydrodynamic regime of the fluidized bed. Major changes can occur to bed expansion, porosity, liquid and gas hold-up and solid hold-up. These changes can affect biofilm detachment and substrate removal. This improves liquid/solid mass transfer and controls biofilm thickness[17]. Three-phase FBBRs can be divided according to flow direction into up-flow and down-flow reactors.

In an up-flow fluidized bed the granular density of medium is higher than 1 g/ml, often 2.5–3.0 g/ml. Compared to previous reactors, they do not require any equipment to control biofilm thickness. However, there are problems with controlling bed expansion and homogeneous flow distribution[18,19]. According to Elmaleh[18], the energy consumption in these reactors is similar to that in activated sludge processes and lower than in two-phase FBBR with pre-oxygenation. The major drawback of these processes is the difficulty to produce fluidization and air injection simultaneously (coalescence of air bubbles, bed turbulence, etc.). These problems were overcome by Fan[20] and Tang[21] who developed a draft-tube reactor by replacing the driving force created by the liquid velocity with the driving force of air-lift type gas injection.

Down-flow (inverse) FBBRs are more reliable. Low-density granular medium (floating medium) of density varying from 0.05 to 0.9 g/ml is made of plastic. This medium can be fluidized by the driving force of the liquid (column type reactor) or gas (draft-tube reactor). Column type bioreactors are recommended for denitrification processes. However they require the homogeneous distribution of liquid flow and effluent recirculation. The rate of expansion of the floating bed is difficult to control plus there is a high risk of bed clogging[10]. In the draft-tube reactors, liquid circulation carried out using the air-lift principle expands the flotation bed to the bottom of the reactor. During the development of the biofilm the density of bioparticles increases and they are led into the internal column where the overgrowth of the biofilm is stripped by shear stress. These reactors ensure complete mixing, tolerating load, concentration and temperature variations. They are characterized by high carbon and nitrogen removal rate. Their efficiency is many times higher than that of the conventional processes[22,23].

AIR-LIFT BIOFILM REACTORS (ALBRs)

Air-lift mobile beds are a new generation of reactors. The driving force of ALBRs is gas injection which gives it some advantages such as those of the three-phase fluidized bed reactors (high removal efficiency, high nitrification rate, no clogging problems) but with more simple design and operation. Several studies have demonstrated the ALBR's ability to provide high mass transfer at low power consumption[24,25,26]. These advantages have the effect of lowering investment and operating costs.

As a result of research an aerobic biofilm air-lift suspension reactor was designed and patented as CIROX® technology[27,28,29] and a circulating bed biofilm reactor was patented as Turbo N® technology[30,31].

The CIRCOX® process is carried out in the air-lift reactor with internal circulation and integral settler (Figure 1). The carriers are recycled into the reactor not by pumping but by the action of the secondary air lift. Basalt (density of 1.6 kg/l) is used as a carrier for the biomass, the diameter of the basalt particles ranged from 0.09 to 0.30 mm. The amount of the carrier is limited by the drop in the oxygen transfer at high solid contents. Therefore, the amount of 10–15% (maximum 20%) v/v carrier material is assumed. Theoretical specific biofilm surface area for that percentage of filling varies from 3000 m^2/m^3 to 6000 m^2/m^3. Biomass concentration of 10–40 kg VSS/ m^3 in the reactor is obtained, typical biofilm thickness is 50–150 μm[27,28,29].

For the pilot scale[27] and for the following operating conditions: the wastewater residence time of 2 hrs and the superficial air velocity of 0.05 m/s., the rate of nitrification was 1.5–2 kg N-NH_4^+/m^3d for high volumetric loading COD (4–10 kg COD/ m^3d) and nitrogen (1–2 kg N-NH_4^+/ m^3d). Full ammonia conversion to nitrate was possible for loading rates up to 5 kg N-NH_4^+/ m^3 reactor per day[32]. The sludge on the basalt particles has high settling capacities (50–200 m/h).

The CIRCOX® air-lift technology with integrated denitrification is suitable for treating municipal sewage. The denitrifying air-lift reactor was tested on pilot scale in two reactors, with an oxic volume of 2.28 m^3 each. One reactor was extended with an anoxic compartment of 1.16 m^3 (34% of the total reactor volume)[29]. In this reactor, the wastewater/sludge mixture was circulated alternately over the oxic (BOD removal and nitration process) and anoxic (denitrification process) compartments by means of an airlift pump. For municipal wastewater at a COD conversion rate of 4.2 kg COD/m^3d, the nitrification efficiency was over 90%, nitrification rate was 1,3 kg N-NH_4^+/ m^3d and denitrification rate was 0.6 kg N-NH_4^+/ m^3d. The mean soluble COD and soluble BOD_5 concentration in the effluent of both reactors were 55 and 7 mg/l respectively.

Recent experimental observations[32,33] have shown that it is possible to obtain full ammonium conversion with approximately 50% nitrate and 50% nitrite in the effluent of

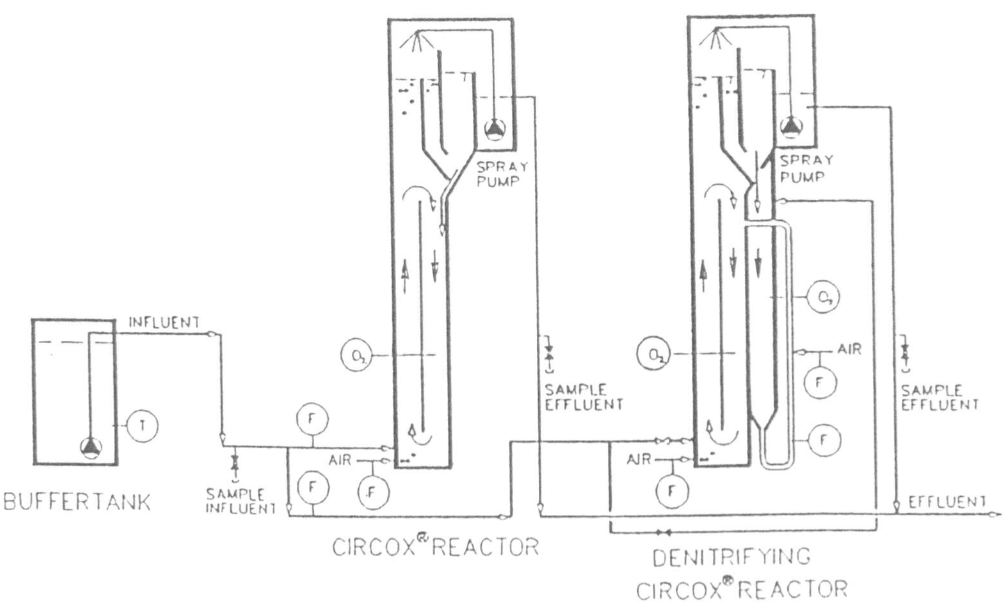

Figure 1. Schematic diagram of pilot plant with air-lift suspension reactor ALSR (according to Frijters et al. 1996).

the biofilm air-lift suspension reactor. For an integrated nitrification-denitrification processes, nitrite formation in the aerobic stage leads to substantial savings. Lower oxygen consumption (25%) in the aerobic stage results in a 60% energy savings. In the anoxic stage the electron donors requirement is lower, and the nitrite denitrification rates are 1.5 to 2 times higher than with nitrate.

Garrido[33] reported that during the normal operation of nitrifying biofilm air-lift suspension reactor in the presence of organic matter coming from the biomass hydrolysis or that added in the feeding medium (formaldehyde), nitrous oxide production was detected thus implying simultaneous nitrification-denitrification in the system. Denitrification rate of nitrite was 3 times higher than the with nitrate.

To sum up, the major advantages of CIRCOX® technology are: (1) high efficiency, (2) low biomass production and high biomass age, (3) highly compact system with integrated settler and (4) small requires area. The major disadvantages are: (1) limited reactor's filling caused by the negative influence of carriers on hydrodynamics and (2) a more complex operation.

The circulating floating bed biofilm reactor (CFBBR) is a new concept of ALBR. This new bioprocess called TURBO N® has been developed by the Degrémont company. The solid medium used is a rough, plastic, granular product on the basis of polyethylene with a diameter between 0.5–3.0 mm and a density slightly below 1.0 kg/l. The reactor is filled with these carriers and an average solid hold-up of 15–20% (up to 40% v/v). Theoretical specific biofilm surface area for that percentage of filling varies from 500 m^2/m^3 to 1200 m^2/m^3. The total amount of fixed biomass in the reactor was 7.5±1.5 kg SS/ m^3. The calculated biofilm thickness is below 50–80 μm[30,31]. The CFBBR is characterized by the dividing of the reactor into two sections :an up-flow aerated section and a down-flow non aerated section (Figure 2). The floating medium, air and water form a homogeneous three phase mixture circulating in both sections of the reactor.

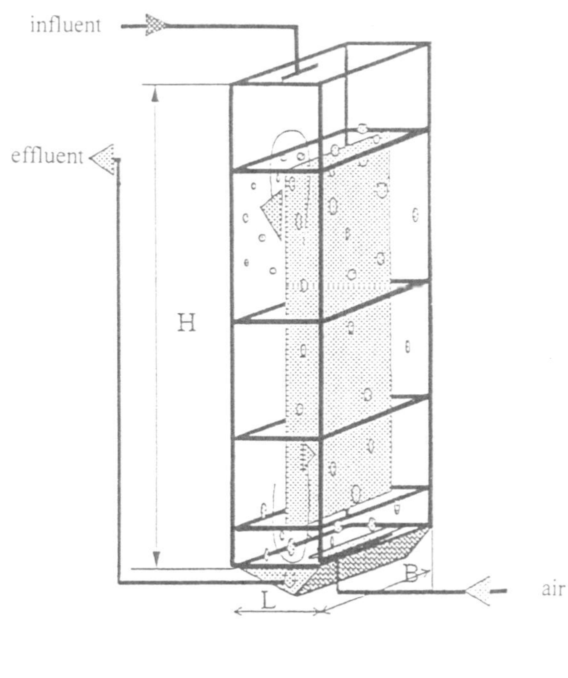

Figure 2. Schematic diagram of circulating bed reactor (according to Lazarova et al. 1996).

The optimum hydrodynamic characteristics of the CFBBR (hydraulic residence time HTR from 0.5 to 4 hours, gas velocity 0.01–0.035 m/s, liquid circulation velocity 0.3–0.4 m/s, mass transfer coefficient k_La 50–300 h^{-1}, average mixing time 85 s) are not deteriorated by the high solid hold-ups (up to 40% v/v). On the contrary three-phase operation improves the local gas hold-up in the downcomer. A good mass transfer in the non-aerated down-flow section could be ascribed to the lower density of particles which is close to the density of gas-liquid fluid. Improved mass transfer and mixing in the CFBBR guarantee high nitrification rates and operation stability[30].

The experiments on an industry scale[31] showed that tertiary nitrification rate was up to 2 kg $N\text{-}NH_4^+/m^3d$ (for COD/N = 0.3–1 and HRT = 0.6–1.2 h), and secondary nitrification rate was up to 0.6 kg $N\text{-}NH_4^+/m^3d$ (for COD/N = 4–10 and HRT = 1.5–4 h). Total carbon elimination was observed with removal rate up to 5 kg COD/ m^3d. The mean value of the effluent COD concentration was about 50 mg/l (BOD_5 <5 mg/l). The system is compact, simple and there is no need for back washing.

The TURBO N® technology is simple without any sophisticated devices (easy effluent and air-flow distribution, no primary settling, no back-washing). Therefore this innovative and more compact technology is an attractive solution for intensive nitrogen and carbon removal both in new compact works and in the existing plants which need upgrading.

The described processes are based on natural basalt and plastic carriers. However research is being done on the application of other carriers. One of the recent investigations concerns an "Aquacel" carrier , made of foam cellulose with continuous macropores[34]. This carrier has a cubic form of particles, size 1–5 mm, mean pore size 100–1260 μm and density 1.05 kg/ m^3, a void fraction of 97%. Broad pore size distribution and the large pores of the carrier prevented its surface from being fully covered with attached cells even after a long-term cultivation under high ammonium loading rate. It was important to enhance the oxygen transfer inside the carrier. The macro porous cellulose carrier was applied for nitrification to develop a compact and rapid nitrogen removal system. The ammonium oxidation was highly influenced by dissolved oxygen, and the maximum value was attained at 6g/l O_2 in air-lift reactor with moderate mixing. For 1mm cubic carrier the maximum ammonium loading rate was 12 kg $N\text{-}NH_4^+$/ m^3 carrier d.

Another modification of the ALBRs was a sequencing batch wastewater treatment process[35]. A modification of the SBR process was introduced to exploit the ALBRs' geometric structure. This unique treatment process combining the advantages of the ALBRs with the flexibility and simplicity operation of the SBRs demonstrated consistent and highly efficient removal of COD. The advantages of the periodically operated biofilm reactors for the treatment of highly variable wastewater were pointed by Woolard[36]. The continuous flow operation of biofilm reactors caused stratification and uneven biomass distribution which limits the efficiency of these reactors. Periodic biofilm reactors achieve better biomass distribution. Periodic operation imposes regular variation in substrate concentration in the biofilm. The biofilm achieves maximum growth rates which can result in improved reaction potential, stable and robust systems.

A very interesting direction of research is a hybrid concept with nitrogen removal using nitrifying biofilm and denitrifying suspended growth in a biofilm airlift suspension reactor[37]. This solution decreases the limitation of biomass transport and increases an efficiency of the process. Another test of hybrid system was also carried out in a fluidized reactor for the dyeing industry[38].

MOVING BED BIOFILM REACTORS (MBBRs)

The third group of MbBBRs consist of the reactors in which the movement of the medium is caused by aeration (turbulent beds) in the aerobic processes or by a mechanical stirrer in the anoxic/anaerobic processes (stirred beds).

The CAPTOR® process uses a combination of attached and suspended biomass by adding polyurethane foam media to the activated sludge process[39,40]. The biological reactor is filled with reticulated polyurethane foam media having about 97% void space with internal and external surfaces for biomass attachment and/or entrapment. The media are of 1" × 1" × 0.5" dimensions having a specific gravity of about 1.0 g/l when filled with biomass. The CAPTOR® system was designed as a separate front portion of the activated sludge process. The suspended biomass in the Captor basin is around 0.9–1.7 g/l and the equivalent attached biomass is maintained at 6.5–12 g/l. The media are retained in the tank by a wire mesh screen that is placed perpendicularly to the direction of flow. It has been found that soluble BOD_5 removal (95%), nitrification (70–90%) and partial denitrification (40–60%) occur simultaneously (at the HRT of 50–90 minutes) in the Captor chamber. Disadvantages of the CAPTOR® system are: (1) the necessity of removing the biomass overgrowth, (2) transport of medium to the beginning of the system, (3) low sludge age and (4) high surplus sludge production.

The Kaldnes process is the most thoroughly tested process from mobile bed biofilm reactors. This technology has been developed in Norway[41,42,43,44]. A sewage inflow to the Kaldnes reactor should not contain suspended matter, therefore preliminary sedimentation is necessary. The MBBR system consists of 4–8 reactors in a series, separated by simple wall constructions (Figure 3). That is why in each reactor specific biofilm is growing on carriers elements, depending on the reactor's function and its load. To keep the biofilm carriers in the reactors, sieves (with 7 mm openings) are placed at the each reactor outlet. The agitation is arranged so that the carrier elements are constantly being moved upwards

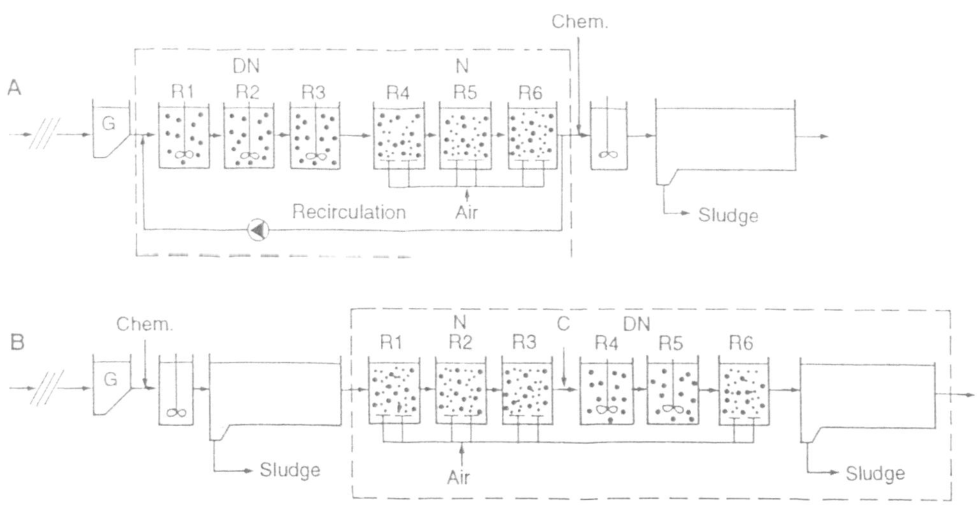

Figure 3. Schematic diagram of treatment plants with moving bed biofilm reactors. (A) Predenitrification and combined precipitation. (B) Preprecipitation and post-denitrification (according to Rusten et al. 1995a). C: External carbon source; DN: denitrification; G: grit chamber; N: nitrification; R1–R6: biofilm reactors.

over the surface of the sieve. This creates a scrubbing action that prevents clogging. The carriers are made of polyethylene (density 0.92–0.96 g/l) and are shaped like small cylinders (about10 mm in diameter and in height) with a cross inside the cylinder and longitudinal fins on the outside The filling of carrier elements in the reactor may be decided for each case, giving considerable flexibility in the specific biofilm area. A maximum filling of about 70% corresponds to the specific potential growth area of biofilm of about 500 m^2/m^3. Because the growth is much less on the outside of the cylindrical carrier than on the sheltered inside area, the maximum practical specific growth area is expected to be about 350 m^2/m^3. Biomass concentration in the reactor reaches 5.2 kg/m^3, average 4.0 kg/m^3, of which the suspended biomass is about 2%. Average concentration of the suspended matter in the inflow to a secondary settling tank varies from 80–85 mg/l. The thickness of the biofilm reaches 700 μm[46,47,48]. The studies show that Kaldnes process as a highly loaded stage combined with an activated sludge forms an efficient, stable and competitive combination process which can be applied both to municipal[42,43,44,46,48] and industrial[45,49,50,51] wastewater treatment. Biological pretreatment of dairy wastewater in an aerobic pilot-plant with two MBBRs showed 85% and 60% COD removal at volumetric organic loading rates of 12 kg COD/m^3d and 21.6 kg COD/m^3d, respectively[45]. At an organic load of 25 kg total COD/m^3d coming from fiber mill, to the first Kaldnes stage, the reduction of soluble COD and BOD in these stages was 40–50%. With an organic load of 12 kg total COD/m^3d based on both stages of the Kaldnes process, the total reduction of soluble COD was 70–80%[49]. Treatment of integrated newsprint mill wastewater in MBBRs achieved a removal efficiency of 65%-82% COD for organic load of 16 kg COD/m^3d, at HRT of 4–5 hrs[50].

The application of MBBRs to nutrient removal in municipal wastewater treatment was presented by Odegard[43]. The results showed that predenitrification/post-denitrification alternative required more than 6 hrs HRT to achieve the best removal efficiency (70% N-removal at a circulation ratio 2). The pre-precipitation/post-denitrification alternative, however, could meet a higher N-removal efficiency (>85%) at a HRT of less than 3 hrs. With low organic removal rates, wastewater temperature of 10 C and 10 mg O_2/l in the reactors, nitrification rate for the predenitrification system was 313 g $N\text{-}NH_4^+/m^2d$, and for the post-denitrification system was 386 g $N\text{-}NH_4^+/m^3d$[48]. Maximum denitrification rate for the predenitrification process, measured over the entire anoxic zone was 130 g $N\text{-}NO_3^-/m^3d$, or 210 g $N\text{-}NO_3^-/m^3d$ measured over the first anoxic reactor. In the post-denitrification process, with addition of external source of carbon, denitrification rates up to700 g $N\text{-}NO_3^-/m^3d$ were obtained in the first anoxic reactor. Specific sludge production for these nutrient removal systems varies slightly (0.36–0.38 kgTSS/kg COD) depending on HRT (2.5–5 hrs)[46]. These values are comparable to the surplus low loaded activated sludge, and are lower when compared to the similar as Kaldnes system activated sludge load.

Similar to ALBRs, organic carbon and nitrogen removal in a heterotrophic sequencing batch MBBR for denitrification and in a continuous-flow autotrophic MBBR for nitrification was tested[52,53]. Denitrification tests performed without external carbon sources yielded denitrification rate that varied from 100–200 g $N\text{-}NO_3^-/m^3d$.

The MBBRs have a simple construction and are easy to operate. In many cases the existing non-nitrifying activated sludge plants may easily be upgraded to nitrogen removal without expanding the existing reactor volumes. There is no need for recycling the biomass and the capacity of the reactor may be altered by simply changing the degree of filling[45,46].

There are many of advantages of MBBR technology compared to the conventional methods. The MBBR technology is as simple as the activated sludge technology, and re-

Table 1. Parameters of mobile bed biofilm reactors

Parameter	Unit	Fluidized reactors	Air-lift reactors			Moving bed reactors		Conventional activated sludge
			Circox	Turbo N	Aquacel	Captor	Kaldnes	
Carrier material	—	Sand, anthracite	basalt	plastic	foamed cellulose	Polyure-thane	Polyethyle-ne	—
Density of material	kg/l	>1	1,6	<1,0	1,05	<1,0	0,92–0,96	—
Size	mm	0,1–0,3	ϕ0,09–0,3	ϕ < 5	1´1´1–3´3´3	25´25´12	ϕ 9, high 7	—
Porosity	%	100	100	100	97	97	95	—
Potential surface	m^2/m^3	3000–6000	3000–6000	500–1200	—	—	360–500	—
Carrier hold-up	% v/v	10–15	10–15	15–40	10	80	50–70	—
Biomass concentration	kg VSS/m^3	2–22	10–40	7,5 ± 1,5	—	6,0–9,0	3–5,2	3,0–6,0
Biofilm depth	µm.	50–150	50–150	50–80	—	—	700	—
HRT	hr	—	2,0	0,5–4,0	3	0,85–1,5	3,0–6,0	7–18
Nitrification rate	kg N/m^3d	—	< 1,3	< 2,0	1,2	—	< 0,40	—
Volumetric loads	kg $/m^3d$	6–24 (COD)	4–10 (COD)	<20(COD)	—	1,4–4 (BOD)	1–25 (COD)	5

quires much smaller area. There is no need to wash carrier. Total volume of the reactor is effectively used. The volumetric efficiency is the same as for submerged biofilters. Reaction rate is higher for the same biomass concentration as for the activated sludge concentration caused by higher activity of the biomass. The main drawback is higher cost of Kaldnes elements and the necessity to maintain high oxygen concentration for economical effectiveness.

SUMMARY

Increasing quality standards concerning industrial and municipal wastewater stimulate the need for introducing novel, new and innivative technologies for their treatment. Currently, one of the new directions in sewage treatment technology which is dynamically evolving are bioprocesses based on biomass attached on a movable carrier. In the 1990s this technology were developed on a full technical scale in several processes: CIRCOX, TURBO N , KALDNES.

In all these processes, there is no need for biomass recirculation. They are more stable then the activated sludge technology. Another advantage is an effective self-regulation, which eliminates washing and prevents clogging. Better phase mixing and larger surface for mass transfer which leads to high speed of reaction. High reaction rates of liquid phase, thickness of biofilm and turbulence result in smaller diffusion limits from liquid to biofilm and in the biofilm itself. This facilities mass transfer. Moreover, in Kaldnes technology, high reaction speed is the effect of high biomass activity achieved through biomass "specialization" i.e. its ability to adapt both to substrate loads and the kind of process carried out in a given reactor. High airing efficiency is obtained in column reactors.

These processes are applied both for industrial and municipal wastewater discharges. Kaldnes technology which was the earliest developed, is applied to discharges from forest, food and paper industries as well as to municipal sewage. It is used for removal of organic matter as the first highly loaded stage of biological treatment, proceeding in the second stage (activated sludge). It is also used for nitrogen removal, most often combined with post-denitrification. Kaldnes technology is useful for modernizing the existing, overloaded treatment activated sludge plants. CIRCOX technology has been successfully applied to treating discharges from pharmaceutical plants. Combined with integrated denitrification it could be used for municipal sewage purification. Similar applications of TURBO N process depended on its ability to remove nitrogen.

Recently research is being conducted on airlift reactors. Concentrating on biomass carrier, reactor's geometry and its modifications aiming at nitrogen removal, simultaneous nitrification and denitrification, phosphorus removal[54,55,56]. These reactors combined with biomass attached on movable carrier are an interesting form of wastewater treatment, due to their good hydrodynamics properties, facilitated biomass transfer and other advantages. One drawback is the lack of phosphorus removal.

REFERENCES

1. Pascik, I., Mann, T., The two step nitrification of ammonia rich waste water. *Wat. Sci. Tech.*, 1984, 16 (10–11), pp. 215–223.
2. Lazarova, V., Manem, J., Advances in biofilm aerobic reactors ensuring effective biofilm control. *Wat. Sci. Tech.*, 1994, 29 (10/11), pp. 319–327.

3. Audic, J.M., Faup, G.M., Navarro, J.M., Specific activity of Nitrobacter through attachment on granular media. *Wat. Res.*, 1984, 18 , pp. 745–750.
4. Klein, J., Ziehr, H., Immobilization of microbial cells by adsorption. *J. Biotechn.*, 1990, 16, pp. 1–16.
5. Broch-Due, A., Andersen, R., Kristoffersen, O., Pilot plant experience with an aerobic moving bed biofilm reactor for treatment of NSSC wastewater. *Wat. Sci. Tech.*, 1994, 29 (5–6), pp. 283–294.
6. Nicolov, L. and Karamanev, D., Change of microbial activity after immobilisation of microorganisms, in: *Physiology of Immobilised Cells*, ed. by J.A.M. DeBont et al., Elsevier 1990, pp. 4461–466.
7. Cooper, P.F., The use of biological fluidized beds for the treatment of domestic and industrial wastewater, 1981, *The Chem. Eng*, (8/9), 373–376.
8. Sutton, P.M., Mishra, P.N., Fluidized bed biological wastewater treatment; effects of scale-up on system performance, 1990, *Wat. Sci. Tech.*, 1990, 22 (1–2), pp.419–430.
9. Hosaka, Y., Minami, T., Nasuno, S., Fluidized-bed biological nitrogen removal, 1991, *Wat. Environ. Tech.* (8), 48–51.
10. Turner, C.D., Koehmstedt, M.J., Gallagher, J.R., A comparison of coupled upflow and coupled downflow fluid bed reactors treating synfuel wastewater, 44th Purdue Ind. Waste Conf. Proc., Lewis Publ., Inc., Chelsea, Michigan, 1990, pp. 475–484.
11. Cooper, P.F., Williams, S.C., High-rate nitrification in a biological fluidized bed. *Wat. Sci. Tech.*, 1990, 22 (1–2), pp. 431–442.
12. Vredenbregt, L., Nielsen, K., Potma, A., Kristensen, G., Sund, Ch., Fluid bed biological nitrification and denitrification in high salinity waste water. *Proceedings of Third International IAWQ Special Conference on Biofilm Systems*, 28–30.08.1996, Copenhagen.
13. Ehlinger, F., La fermentation methanique en lits fluidises en France. *L'eau, L'ind., Les Nuisances*, 1992, (155), pp. 48–50.
14. Semon, J., Sadick, T., Palumbo, D., Santoro, M., Keenan, P., Biological Upflow Fluidized Bed Denitrification Reactor Demonstration Project—Stamford, Ct. *Proceedings of Third International IAWQ Special Conference on Biofilm Systems*, 28–30.08.1996, Copenhagen.
15. Popel, H.J., W Kristeller, W., Post-denitrification at the Frankfurt-Niederrad wastewater treatment plant by fluidized-bed technology. *Wat. Sci. Tech.*, 1997, 35 (10), pp. 95–102.
16. Hirata, A., Meutia, A.A., Denitrification of nitrite in a two-phase fluidized bed bioreactor. *Wat. Sci. Tech.*, 1996, 34 (1–2), pp. 339–346.
17. Trinet, F., Heim, R., Amar, D., Chang, H.T., Rittman, B.E., Study of biofilm and fluidization of bioparticles in a three-phase liquid-fluidized-bed reactor. *Wat. Sci. Tech.*, 1991, 23 (7–9), pp. 1347–1354.
18. Elmaleh, S., Papaconstantinou, S., Rios, G.M., Biological wastewater treatment in a high compacting multiphasic reactor. In: *Recents progres en genie des procedes*, 1992, 6 (20), pp.1–9, Toulouse.
19. Chang, H.T., Rittmann, B.E., Predicting bed dynamics in three-phase, fluidized-bed biofilm reactors. *Wat. Sci. Tech.*, 1994, 34 (10–11), pp. 231–241.
20. Fan, L.S., Fujie, K., Long, T.R., Some remarks on gas-liquid mass transfer and biological phenol degradation in a draft tube gas-liquid-solid fluidized bed bioreactor. *A.I.Ch.E. Symp. Series*, 1984, 80 (241), pp. 102–109.
21. Tang, W.T., Wisecarwer, K., Liang-Shih Fan, Dynamics of draft tube gas-liquid-solid fluidized bed bioreactor for phenol degradation. *Chem. Eng. Sci.*, 1987, 42 (9), pp. 2123–2134.
22. Karamanev, D.G., Nikolov, L.N., Influence of some physicochemical parameters on bacterial activity of biofilm: ferrous iron oxidation by Thiobacillus ferrooxidans. *Biotech. Bioeng.*, 1988, 31, pp. 295–299.
23. Nikolov, L., Karamanev, D., Penev, T., Dimitrov, D., Full-scale inverse fluidized bed biofilm reactor for waste water treatment. *2nd Int. Biotechnology Conf.* may 6–9 1990, Seoul, p.12.
24. Siegel, M.H., Merchuk, J.C., Mass transfer in a rectangular air-lift reactor: Effects of geometry and gas recirculation. *Biotechnol. Bioeng.*, 1988, 32, pp. 1128–1137.
25. Siegel, M.H., Hallaile, M., Merchuk, J.C., Air-lift reactors: Design, operation, and applications. In Mizrahi A. (ed): "*Advances in Biotechnological Processes,*" Vol. 7, New York, 1988, Alan R. Liss, p. 79.
26. van den Heuvel, J.c., Varschuren, P.G., Ottengraf, S., Acceleration of mass transfer in loop reactors. *Proceedings of Third International IAWQ Special Conference on Biofilm Systems*, 28–30.08.1996, Copenhagen.
27. Heijnen, J.J., van Loosdrecht, M.C., Mulder, R., Mulder, A., Weltervrede, R., Development and scale-up an aerobic biofilm air-lift suspension reactor. *Wat. Sci. Tech.*, 1993, 27 (5–6), pp. 253–261.
28. van Benhum, W.A.J., van Loosdrecht, M.C.M., Tijhuis, L.., Heijnen, J.J., Solids retention time in heterotrofic and nitrifying biofilms in a biofilm airlift suspension reactor. *Wat. Sci. Tech.*, 1995, 32 (8), pp. 35–43.
29. Frijters, C.T., Eikelboom, D.H., Mulder, A., Mulder, R., Treatment of municipal wastewater in a CIRCOX® airlift reactor with integrated denitrification. *Proceedings of Third International IAWQ Special Conference on Biofilm Systems*, 28–30.08.1996, Copenhagen.

30. Lazarova, V., Manem, J., An innovative process for waste water treatment: the circulating floating bed reactor. *Wat. Sci. Tech.*, 1996, 34 (9), pp. 89–99.
31. Lazarova, V., Nogueira, R., Manem, J., Melo, L., Control of nitrification efficiency in a new biofilm reactor. *Proceedings of Third International IAWQ Special Conference on Biofilm Systems*, 28–30.08.1996, Copenhagen
32. Picioreanu, C., van Loosdrecht, M.C., Heijnen J.J., Modeling the effect of oxygen concentration on nitrite accumulation in a biofilm airlift suspension reactor. *Proceedings of Third International IAWQ Special Conference on Biofilm Systems*, 28–30.08.1996, Copenhagen.
33. Garrido, JH.M., Campos, J.L., Mendez, R., Lema, J.M., Nitrous oxide production by nitrifying biofilms in a biofilm air lift suspension reactor. *Proceedings of Third International IAWQ Special Conference on Biofilm Systems*, 28–30.08.1996, Copenhagen.
34. Matsumura, M., Yamamoto, T., Wang, P., Shinabe, K., Yasuda, K., Rapid nitrification with immobilized cell using macro-porous cellulose carrier. *Wat. Res.*, 1997, 31 (5), pp. 1027–1034.
35. Brenner, A., Ben-Shushan, N., Siegel, M.H., Merchuk, J. C., Pilot plant performance and model calibration of a sequencing batch air-lift reactor. *Wat. Sci. Tech., 1997,* 35 (1), pp. 121–128.
36. Woolard, C.R., The advantages of periodically operated biofilm reactors for the treatment of highly variable wastewater. *Wat. Sci. Tech.*, 1997, 35 (1), pp. 199–206.
37. van Benthum, W.A.J., van Loosdrecht, M., Heijnen, J.J., Nitrogen removal using nitrifying biofilm and denitrifying suspend growth in a biofilm airlift suspension reactor. *Proceedings of Third International IAWQ Special Conference on Biofilm Systems*, 28–30.08.1996, Copenhagen.
38. Park, Y., Lee, C., Dyeing wastewater treatment by activated sludge process with a polyurethane fluidized bed biofilm. *Wat. Sci. Tech.*, 1996, 34 (5–6), pp. 193–200.
39. Golla, P.S., Lin, J.N., Cold Temperature Nitrification Using the Captor Process. *Proceedings of the 46th Industrial Waste Conf.* Lewis Publishers, 1991, pp. 631–640.
40. Golla, P.S., Reddy, M.P., Simms, M.K., Laken, T.J., Three years of full-scale Captor® process operation at Moundsville WWTP. *Wat. Sci. Tech.*, 1994, 29 (10–11), pp. 175–181.
41. Hem, L.J., Rusten, B., Odegaard, H., Nitrification in a moving bed biofilm reactor. *Wat. Sci. Tech.*, 1994, 28 (6), pp. 1425–1433.
42. Odegaard, H., Rusten, B., Badin H., Small wastewater treatment plants based on moving bed biofilm reactors. *Wat. Sci. Tech.*, 1993, 28 (10), pp. 351–359.
43. Odegaard, H., Rusten, B., Westrum, T., A new moving bed biofilm reactor—applications and results. *Wat. Sci. Tech.*, 1994, 29 (10–11), pp. 157–165.
44. Rusten, B., Kolkin, O., Odegaard, H., Moving bed biofilm reactors and chemical precipitation for high efficiency treatment of wastewater from small communities. *Wat. Sci. Tech.*, 1997, 35 (6), pp. 71–79.
45. Rusten, B., Odegaard, H., Lundar, A., Treatment of dairy wastewater in a novel moving bed biofilm reactor. *Wat. Sci. Tech.*, 1992, 24 (3–4), pp. 703–711.
46. Rusten, B., Siljudalen, J.G., Nordeidet, B., Upgrading to nitrogen removal with the KMT moving bed biofilm process. *Wat. Sci. Tech.*, 1994, 29 (12), pp. 185–195.
47. Rusten, B., Hem, L.J., Odegaard, H., Nitrogen removal from dilute wastewater in cold climate using moving-bed biofilm reactors. *Wat. Envir. Res.*, 1995, 67 (1) 65–74.
48. Rusten, B., Hem, L.J., Odegaard, H., Nitrification of municipal wastewater in moving-bed biofilm reactors. *Wat. Envir. Res.*, 1995, 67 (1), pp. 75–86.
49. Dalentoft, E., Thulin, P., The use of the Kaldnes suspended carrier process in treatment of wastewater from the forest industry. *Wat. Sci. Tech.*, 1997, 35 (2–3), pp. 123–130.
50. Broch-Due, A., Andersen, R., Opheim, B., Treatment of integrated newsprint mill wastewater in moving bed biofilm reactors. *Wat. Sci. Tech.*, 1997, 35 (2–3), pp. 173–180.
51. Welander, U., Henrysson, T., Welander, T., Nitrification of landfill leachate using suspended carrier biofilm technology. *Wat. Res.*, 1997, 31 (9), pp. 2351–2355.
52. Pastorelli, G., Andreottola, G., Canziani, R., de Fraja Frangipane, E., De Pascalis, F., Gurrieri, G., Rozzi, A., Pilot-plant experiences with moving-bed biofilm reactors. *Proceedings of Third International IAWQ Special Conference on Biofilm Systems*, 28–30.08.1996, Copenhagen.
53. Pastorelli, G., Andreottla, R., Canziani, C., Darriulat, E., de Fraja Frangipane, Rozzi, A., Organic carbon and nitrogen removal in moving-bed biofilm reactors. *Wat. Sci. Tech.*, 1997, 35(6), pp. 91–99.
54. Garzon-Zuniga, M.A., Gonzalez-Martinez, S., Biological phosphate and nitrogen removal in a biofilm sequencing batch reactor. *Wat. Sci. Tech.*, 1996, 34 (1–2), pp. 293–301.
55. Liu, J.X., van Groenstijn, J.W., Doddema, H.J., Wang, B.Z., Removal of nitrogen and phosphorus using a new biofilm-activated-sludge system. *Wat. Sci. Tech.*, 1996, 34 (1–2), pp. 315–322.
56. Munoz-Colunga, A., Gonzalez-Martinez, S., Effects of population displacements on biological phosphorus removal in a biofilm SBR. *Wat. Sci. Tech.*, 1996, 34 (1–2), pp. 303–313.

14

RETENTION MECHANISMS IN NANOFILTRATION

Johan Schaep,* Bart Van der Bruggen, Carlo Vandecasteele, and Dirk Wilms

University of Leuven
Department of Chemical Engineering
W. de Croylaan 46, B-3001 Heverlee, Belgium

ABSTRACT

Nanofiltration is a pressure-driven membrane separation process that falls in between reverse osmosis and ultrafiltration in its separation characteristics. In general, even low molecular weight organics (> 200 g/mol) and multivalent ions are highly retained, while monovalent ions are retained to a smaller extent. Nanofiltration can be used as well for the production of drinking water, as for the treatment of process and waste waters. The mechanism for separation can be explained in terms of charge effects and/or size effects. Experimental evidence is given for several salt solutions and saccharide solutions. Three commercial nanofiltration membranes are used. The results are interpreted qualitatively by the Donnan exclusion mechanism, which takes the charge effects into account. The steric hindrance pore model is used to determine the membrane pore radii. To predict the results for multicomponent mixtures quantitatively the extended Nernst-Planck model could be used.

Keywords: nanofiltration, salt retention, Donnan exclusion, retention mechanisms.

1. INTRODUCTION

Because of a higher pollution of ground and surface waters and because of stronger regulations for the production of drinking water and the evacuation of waste waters, water treatment is becoming more and more important. The need for advanced separation techniques is therefore increasing. In that way, there is a growing interest in membrane separation processes. A better technological knowledge with regard to membrane development

* Corresponding author Tel: +32 16 322340 Fax: +32 16 322991 E-mail: johan.schaep@cit.kuleuven.ac.be

and production, leads to a wide variety of membranes and causes a decrease in membrane cost price. Therefore, membrane processes are more and more being used in practice.

Well-known membrane processes, like reverse osmosis, ultrafiltration, microfiltration and electrodialysis, are frequently used for water treatment. For a long time these processes are used in commercial applications. In the eighties a new membrane process was developed, called nanofiltration because of its selectivity for components of about 1 nanometer in size. Nanofiltration certainly extends the applications of membrane processes and is a promising technique for future water treatment. The first major application of nanofiltration was in Florida (USA) for the treatment of ground and surface waters, where 85–95% of the hardness is removed. In general, nanofiltration can be used in the production process of drinking water for softening, removal of color and removal of dissolved organics. Nanofiltration can also be used for the treatment of process and waste waters. Commercial applications are found in the paper and the textile industry (removal of color) , in the food industry (production of cheese, fruit juices,...) and in the pharmaceutical industry (reuse of amino acids and antibiotics)[1].

In this paper, the nanofiltration process is situated among the other membrane processes, and the separation characteristics are given. The separation mechanisms will be discussed using experimental results on several salt solutions and saccharide solutions.

2. MEMBRANE PROCESSES

Membrane processes are separation processes where certain components are removed from a feed solution. The driving force for membrane separation can be a pressure difference, an electrical potential difference, a concentration difference,... The principle of pressure-driven processes is schematically shown in Fig. 1.

A feed solution flows along the membrane. As a consequence of an applied pressure difference the solvent (water) is forced through the membrane, where the solutes are more or less retained.

Table 1 gives an overview of the pressure-driven membrane processes and their major applications. Going from microfiltration (MF) to ultrafiltration (UF), nanofiltration (NF) and reverse osmosis (RO) the membranes have smaller pores and therefore it is possible to remove smaller components from the water. Higher pressures are used to obtain sufficiently high water fluxes.

Table 1. Overview of pressure-driven membrane processes and their major applications

	Pressure (bar)	Pore radius (nm)	Mechanism	Separation of	Application
Micro-filtration	0.1–2	50–10,000	sieving	particles	sterilisation (food, pharmaceutical industry)
Ultra-filtration	1–5	1–50	sieving	high MW organics (10,000–100,000 g/mol)	dairy industry; food; pharmaceutical (enzymes, antibiotics)
Nano-filtration	5–20	ca. 1	sieving ; Donnan-exclusion	low MW organics (200–1,000 g/mol) ; salts	removal of colour (textile); removal of micropollutants; softening
Reverse osmosis	10–100	not porous	solution-diffusion	organics; salts	desalting; ultrapure water; concentration of liquids (food, dairy)

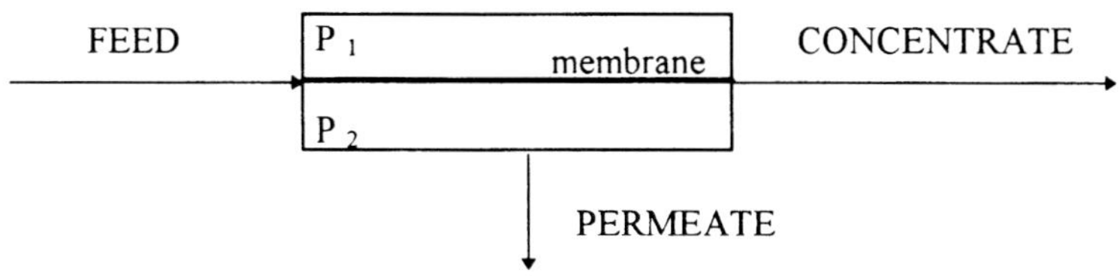

$$\text{driving force} = P_1 - P_2$$

$$\text{Retention (-)} = 1 - \frac{\text{permeate concentration}}{\text{feed concentration}}$$

$$\text{Flux } (l/hm^2) = \text{permeate flow}$$

$$\text{Recovery (-)} = \frac{\text{permeate flow}}{\text{feed flow}}$$

Figure 1. Diagram of pressure-driven membrane processes.

MF is applied for the removal of particles having a diameter of 0.1 to 10 μm and the membranes are characterized by the size of the particles that are removed. High molecular organic components can be removed with UF while small molecules and ions can move through the membrane. UF membranes are characterized by the molecular weight cut-off (MWCO) value, i.e. the molecular weight of a component that is retained for 90%; components with a higher molecular weight are then retained for more than 90%.

NF has to be situated between UF and RO. Membranes for RO are dense membranes with a high hydrodynamic resistance. Even small molecules and ions are removed to a high extent. Because of the high resistance and the osmotic pressure of the solution (e.g. 25 bar for sea water) high pressures have to be applied. Membranes for NF are not so dense and show pores of about 1 nm. Molecules with a molecular weight above 200 are highly retained. The membranes mostly show a high retention for multivalent ions (more than 90%) while monovalent ions have a lower retention (10–80% depending on the membrane). Therefore, the osmotic pressure of the solution is much lower and high pressure differences do not have to be applied as in RO applications. NF can be used when there is no need to remove all monovalent ions.

NF membranes are usually polymeric. The membranes consist of several layers with the top layer, where the separation is achieved, as thin as possible (less resistance). The thickness of the top layer (< 1 μm) is dependent on the preparation technique. This top layer is supported by one or more support layers (150 μm). Not only polymeric but also ceramic membranes based on alumina (Al_2O_3) or zirconia (ZrO_2) can be used for high temperature applications or applications in an aggressive environment (pH, organic solvents). Promising results have been published for NF using such membranes but the production of ceramic NF membranes is still in the developing phase[2].

3. RETENTION MECHANISMS IN NANOFILTRATION

Both charge and size effects play a role in the separation process. As NF membranes are porous, the membrane pore size is an important characteristic to evaluate the effect of the solute size. NF membranes are also characterized by the presence of ionic groups like $-SO_3H$ and -COOH on the polymer structure. Therefore the membrane becomes charged when placed in a solution and an interaction takes place between the membrane and a charged component.

In NF, components can be retained by their size. To describe this size effect, it is necessary to determine the membrane pore size. The methods frequently used for UF membranes (gas adsorption-desorption, permporometry) are not applicable to determine the pore size of NF membranes as the pores are too small. Advanced microscopic techniques (FESEM, AFM) can give an image of the membrane surface. However, imaging the outside of a pore does not tell anything about the structure of the pore itself. Therefore in practice, filtration experiments are carried out to evaluate the pore size of NF membranes.

Normally experiments are carried out with saccharides, but other uncharged molecules can also be used. For a solute with a known solute radius the percentage of retention is determined in a filtration experiment. Using the Spiegler-Kedem equation[3] and the steric hindrance pore model[4] the pore radius can be evaluated (see Fig. 2).

The models make use of the so called reflection coefficient, the solute retention at infinitely high fluxes. Fig. 2 clearly shows the effect of the size of a solute on its reflection coefficient. The experimental results for 4 commercially available NF membranes are shown in Table 2. The pore radii of the membranes varied between smaller than 0.37 nm to 1.12 nm.

With regard to charged components not only size effects are of importance. When membranes having ionic groups on their polymer structure are placed in a salt solution, an equilibrium occurs between the membrane and the solution. Because of the presence of the fixed membrane charge, the ionic concentrations in the membrane are not equal to those in the solution. The counter-ion (opposite sign of charge of the fixed charge in the membrane) concentration is higher in the membrane phase than in the bulk solution, while the co-ion (same sign of charge as the fixed membrane charge) concentration is lower in the membrane phase. A potential difference at the interphase, called Donnan potential, is created to counteract the transport of counter-ions to the solution phase and of co-ions to the membrane phase. When a pressure gradient across the membrane is applied, water is transported through the membrane. The effect of the Donnan potential is to repel the co-

Table 2. Pore radii and reflection coefficients of saccharide solutions for 4 commercial NF membranes

	NITTO 7450	PES 10	UTC 20	NF 70
Pore radius (nm)	0.80	1.12	0.40	< 0.37
Galactose (0.37 nm)	32 %	16 %	95 %	100 %
Maltose (0.47 nm)	50 %	29 %	/	/
Raffinose (0.57 nm)	67 %	34 %	/	/

SPIEGLER - KEDEM equation

$$R = \frac{\sigma(1-F)}{1-\sigma F} \quad \text{with} \quad F = \exp(-\frac{1-\sigma}{P}J)$$

R = retention (-)
J = flux (l/hm²)
σ = reflection coefficient (-)
P = permeability (l/hm²)

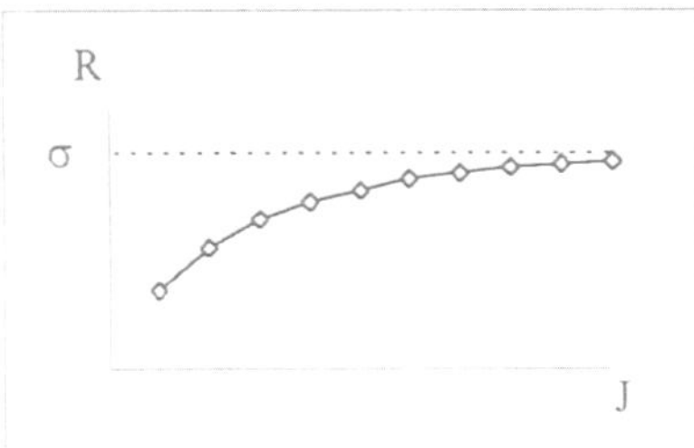

At higher pressures, higher fluxes J are reached.
R is limited by σ.

MODEL: STERIC HINDRANCE PORE MODEL (SHP)

$$\sigma = 1-(1+\frac{16r_s^2}{9r_p^2})(1-\frac{r_s}{r_p})^2(2-(1-\frac{r_s}{r_p})^2)$$

r_s = solute radius
r_p = pore radius

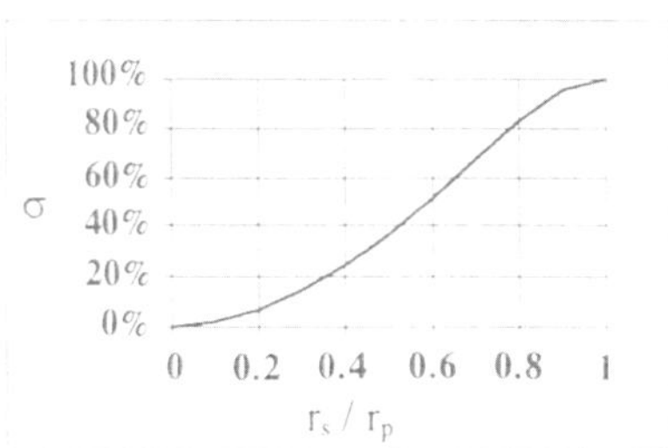

A higher ratio r_s/r_p means more steric hindrance.
For a solute with r_s larger than r_p 100% is retained.

The retention - flux curve is experimentally determined for a solute with known solute radius r_s. Spiegler-Kedem gives σ, SHP gives r_s/r_p so r_p is known.

Figure 2. Determination of the membrane pore size.

ion from the membrane. Because of electroneutrality requirements, the counter-ion is also rejected and salt retention occurs. This mechanism is called Donnan exclusion.

Electroneutrality is also required for the membrane phase. Simple calculations for single salt solutions show how the concentrations in the membrane phase differ from the feed concentrations as a consequence of the existence of the Donnan potential. Some examples for a negatively charged membrane are shown in Fig. 3. Because of the negative

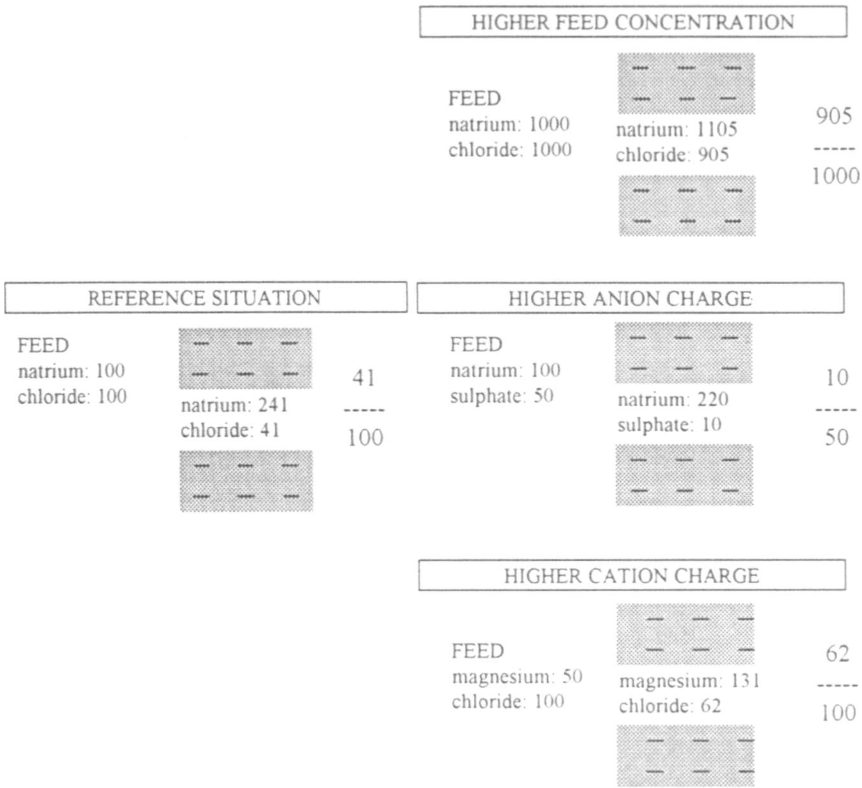

Figure 3. Theoretical calculation of the influence of some parameters on the ion concentration in the membrane phase. All values in mmol/l. Membrane charge density = −200 mmol/l.

membrane charge, the concentration of the anion is lower in the membrane phase than in the feed and the concentration of the cation is higher in the membrane phase. The concentrations are calculated using chemical equilibrium conditions between the feed solution and the membrane phase and taking into account electroneutrality requirements. Activity coefficients were not taken into account and the ions are considered as point charges, so that the effect of ion size is neglected.

From Fig. 3, it is clear that for a negatively charged membrane a higher anion charge leads to a higher exclusion and thus, a higher salt retention. A high cation charge has the opposite effect. When a higher feed concentration is used, the effect of the membrane charge is less important and this leads to relatively high ion concentrations in the membrane.

These theoretical considerations can now be experimentally verified. Experiments were carried out on 3 commercial NF membranes. Single salt solutions of NaCl, $MgCl_2$, Na_2SO_4 and $MgSO_4$ were used. The experimental results for the NTR 7450 membrane are presented in Fig. 4 as a function of the feed concentration.

From Fig. 4 it can be seen that the performance of the NTR membrane is in accordance with the Donnan exclusion theory. A higher anion charge (of sulphate to chloride) leads to a higher salt retention while a higher cation charge (of magnesium to sodium) leads to a lower salt retention. The effect of a higher feed concentration is obvious. All results can be explained by the effect of the Donnan potential.

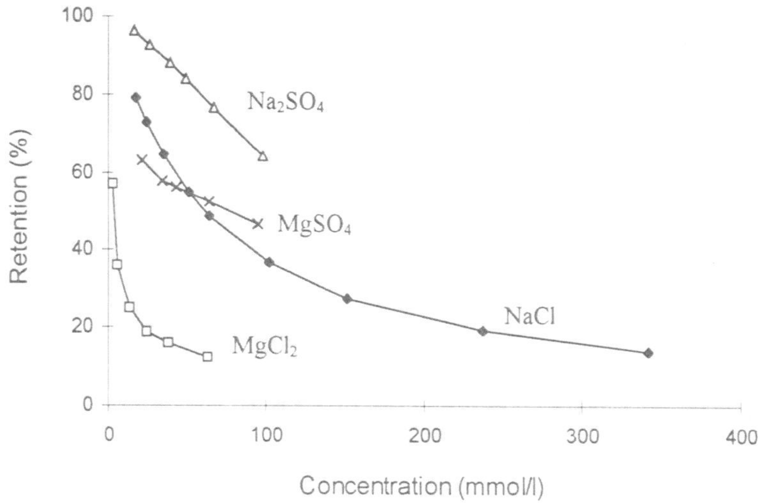

Figure 4. Retention of single salt solutions as a function of the feed concentration. Pressure: 10 bar. Membrane: NTR 7450.

This is not the case for the NF 70 and the UTC 20 membrane (see Table 3).

For NF 70 (negatively charged) the retention for $MgCl_2$ is expected to be lower than for NaCl if based on charge effects. For UTC 20 (positively charged), the Donnan exclusion predicts a lower retention for Na_2SO_4 than for NaCl. The high retentions for respectively $MgCl_2$ and Na_2SO_4 cannot be ascribed to charge effects alone. Comparing the hydrated radii of ions (Table 4) to the membrane pore radius (see Table 2), it is obvious that for both NF 70 and UTC 20 the effect of the ion size cannot be neglected. Therefore the Donnan exclusion theory cannot be used.

Table 3. Retention of single salt solutions for two commercial NF membranes

Membrane charge	NF 70 negative (%)	UTC 20 positive (%)
NaCl	93	38
Na_2SO_4	97	79
$MgCl_2$	97	88
$MgSO_4$	98	95

Table 4. Hydrated ionic radii[5]

Ion	Radius (nm)
Chloride	0.33
Sodium	0.36
Sulphate	0.38
Magnesium	0.43

Table 5. Ion retention in a salt mixture. Pressure: 10 bar. Concentration: 11 meq/l

Ion	NTR 7450 (%)	NF 70 (%)	UTC 20 (%)
Na^+	26	94	77
Mg^{2+}	47	98	96
Cl^-	19	94	76
SO_4^{2-}	59	98	99

Experiments were also carried out for a mixture of NaCl and $MgSO_4$ to study the effect on the salt retention. It can be seen from Table 5 that for both NF 70 and UTC 20 high retentions were found, with the retention of multivalent ions close to 100%. For the NTR 7450 membrane lower retentions were found. Again multivalent ions are better retained than monovalent ions. As magnesium has a higher charge than sodium, it is more strongly retained by the charged groups of the membrane and thus shows a higher retention.

To be able to predict the separation performance of a NF membrane, the transport of a solute through the membrane has to be described. For that purpose the extended Nernst-Planck equation could be used:

$$J_i = -D_i^m \frac{dc_i^m}{dx} - \frac{z_i c_i^m D_i^m}{RT} F \frac{d\psi}{dx} + K_{i,c}\, c_i^m J_v$$

J_i = ion flux
D_i^m = hindered diffusion coefficient
c_i^m = concentration in membrane
x = distance normal to membrane
z_i = valence of ion
F = Faraday constant
R = gas constant
T = absolute temperature
ψ = electric potential
$K_{i,c}$ = hindrance factor for convection
J_v = solvent flux

This equation describes the solute transport in terms of convection, diffusion and electromigration. Experiments with single salt solutions and saccharide solutions are used to determine the membrane pore radius, the membrane charge density and the ratio of membrane porosity to membrane thickness. These membrane characteristics can then be used to predict the membrane performance at other operating conditions.

The extended Nernst-Planck equation has already been used for NF membranes with very good results[6,7]. Further attention will be paid to the optimization of the modelling of the transport through NF membranes by using the extended Nernst-Planck equation.

4. CONCLUSIONS

The retention mechanisms were studied for commercially available NF membranes. Both size and charge effects play a role in the separation of salt solutions. It can be concluded that the pore size is an important membrane characteristic. The pore radii of both UTC 20 and NF 70 were estimated at respectively 0.40 nm and lower than 0.37 nm. Both membranes showed very high ion retentions, as well in single salt solutions as in salt mixtures. The membrane charge seemed to be of no importance.

For the NTR 7450 membrane which was found to have larger pores (r_p = 0.80 nm), the Donnan exclusion theory can be used and salt retention is determined by charge effects.

ACKNOWLEDGMENT

This research was financed with two fellowships from "het Vlaams Instituut voor de bevordering van het wetenschappelijk-technologisch onderzoek in de industrie (IWT)" and supported by the FWO (Fonds Wetenschappelijk Onderzoek - Vlaanderen).

REFERENCES

1. L.P. Raman, M. Cheryan, N. Rajagopalan, Consider nanofiltration for membrane separations, *Chemical Engineering Progress*, *90* (1994) 68–74
2. S. Alami-Younssi, A. Larbot, M. Persin, J. Sarrazin, L. Cot, Gamma alumina nanofiltration membrane. Application to the rejection of metallic cations, *Journal of Membrane Science*, *91* (1994) 87–95
3. K.S. Spiegler, O. Kedem, Thermodynamics of hyperfiltration (reverse osmosis): criteria for efficient membranes, *Desalination*, *1* (1966) 311–326
4. X.L. Wang, T. Tsuru, M. Togoh, S. Nakao, S. Kimura, Evaluation of pore structure and electrical properties of nanofiltration membranes, *Journal of Chemical Engineering of Japan*, *28* (1995) 186–192
5. E.R. Nightingale, Phenomenological theory of ion solvation. Effective radii of hydrated ions, *The Journal of Physical Chemistry*, *63* (1959) 1381–1387
6. W.R. Bowen, H. Mukhtar, Characterisation and prediction of separation performance of nanofiltration membranes, *Journal of Membrane Science*, *112* (1996) 263–274
7. W.R. Bowen, A.W. Mohammed, N. Hilal, Characterisation of nanofiltration membranes for predictive purposes - use of salts, uncharged solutes and atomic force microscopy, *Journal of Membrane Science*, *126* (1997) 91–105

NANOFILTRATION FOR REMOVAL OF ORGANIC SUBSTANCES FROM WASTE WATER

Application in the Textile Industry

B. Van der Bruggen,[1] J. Schaep,[1] D. Wilms,[1] C. Vandecasteele,[1] and M. Van den Bosch[2]

[1]Department of Chemical Engineering
University of Leuven
W. de Croylaan 46, B-3001 Heverlee, Belgium
[2]Centexbel
Technologiepark–Zwijnaarde 7, B-9052 Zwijnaarde, Belgium

1. ABSTRACT

The possibility of using nanofiltration as a treatment method for waste water in the textile industry was investigated. An effluent from a textile factory, after biological treatment and sand filtration, was taken as feed solution for a lab-scale nanofiltration unit. The retention of organic and inorganic compounds was tested for three nanofiltration membranes: NF70 (FilmTec), Nitto NTR 7450 (Nitto-Denko) and UTC-20 (Toray). As expected, considerable differences in retention were observed. The highest retentions were obtained for NF70. For UTC-20 and Nitto 7450, retentions were slightly lower. Colour removal was nearly complete for the three membranes.

Depending on the choice of the membrane, the permeate can meet the quality standards for process waters imposed by textile factories and reused as process water. If the permeate is recycled, a considerable amount of water can be saved.

The experimental results were also used to obtain information about the membranes. Conclusions were made on the shape of the retention curves of the different membranes as a function of molecular size.

2. INTRODUCTION

Membrane processes in general, and nanofiltration in particular, have become more and more important in environmental technology[1,2]. Nanofiltration membranes have the

Chemistry for the Protection of the Environment 3, edited by Pawłowski *et al.* Plenum Press, New York, 1998

possibility to remove multivalent ions as well as organic molecules with a molecular weight of about 300 and higher. Energy consumption is low in comparison with reverse osmosis, because the membrane pressure is much lower (generally 5 to 15 bar).

Typical applications are the simultaneous removal of hardness and colour from ground or process water[2,3] and the removal of pesticides from ground water[4].

However, the retention mechanism for organic as well as inorganic molecules is only partly understood. Due to the presence of pores in the membrane, a sieving process takes place in which large molecules are separated from smaller ones. It is not clear which factors are important in this sieving process, and, consequently, how this separation can be predicted for a given feed solution. Therefore, the study of the application of nanofiltration should be combined with theoretical considerations about retention and membrane structure. Particularly, very little is known about the retention behaviour of organic molecules. The molecular weight cut-off (MWCO), a widely used parameter that represents the molecular weight of a component which is retained for 90 %, can only provide a rough estimation of membrane performance[5,6]. Although molecules with a higher (lower) molecular weight than the MWCO will show in principle a higher (lower) retention than 90 %, nothing is known about the pace at which the retention increases or decreases below or above the MWCO. More details about the retention curves as a function of a measure of molecular size is needed. Moreover, the retention of an organic molecule may also depend on chemical structure. It can be concluded that a better understanding of the retention of organic molecules is needed in order to find new applications in this field and for better control of the present applications.

In this work, nanofiltration is used for treatment of the effluent of a textile factory after biological treatment. Sand filtration was also used as a pretreatment before nanofiltration. The feed water contained mainly organic waste products, as well as inorganic ions from textile dyeing processes. The retention of organic compounds is related to retention curves for the nanofiltration membranes used in the experiments.

3. EXPERIMENTAL

Biological Treatment

The feed water, taken from the waste water of a textile factory, was stored in a receptor where it was aerated and continuously stirred. After buffering it was pumped to an active sludge system with a retention time of 24 hours. The load was 0.05 kg Biochemical Oxygen Demand (BOD) / kg sludge / day and the Mixed liquor Suspended Solids (MLSS) was 3 g/l. The sludge was then removed in a secondary settling tank.

Sand Filtration

In order to remove suspended and settleable solids, a sand filter was placed after the biological treatment.

3.3. Nanofiltration

After filtration over a glass fiber filter (Whatman GF/C), the biologically treated waste water was brought to a lab-scale nanofiltration unit (Amafilter, Test Rig PSS1TZ). Figure 1 is a schematic representation of the unit.

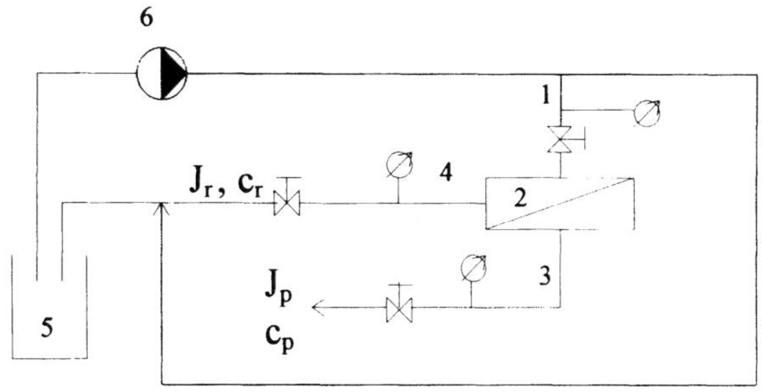

Figure 1. schematic representation of the nanofiltration process (1: feed, 2: module, 3: permeate, 4: retentate, 5: feed tank, 6: pump).

The feed solution was pumped to the module by a three step membrane pump, where it flowed past a flat sheet membrane. Although spiral wound membranes are mostly used for industrial applications, flat sheet membranes are more favorable for laboratory use, because they are easily replaceable. The total effective area of the membrane in the module was 0.0044 m^2.

The pressure over the membrane can be varied from 0 to 60 bar. Pressures and flows were set by manual valves. The feed flow varied from 0 to 1000 l/h, corresponding to a velocity over the membrane of 0 to 7.5 m/s. The temperature was fixed by an electronic circuit.

The retentate flow J_r, which was not passed through the membrane, was recycled to the feed. The permeate flow J_p was removed or recycled to the feed tank. Sampling was possible in the feed tank and from the permeate flow.

3.4. Analysis

Metal concentrations in solution were measured by Inductively Coupled Plasma Mass Spectrometry (ICP-MS). Anions in solution were measured with Capillary Ion Analysis (CIA). Total phosphorus was determined using the *Molybdenum-blue* spectrophotometric method, after destruction to orthophosphate. COD (Chemical Oxygen demand) and BOD (measured after 5 days), which are measures for organic compounds, were determined according to their respective definitions. Suspended solids were measured by filtration over a glass fiber filter (Whatman GF/C 0.45 μm).

A Shimadzu UV-210A double beam spectrophotometer was used for absorbance measurements.

4. RESULTS

Activated Sludge Treatment and Sand Filtration

In Table 1, the concentrations in the raw water are compared to those after activated sludge treatment and those after sand filtration. All values were averaged over a period of four weeks.

During the biological treatment, the BOD and the COD were strongly reduced. For suspended solids and total phosphorus there was also an important reduction.

The sand filtration did not change the composition of the water significantly.

Table 1. Characteristics of raw water, biologically treated water and sand filtered water

Parameter	Raw water	After biological treatment	After sand filtration
BOD (mg/l)	454	5	3
COD (mg/l)	1690	277	213
Suspended solids (mg/l)	188	81	61
Cr (µg/l)	306	300	200
Fe (µg/l)	533	300	300
Total P (mg/l)	3.9	2.3	1.8

4.2. Nanofiltration

A single fraction of the effluent of the sand filtration (25 l) was used for nanofiltration experiments. Three nanofiltration membranes were used: NF70 (Film-Tec), Nitto NTR 7450 (Nitto-Denko) and UTC-20 (Toray). The latter membrane has a positive charge, whereas the other membranes have a negative charge. The temperature was set at 25 °C; membrane pressure was 15 bar. The feed flow was 800 l/h.

Analytical results are given in Table 2.

The colour removal results are shown in Figure 2, which gives the absorbance as a function of wave length. In comparison with the flux with distilled water, the water flux was only slightly lower for NF70, Nitto 7450 and UTC-20. Colour removal is very good for the three membranes. None of the permeates was still visibly coloured. This corresponds with a high retention for COD (88% for UTC-20, 85% for Nitto 7450 and 92% for NF70. No membrane fouling occurred during the experiments.

A high sulphate retention was obtained for the three membranes. This retention is caused by charge interactions between sulphate and the charged membrane surface. For NF70 and UTC-20, size effects could also play a role in the retention of sulphate. This explains why sulphate retention was over 99 % with NF70 and UTC-20, and only 71 % for Nitto 7450. For chloride, a somewhat smaller molecule, the size effect was only important

Table 2. Nanofiltration results for NF70, Nitto 7450 and UTC-20

Parameter	Feed	Glass fiber filtration	NF70	Nitto 7450	UTC-20
pH	8.3	8.3	9.3	8.6	8.6
Suspended solids (mg/l)	18.8	(–)	(–)	(–)	(–)
Ca (mg/l)	85.6	86.6	0.6	27.8	9.6
Cr (µg/l)	140	110	< 10	10	< 10
Fe (µg/l)	360	210	< 10	10	< 10
Zn (µg/l)	610	550	13	149	17
COD (mg O_2/l)	231.8	187	17.7	31.4	33.4
BOD (mg O_2/l)	5.2	4,1	< 1	< 1	(–)
Conductivity (mS/cm)	3.8	3.81	0.11	2.29	1.78
SO_4^{2-} (mg/l)	313	313	0,9	93	N.D.
Cl^- (mg/l)	740	710	17,4	587	496
Total P (mg/l)	1.0	0.8	0.2	0.34	0.17
Water flux (l/m².h.bar)			8.1 (85*)	13.3 (85*)	15.7 (90*)

*percentage of the water flux with distilled water as feed.
(–) = no measurements available

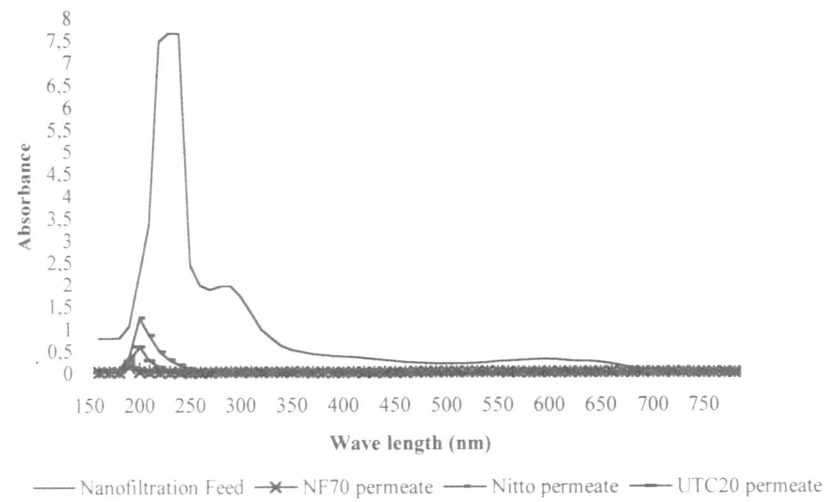

Figure 2. Colour removal with nanofiltration membranes NF70, Nitto 7450 and UTC-20.

for NF70, the membrane with the tightest pores. The pores of the UTC-20 membrane are not small enough to retain chloride. Charge effects are small for both membranes, because the chloride charge is only one unit. In this way, a good separation between sulphate and chloride was obtained for UTC-20, whereas NF70 retained both components almost equally.

Metal retention was generally high. Chromium (+III) and iron (+III) retentions are higher than calcium (+II) and zinc (+II) retention, due to their higher charge. For NF70 and UTC-20, which are negatively charged membranes, this effect remains unexplained.

5. DISCUSSION

In Table 3, the COD retention for the three membranes used in the experiments is compared to retention of three organic components.

For NF70 and UTC-20, retention of raffinose and galactose was higher than the COD retention in this experiment. For Nitto 7450, raffinose and galactose retention was lower. This can be explained by the assumption that the major part of the organic material in the waste water has a relatively high molecular weight (and, correspondingly, a large molecular size). Thus, the waste water contains a large amount of molecules with a mo-

Table 3. COD retention compared to retention of benzoïc acid, galactose and raffinose, and estimated MWCO

	COD	Benzoic acid (MW 122)	Galactose (MW 180)	Raffinose (MW 504)	Estimated MWCO
NF70	92	69	95	99	250
UTC-20	88	7	95	97	180
Nitto 7450	85	20	29	67	600–800

lecular size that corresponds approximately with a molecular weight of over 800, the estimated molecular weight cut-off of the Nitto 7450 membrane.

NF70 and UTC-20 both have tight pores; only very small molecules like ethanol can permeate through these membranes. Because galactose retention is higher than the obtained COD retention, the waste water must also contain a fraction with a relatively small molecular size.

Furthermore, the experimental results confirm the finding that the retention curve for UTC-20 is much sharper than for Nitto 7450[7]. UTC-20 has a higher raffinose retention than Nitto 7450, but retention for benzoic acid (molecular weight 122) is higher with Nitto 7450 (20%) than with UTC-20 (7%). In this experiment, the COD retention is lower for Nitto 7450 because the gradual retention curve allows more organic molecules to permeate partly through the membrane.

This illustrates the importance of a more detailed description of the retention curve as a function of molecular size.

6. CONCLUSIONS

The composition of the permeate can be compared to the composition of treated ground water or surface water. When the permeate has an equal or better quality, reuse as process water can be considered. In this way, a useful product can be obtained from the waste water, using the treatment process.

From Table 2, it can be seen that the possibility of permeate reuse depends greatly on the nature of the nanofiltration membrane. With the Nitto 7450 membrane, the permeate cannot be applied as a water supply source, although a considerable reduction of contaminants is obtained. UTC-20 provides better results, especially for removal of metals. The COD is also acceptable. Chloride retention may be too low for this membrane.

For the NF70 membrane, reuse of the permeate is possible, on condition that the permeate quality remains constant over a long period. COD, concentrations of calcium and other metals and concentrations of anions are lower in the NF70 permeate than in process waters used in the textile industry. Depending on the specific process for which the water will be used, pH and total dissolved salts can be adjusted.

It can be concluded that the permeate obtained from the nanofiltration unit can be reused as a process water, on condition that the membrane is well chosen. The major part of the waste water can thus be recycled, with addition of a smaller part of fresh water to compensate for the retentate.

The further treatment of the retentate must also be considered. Since concentrations in the retentate can be five times higher as in the feed, these may be too high for discharging according to legal regulations. The water volume, however, is much smaller. Other techniques, such as ozonation, could solve this problem.

Furthermore, the permeation results should also be studied without recirculation of the permeate. Indeed, the permeate concentrations will increase with feed concentrations.

ACKNOWLEDGMENT

This research was financed with a fellowship from IWT (Vlaams Instituut voor de bevordering van het wetenschappelijk-technologisch onderzoek in de industrie).

REFERENCES

1. Owen J., Bandi M., Howell J.A., Churchouse S.J., *Economic assessment of membrane processes for water and waste water treatment*, Journal of Membrane Science, vol. 102, 77–91 (1995)
2. Rachwal A.J., Khow J., Colbourne J.S., O'Donnell J., *Water treatment for public supply in the 1990's: a role for membrane technology?*, Desalination, vol. 97, 427–436 (1994)
3. Schneider B.M., *Membranes - Nanofiltration compared to other softening processes*, Ultrapure water, Oct. 1994, 65–74 (1994)
4. Taylor J.S., Mulford L.A., Chen S.S., Hofman J.A.M., *Membrane Filtration of Pesticides, Proceedings of the Annual Conference of the American Water Works Association*, 593–606 (1995)
5. Meireles M., Bessieres A., Rogissart I., Aimar P., Sanchez V., *An Appropriate Molecular Size Parameter for Porous Membranes Calibration*, Journal of Membrane Science, vol. 103, 105–115 (1995)
6. Rautenbach R., Gröschl A., *Separation Potential of Nanofiltration Membranes*, Desalination, vol. 7, 73–84 (1990)
7. Van der Bruggen B. et al., to be published

16

METAL-ION SELECTIVITY OF PHOSPHORIC ACID RESIN IN AQUEOUS NITRIC ACID MEDIA

Akinori Jyo and Xiaoping Zhu

Department of Applied Chemistry and Biochemistry
Faculty of Engineering, Kumamoto University
Kumamoto 860, Japan

ABSTRACT

Metal ion selectivity of a phosphoric acid resin (RGP), which was derived from addition of phosphoric acid to epoxy groups of macroreticular poly(glycidyl methacrylate-*co*-divinylbenzene), was studied by measuring pH dependency of distribution ratios of various metal ions between RGP and nitric acid solutions. In nitric acid media, RGP exhibits extremely high adsorption ability toward Ti(IV), Mo(VI), and Fe(III). Furthermore, its selectivity sequence for common trivalent and divalent metal ions was quite different from that of conventional cation exchange resins having sulfonic acid groups, suggesting that RGP will be useful for the elimination of some heavy metal ions which are not strongly adsorbed by sulfonic acid resins.

1. INTRODUCTION

Recently, much attention has been paid to phosphonic acid resins.[1–9] However, phosphoric acid resins have not been extensively studied compared with phosphonic ones. In this connection, Jyo *et al.*[10] have reported preparation of a phosphoric acid resin (RGP, Scheme 1) having acid capacities as high as 7–8 meq/g and its behavior in adsorption of various metal ions from aqueous hydrochloric acid media as well. Characteristics of RGP can be summarized as:[10)] (1) it can be easily prepared by heating poly(glycidyl methacrylate-*co*-divinylbenzene) beads in commercially available phosphoric acid (85%) at 140 °C for 2 h without use of any other specific reagent in the functionalization reaction, and (2) it prefers hard Lewis acid cations (Fe(III), Th(IV), U(VI)) to soft ones (Hg(II)). However, much more knowledge about metal ion selectivity of RGP will be required to estimate its role in the elimination of heavy metal ions from water.

Chemistry for the Protection of the Environment 3, edited by Pawłowski *et al.* Plenum Press, New York, 1998

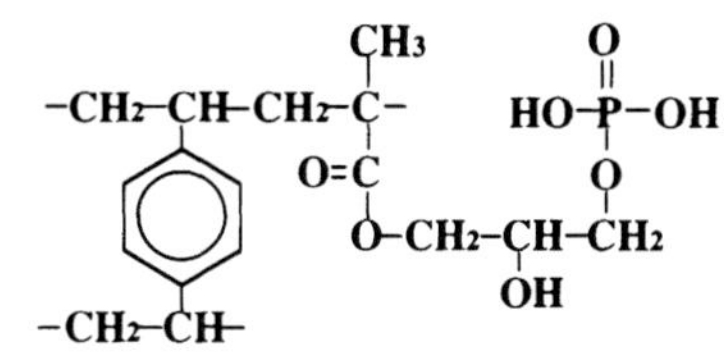

Scheme 1. Phosphoric acid resin (RGP).

It is well known that esters of phosphoric acid, such as tributyl phosphate (TBP) and di-2-ethylhexylphosphoric acid (D2EHPA), selectively extract some metal ions (U(VI), Mo(VI), and Fe(III)) from aqueous nitric acid solutions of high acidity (1–10 M) through the synergistic effect of nitrate ion.[11,12] Thus, of interest is the behavior of RGP in the uptake of various metal ions from nitric acid media, since it can be regarded as a kind of ester of phosphoric acid. This work was planned to investigate the behavior of RGP in the distribution of various metal ions from aqueous nitric acid solutions. For the sake of comparison, a conventional sulfonic acid resin (Diaion PK216) was also used. Eighteen kinds of metal ions were tested, including (Ba(II), Ca(II), Cd(II), Cu(II), Mg(II), Mn(II), Ni(II), Pb(II), Zn(II), Al(III), Cr(III), Fe(III), Gd(III), La(III), Lu(III), Tl(III), Ti(VI) and Mo(VI)).

2. EXPERIMENTAL

2.1. Materials

The resin RGP was prepared by the functionalization of the precursory copolymer beads, which were obtained by the suspension polymerization of glycidyl methacryalte and divinylbenzene (10 mol% in the monomer mixture) in the presence of 140 volume% of porogen (isobutyl acetate) per the monomer mixture.[10] Diaion PK216 was provided from Mitsubishi Chemical Co. Both resins were conditioned as described,[13] and were used in the hydrogen ion form. Properties of both resins are given in Table 1. Metal ion solutions were prepared by dissolving nitrate salt of each metal ion into water. All reagents were of reagent grade, and deionized water was used throughout.

2.2. Distribution Study

A resin (0.04g) and a 0.0001 M metal ion solution (25ml) were placed in a 50 ml Erlenmeyer flask. The flask was sealed with a Parafilm, then shaken for 24 h at 30 °C to ensure establishment of distribution equilibrium. The concentration of metal ion in the supernatant was determined by means of ICP-AES. From a decrease in the metal ion con-

Table 1. Properties of RGP and Diaion PK216 in hydrogen ion form

Resins[a]	Specific surface area (m^2/g)	Acid capacity (meq/g)	Salt splitting capacity (meq/g)	Phosphorus content (mmol/g)
RGP	29.2	6.98	2.77	3.75
Diaion PK216	0	4.71	4.50	—

[a]Particle size of both resins was 32–60 mesh.

centration in the supernatant, adsorption percentages and distribution ratios (D) were calculated. Here, D is designated by the following equation.

$$D = \frac{\text{amount of metal ion adsorbed (mmol / g)}}{\text{amount of metal ion in solution (mmol / mL)}}$$

The acidity of the aqueous phase was adjusted with nitric acid, and $-\log[HNO_3]$ was adopted as pH, when the concentration of nitric acid was higher than 1 M.

3. RESULTS AND DISCUSSION

3.1. Divalent Metal Ions

Figure 1 shows the dependence of extraction percentages of divalent metal ions on pH. Here, results for Pb(II), Ca(II), Ba(II), Mg(II), and Zn(II) are given as examples, and results of Diaion PK216 are also shown for the sake of comparison. Under conditions in the present measurement of distribution ratios, the total amount of a metal ion (0.0025 mmol) is much less than that of functional groups (0.14 mmol for RGP and 0.17 mmol for

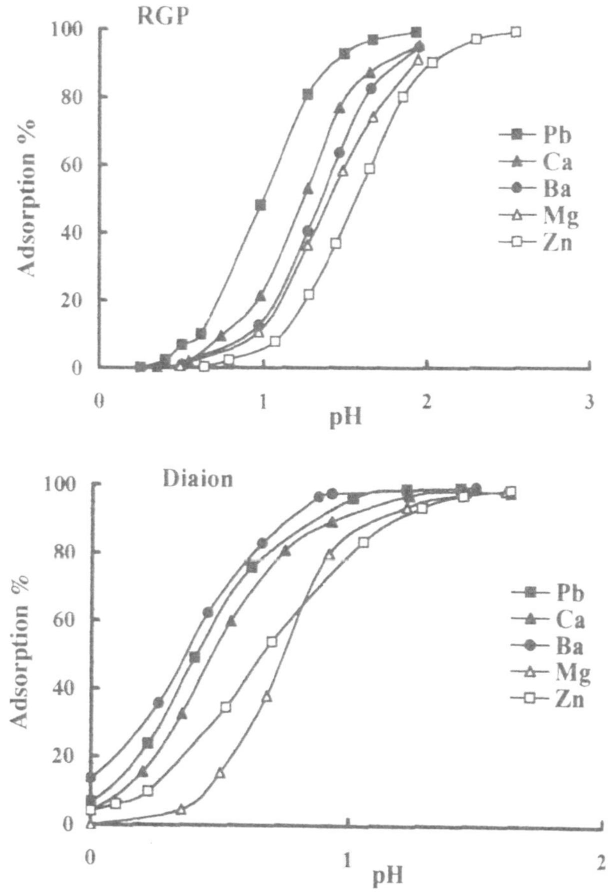

Figure 1. Dependence of adsorption percentages for typical divalent metal ions on pH.

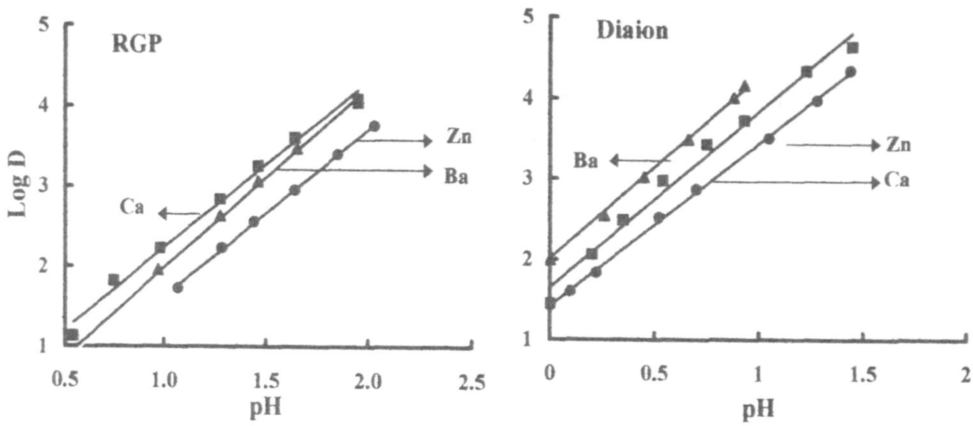

Figure 2. Relationship between log D and pH.

Diaion PK216). The amount of the hydrogen ion in the resin phase can be essentially regarded as constant. Then, the following equation holds between log D and pH, so long as the metal ion is adsorbed through the stoichiometric cation exchange between hydrogen ion and a metal ion.

$$\log D = \text{constant} + n\text{pH}$$

Here, n represents formal charge of the metal ion. In Fig. 2, some examples of plots of log D *vs.* pH are shown. In the distribution of all tested divalent cations, linear relationship was observed between log D and pH, and Table 2 summarizes least square slopes and pH values at half adsorption ($pH_{1/2}$). Slopes of the plots are nearly equal to +2, although they are slightly greater than +2 in the distribution of Mg(II), Cd(II), and Pb(II) into Diaion

Table 2. Slopes of log D vs. pH plots and values of pH at a half adsorption ($pH_{1/2}$) for metal ions giving linear relationship between log D and pH

	RGP		Diaion PK216	
Metal ion	Slope	$pH_{1/2}$	Slope	$pH_{1/2}$
Ni(II)	2.0	1.64	2.0	0.65
Zn(II)	2.1	1.56	2.0	0.68
Mg(II)	2.2	1.43	2.5	0.78
Cu(II)	2.0	1.40	2.2	0.66
Mn(II)	2.1	1.36	2.3	0.57
Ba(II)	2.2	1.36	2.3	0.34
Cd(II)	2.1	1.30	2.6	0.66
Ca(II)	2.1	1.27	2.2	0.52
Pb(II)	2.3	1.04	2.4	0.43
Cr(III)	2.9	1.04	3.2	0.19
La(III)	2.9	0.33	3.0	−0.04
Gd(III)	3.0	0.14	3.5	0.05
Al(III)	3.0	0.02	3.2	0.19
Lu(III)	2.6	−0.15	3.1	0.05
Fe(III)	—	—	3.0	0.25
Tl(III)	—	—	1.0	1.25

PK216. From $pH_{1/2}$ values, it can be concluded that the affinity of both resins toward the tested divalent metal ions decreases in the following order.

RGP: Pb(II) > Ca(II) ≈ Cd(II) > Ba(II) ≈ Mn(II) > Cu(II) ≈ Mg(II) > Zn(II) ≈ Ni(II)
Diaion PK216: Ba(II) > Pb(II) > Ca(II) > Mn(II) > Cu(II) ≈ Ni(II) ≈ Cd(II) ≈ Zn(II) > Mg(II)

Thus the divalent metal ion selectivity sequence of RGP is significantly different from that of Diaion PK216 (sulfonic acid resin). In addition, $pH_{1/2}$ values of RGP for divalent metal ions range from 1.04 to 1.64, but those of Diaion PK216 from 0.34 to 0.78. This means that the interionic selectivity of RGP is higher than that of Diaion PK216, and the elution of divalent metal ions adsorbed on RGP with acid solutions are much more easily achieved compared with the case of the sulfonic acid resin. Indeed, Pb(II) adsorbed on RGP was much more rapidly eluted with aqueous nitric acid solutions compared with that of Pb(II) adsorbed on a sulfonic acid resin, as reported in a previous paper.[14)]

3.2. Trivalent Metal Ions

Figure 3 shows uptake of Al(III), Fe(III), Cr(III), La(III), Gd(III), Lu(III), and Tl(III) by RGP as a function of pH. In the adsorption of Fe(III) and Tl(III), quite opposite tendencies are observed. RGP strongly adsorbs Fe(III) even from concentrated nitric acid solutions. Indeed, extraction percentages of Fe(III) exceed 90% in the tested pH range. On the other hand, Tl(III) was little adsorbed by RGP. In the adsorption of five other kinds of cations, normal sigmoid curves are obtained. Results of Diaion PK216 are shown in Fig. 4. This resin gives normal sigmoid curves for all tested metal ions except for Tl(III). From data shown in Figs. 3 and 4, least square slopes of log D *vs.* pH plots and values of $pH_{1/2}$ were calculated (Table 2). Since RGP did not give sigmoid curves for Fe(III) and Tl(III), these were omitted from calculations. Slopes for trivalent cations are nearly equal to +3 except for the distribution of Tl(III) into Diaion PK216. Tl(III) is susceptible to hydrolyze, resulting in the species $Tl(OH)_2^+$.[11] This may be the reason why Tl(III) gives a lower slope of +1. From $pH_{1/2}$ values in Table 2 as well as from data in Figs. 3 and 4, it can be concluded that the decreasing order of the affinity to trivalent cations is as follows:

RGP: Fe(III) >> Lu(III) > Al(III) > Gd(III) > La(III) > Cr(III) >> Tl(III)
Diaion PK216: La(III) > Gd(III) ≈ Lu(III) > Al(III) ≈ Cr(III) > Fe(III) > Tl(III)

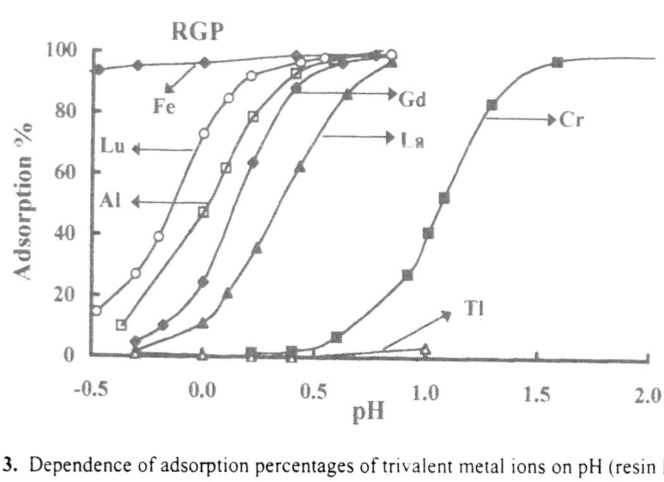

Figure 3. Dependence of adsorption percentages of trivalent metal ions on pH (resin RGP).

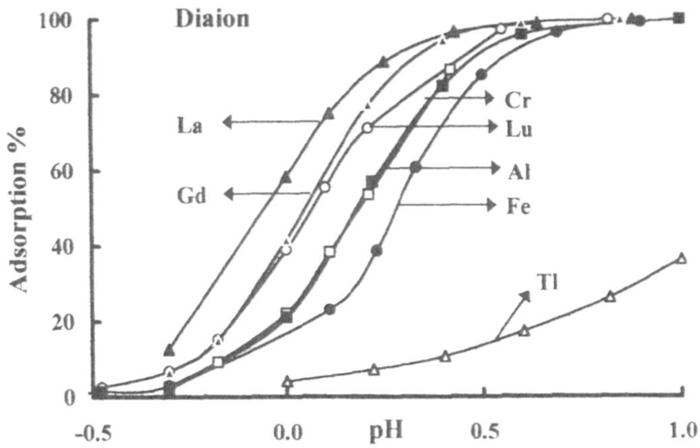

Figure 4. Dependence of adsorption percentages of trivalent metal ions on pH (resin Diaion PK216).

Thus, the selectivity sequence of RGP toward trivalent metal ions is also quite different from that of Diaion PK216. As in the case of divalent cations, RGP also exhibits much higher interionic selectivity toward trivalent cations than does Diaion PK216,

From the results mentioned above, characteristics of RGP in the adsorption of trivalent metal ions from nitric acid media can be summarized as follows: (1) RGP prefers heavy lanthanide ions to light ones, whereas Diaion PK216 exhibits the reversed tendency. (2) The distribution of Fe(III) into RGP from concentrated nitric acid media is extraordinarily high. (3) RGP does not exhibit appreciable adsorption ability toward so called soft Lewis acid cations like Tl(III). These characteristics of RGP can be explained by coordination of oxygen atoms in phosphoric acid groups to some specific cations, such as Fe(III), Al(III), and lanthanide ions.[12] Of particular interest is the extraordinarily high adsorption of Fe(III) from concentrated nitric acid solutions by RGP, since the distribution ratio of Fe(III) between RGP and hydrochloric acid media decreases with an increase in hydrochloric acid concentrations as already reported.[10] Then, synergistic effect of nitrate ion occurs in the adsorption of Fe(III) from concentrated nitric acid media.[12]

3.3. Mo(VI) and Ti(VI)

Figure 5 compares uptake of Mo(VI) and Ti(IV) by RGP with that by Diaion PK216. RGP almost quantitatively adsorbs Ti(IV) even from 6 M nitric acid, but not Diaion PK216. As already clarified in a previous work,[10] RGP exhibits high affinity towards uranyl ion, and then the high affinity of RGP toward Ti(IV) is explained by the fact that Ti(IV) exist as titanyl ion (TiO^{2+}) in the tested pH range, and its properties are very close to those of uranyl ion. Indeed, Ti(IV) and U(VI) bind strongly with phosphate or hydrogen phosphate anions, resulting in sparingly soluble salts.[11,15]

As well known, Mo(VI) also strongly interacts with phosphoric acid forming molybdophosphoric acid in the acidic pH region.[11] In addition, Mo(VI) is extracted from nitric acid media into organic phase containing oraganophosphrous compounds such as tributyl phosphate (TBP) acccording to following synergistic effect of nitrate ion:[11]

$$(MoO_2^{2+})_{aq} + 2(NO_3^-)_{aq} + 2(TBP)_{org} \rightleftharpoons (MoO(NO_3)_2TBP_2)_{org}$$

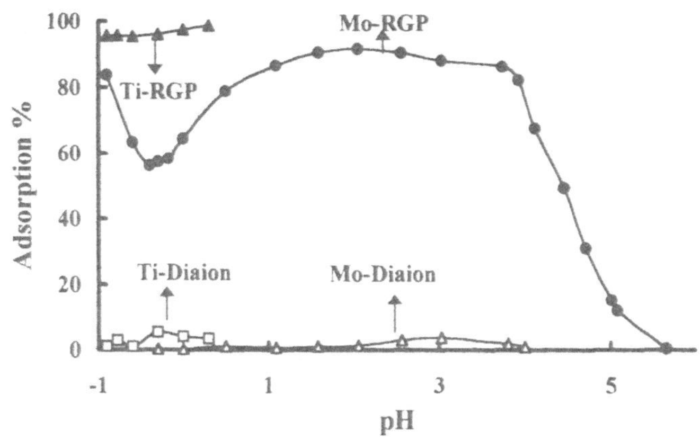

Figure 5. Adsorption of Ti(IV) and Mo(VI) by RGP and Diaion PK216.

It is expected that RGP exhibits high affinity toward Mo(VI) in strongly acidic media. As is shown in Fig. 5, no Mo(VI) was adsorbed in above pH *ca.* 6, and the adsorption of Mo(VI) became maximun around pH 2. With a further increase in the acidity, adsorption percentages decrease until pH=–0.3, and then increase again with an increae in the acidity. This pH dependency clearly suggests that there are two different mechanisms in the adsorption of Mo(VI). One is the adsorption through cation exchange of MoO_2^{2+} with hydrogen ion. This mechanism occurs mainly between pH –0.3 and 6. The marked decrease in adsorption percentages of Mo(VI) above pH 4 can be explained by change of MoO_2^{2+} species into $HMoO^{4-}$, $H_2Mo_7O_{24}^{4-}$, $Mo_7O_{24}^{6-}$, MoO_4^{2-}, and so on with an increase in pH.[16] On the other hand, decrease in the adsorption percentages from pH 2 to pH –0.3 is ascribable to replacement of MoO_2^{2+} with hydrogen ion. The increase in the adsorption from pH –0.3 to pH 0.8 can be explained by the synergistic effect of nitrate ion. Below pH –0.3, phosphoric acid groups exist as protonated neutral species (R-PO_4H_2), since pKa_1 of phosphoric acid groups is nearly equal to 2.[10] Then, Mo(VI) is adsorbed by the similar mechanism in the solvation extraction of MoO_2^{2+} and UO_2^{2+} with TBP from concentrated nitric acid media (above pH –0.3). This kind of synergistic effect of nitrate also occurs in the adsorption of Fe(III), since Fe(III) is also extracted into organic phases containing oraganophosphorus acid from concentrated nitric acid solutions.[12]

4. CONCLUSIONS

Among the nine kinds of divalent metal ions tested in nitric acid media, the phosphoric acid resin RGP exhibits the highest selectivity toward Pb(II), and its selectivity sequence toward these divalent metal ions is quite different from that of Diaion PK216. In the adsorption of trivalent metal ions, RGP also exhibits the characteristic selectivity pattern different from that of the sulfonic acid resin; RGP exhibits extraordinarily high selectivity toward Fe(III), and it prefers heavy lanthanide ions to light ones. Drastic difference in adsorption ability between RGP and Diaion PK216 was observed in the adsorption of Ti(IV) and Mo(VI). RGP strongly adsorb both Ti(IV) and Mo(VI), but not the sulfonic acid resin at all. Because of these characteristics in the metal ion selectivity of RGP, it will be useful for the elimination or separation of heavy metal ions, *e.g.*, the elimination of Pb(II) in the presence of Ba(II), Ca(II), and the separation of Fe(III) from other divalent and trivalent metal ions. Furthermore, we would like to emphasize that RGP exhibits high adsorption ability toward some specific metal ions such as Mo(VI) and Ti(VI) even from concentrated nitric acid media. This behavior of RGP is very close to that of organophosphorous compounds in the solvent extraction of these metal ions from concentrated nitric acid media. Thus, RGP also has significant application in nuclear fuel industries, since it can adosorb Mo(VI), Ti(VI), and U(VI) from concentrated nitric acid solutions without use of flamable and toxic organic solvents.

REFERENCES

1. C. Kantipuly, S. Katragadda, A. Chow, and H. D. Gesser, *Talanta*, **37**, 491–517 (1990).
2. R. Chiarizia and E. P. Horwitz, *Solvent Ext. Ion Exch.*, **12**, 847–871(1994).
3. A. W. Trochimczuk and S. D. Alexandratos, *J. Appl. Polym. Chem.*, **52**, 1273–1277(1994).
4. R. Chiarizia, J. R. Ferraro, K. A. D'Arcy, and E. P. Horwitz, *Solvent Ext. Ion Exch.*, **13**, 1063–1082(1995).
5. R. Chiarizia, K. A. D'Arcy, and E. P. Horwitz, S. D. Alexandratos, and A. W. Trochimczuk, *Solvent Ext. Ion Exch.*, **14**, 519–542(1996).
6. R. Chiarizia, E. P. Horwitz, K. A. D'Arcy, S. D. Alexandratos, and A. W. Trochimczuk, *Solvent Ext. Ion Exch.*, **14**, 1077–1100 (1996).
7. S. D. Alexandratos and D. W. Crick, *Ind. Eng. Chem. Res.*, **35**, 635–644(1996).
8. A. Jyo, K. Yamabe, and H. Egawa, *Sep. Sci. Technol.*, **31**, 513–522(1996).
9. A. Jyo, K. Yamabe, and H. Egawa, *Sep. Sci. Technol.*, **32**, 1099–1105(1997).
10. A. Jyo, S. Matsufume, H. Ono, and H. Egawa, *J. Appl. Polym. Sci.*, **63**, 1327–1334(1997).
11. L. G. Sillen and A. E. Martell, *Stability Constants of Metal-Ion Complexes* (Special Publication 25), The Chemical Society London, 1971.
12. E. S. Stoyanov, V. A. Mikhailov, O. M. Petrukhin, E. V. Shipulo, and G. A. Yagodin, *Solvent Ext. Ion Exch.*, **9**, 787–831 (1991).
13. F. Helfferich, "Ion Exchange", McGraw-Hill, New York, 1962, p230.
14. A. Jyo, K, Yamabe, and H. Egawa, *Chemistry for the Protection of the Environment 2*, Ed. by L. Pawlowski, W. J. Lacy, C. G. Uchrin, and M. R. Dudzinska, Plenum Press, London and New York, pp. 121- 129.
15. M. Kakuda Ed., *Saishin Kyucyaku Gijutsu (The Newest Adsorption Techiniques)*, Sogogijutsu Senta, Tokyo, 1993, pp.271–286.
16. Y. Sato, F. Valenzuela, T. Tsuneyuki, K. Kondo, F. Nakashio, *J. Chem. Eng. Jpn.*, **20**, 317–321(1987).

17

CATALYTIC OXIDATION OF 1,2-DICHLOROPROPANE ON COPPER-ZINC CATALYST

Zbigniew Gorzka, Marek Kaźmierczak, and Andrzej Żarczyński

Institute of General and Ecological Chemistry
Technical University of Łódź
90-924 Łódź, 36 Żwirki Str., Poland.

ABSTRACT

Study on thermocatalytic oxidation of 1,2-dichloropropane (DCP) on the copper-zinc were carried out. Special attention was paid to determination of the lowest temperature required to provide complete combustion. Tests included oxidation of DCP in temperature range from 200 to 700°C. The concentration of DCP in a initial tested solution and in combustion gases was determined by chromatographic method. It was found that DCP was completely oxidized in the presence of an oxide copper-zinc catalyst at temperature of 400°C and contact time of 0.36 s or at temperature of 600°C and contact time of 0.036 s.

Key words: copper-zinc catalyst, 1,2-dichloropropane, catalytic oxidation.

1. INTRODUCTION

Combustion is one of the treatment methods applied for destruction of chloroorganic compounds contaminating off-gases exhausted from different production processes . Based on literature oxidation of chloroorganic compounds without catalyst has to be carried out in the temperature exceeding 1200°C. Reaction time longer than 2 s, oxygen concentration in combustion gases 3 to 6% and very good mixing of reagents has to be provided. Too low temperature of combustion or not fulfillment of other above mentioned conditions can cause formation of very toxic polychlorinated dibenzo-p-dioxins and dibenzofurans (PCDD/Fs), and emission of formaldehyde and carbon monoxide. Recombination of PCDD/Fs is prevented by quick cooling of combustion gases and elimination of dust from waste gases[1].

Application of a proper catalyst allows to decrease the temperature in the reaction zone. It is important that decrease in temperature cannot cause increase in emissions of toxic substances in exhaust gases. The temperature of combustion must be low enough to

Chemistry for the Protection of the Environment 3, edited by Pawłowski *et al.* Plenum Press, New York, 1998

prevent the synthesis of PCDD/Fs and high enough to enable total oxidation of organic substances on a catalyst.

Catalysts composed of metal oxides e.g. oxides of manganese, iron, copper and multicomponent mixtures spread on active aluminium oxide are often applied to oxidation of chloroorganic compounds[2–5]. Platinum and other catalysts e.g. copper-zinc-chromium spread on aluminum oxide have also been successfully used[4–10].

Investigations were carried out on thermocatalytic oxidation of chloroorganic compounds to determine the lowest temperature required to achieve complete combustion. Experiments of oxidation of 1,2-dichloropropane (DCP) with copper-zinc oxide catalyst were performed. A method of thermocatalytic oxidation is applicable to industrial wastewaters with high content of toxic organic compounds like spent cooling oil-emulsions and some wastes from chemical industrial syntheses[11]. Such wastes often contain organic compounds of chlorine. 1,2-dichloropropane was chosen from these compounds to investigations[12]. Aqueous synthetic solutions of 1,2-dichloropropane were used in experiments in order to determine intermediate products, mechanism and reaction kinetics under conditions of thermocatalytic treatment.

2. EXPERIMENTAL

Experiments were carried out in a tubular quartz reactor with an internal diameter of 1.7 cm and 20 cm long. This reactor contained 0.95–9.5 g copper-zinc catalyst. An aqueous solution of DCP was pumped with air to an evaporator heated to 160°C. Reactor and evaporator were heated electrically. Gaseous mixture was passing through the reactor. Vapours leaving the reactor were cooled and the cooled combustion gases and condensate were analysed.

Parameters of the process were as follows:

- a solution of DCP was prepared by dissolving 0.6 g/l of DCP in deionized water (TOC concentration was 180 ppm),
- flow rate of a solution—32.5 g/h and air—30 L/h,
- the range of process temperature: 200 to 700°C,
- oxide catalyst Cu-Zn (TMC-2) is produced by Zakłady Azotowe in Tarnów. Catalyst was ground to 0.75–1.02 mm.

The following analysis were carried out:

- DCP in oxidized solution and outlet gases using gas chromatograph GCHF 18.3 with a flame ionization detector and a steel column 2 m long and 4 mm in diameter filled with Porapak P. Combustion gases were sampled according to Polish analytical standards[13],
- TOC in solutions and condensate using the TOC 5050A (Shimadzu) analyzer,
- concentration of chlorides in a condensate was analysed by Mohr or turbidimetric method[14],
- concentration of carbon oxide and chlorine in combustion gases using automatic analyzers GA-20 (Madur),
- concentration of formaldehyde in condensate and combustion gases using colorimetric method with phenylhydrazine hydrochloride[15].

3. RESULT AND DISCUSSION

Results of experiments are presented in Figures 1–5.

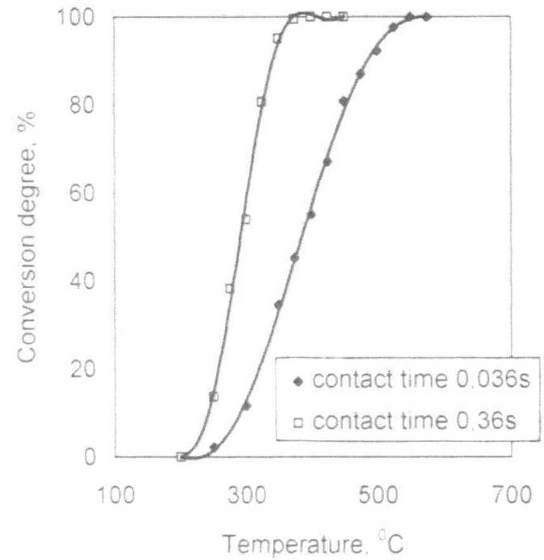

Figure 1. Destruction rate of DCP versus temperature.

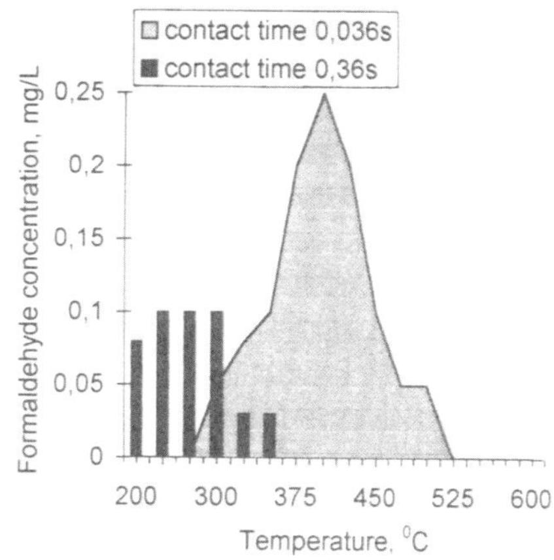

Figure 2. Concentration of formaldehyde in condensate versus temperature.

Figure 1 presents rate of destruction of DCP as a function of temperature. Destruction rate of DCP was determined based on content of DCP in reaction products. Total destruction of the substrate was obtained at a contact time of 0.36 s and temperature of 400°C. The second curve in Figure 1 presents the destruction rate of DCP as a function of temperature for the process carried out at the contact time of 0.036 s. Experiments carried out at lower contact times give a better possibility of investigation of intermediate reaction products. In a temperature range from 275 to 525°C and at contact time of 0.036 s, formaldehyde was detected in the condensate and combustion gases. Formaldehyde was generated by a partial oxidation of DCP.

Figure 2 presents concentration of formaldehyde in the condensate versus temperature of catalyst layer.

Figure 3 shows how the formaldehyde concentration in combustion gases depends on temperature in the reaction zone. At a contact time of 0.036 s, generation of formalde-

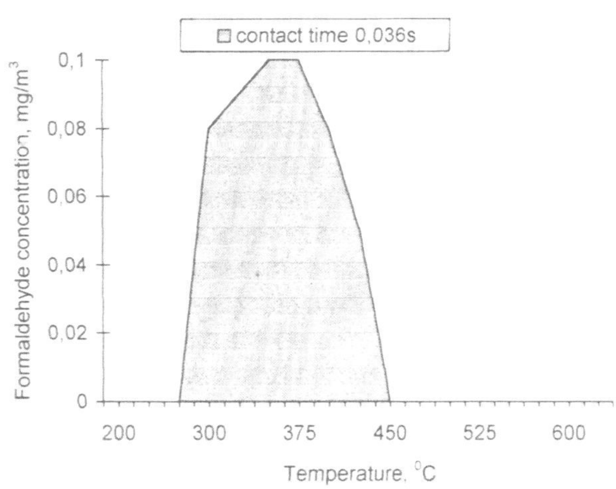

Figure 3. Concentration of formaldehyde in combustion gases versus temperature.

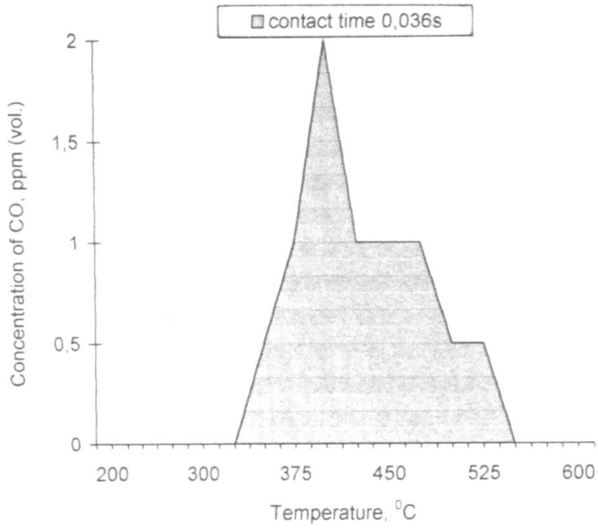

Figure 4. Carbon monoxide concentration in combustion gases versus temperature.

hyde started at the temperature of 260°C. Its concentration reached its peak of 0.1 mg/m^3 at a temperature of 350. At a contact time 0.36 s, combustion gases did not contain formaldehyde.

Carbon monoxide concentration in combustion gases as a function of temperature is shown in Figure 4. The CO concentration was low (2 ppm) for the contact time 0.036 s at about 400°C. At contact time of 0.36 s, carbon monoxide did not occur in reaction products.

Concentration of chloride ions in the condensate as a function of the process temperature is presented in Figure 5. Curve showing destruction rate of DCP calculated based on inorganic chlorine in reaction products is similar to that presented in Figure 1.

Combustion gases did not contain chlorine.

4. CONCLUSIONS

1. Figure 1 shows that 100% destruction was obtained. DCP is oxidized almost at the temperature of 550°C using the copper-zinc catalyst and at a contact time of

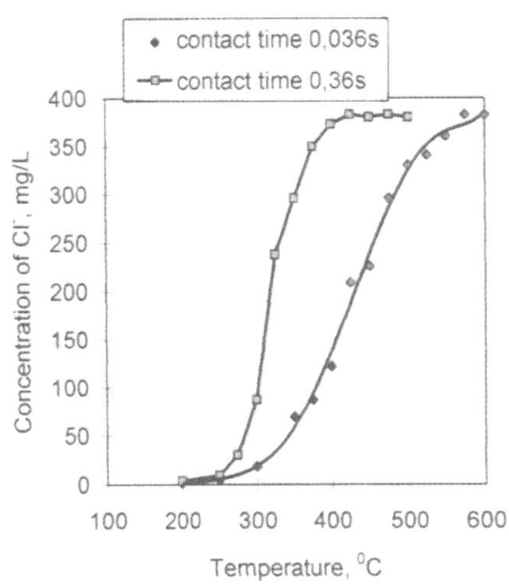

Figure 5. Concentration of chloride ions in condensate as a function of the process temperature.

0.036 s. Intermediate products of this reaction i.e. formaldehyde and carbon monooxide are also oxidized at this temperature. The maximum destruction rate determined by concentration of chloride ions in condensate was obtained at a temperature of 575°C.
2. Maximum concentration of the above mentioned intermediate products was obtained at a temperature of 400°C and at a contact time of 0.036 s.
3. Application of the copper-zinc catalyst and a contact time of 0.36 s leads to almost complete oxidation of DCP at temperatures lower than 400°C. During experiments carried out at this contact time, trace concentrations of the intermediate products were observed.
4. Combustion gases did not contain chlorine.
5. Results of experiments show the possibility of further decrease in temperature of DCP treatment process by an increase in a contact time or application of more active catalyst e.g. platinum. Decrease in temperature of the oxidation is very important because it decreases probability of formation of polychlorinated dioxins and furanes.

REFERENCES

1. Przondo J. and J.Rogala, Przemysłowa instalacja spalania ciekłych odpadów chloroorganicznych w Zakładach Chemicznych "Rokita" S.A., Przem. Chem., 1996,*3*,98.
2. Weisweiler W. and M.Lueck, Katalysierte Umsetzung chlorierter Kohlenwasserstoffe an Manganoxid-Katalysatoren, Chem. Ing. Tech., 1994, *4*, 518.
3. Gervasini A., Bianchi C.L. and V.Ragaini, Low-temperature catalytic combustion of valatile organic compounds, ACS Symp. Ser. 1994, 552 (Environmental Catalysis), 353.
4. Barresi A.A., Distruzione termica e catalitica in fase gassosa di composti organici clorurati, Acqua-Aria, 1991, *7*, 649.
5. Nowicki B., and J.Hetper, Badanie aktywnoœci katalizatorów w reakcji spalania chlorku winylu w powietrzu, Chem. Inż. Ekol., 1995, *3*, 423.
6. Mendyka B., Deactivation of platinum catalysts during combustion of multi-component chlororganic mixture, Environ. Prot. Eng., 1990, *3–4*, 129.
7. Papenmeier D.M. and J.A.Rossin, Catalytic oxidation of dichloromethane, chloroform, and their binary mixtures over platinum aluminia catalyst, Ind. Eng. Chem. Res., 1994, *12*, 3094.
8. Gorzka Z., Żarczyński A, Kaźmierczak M. and G.Miedzianowska, Termokatalityczne utlenianie 1,2-dichloropropanu na katalizatorze platynowym PA-2, Mat. IV Ogólnopol. Symp. "Ochrona powietrza w przemyœle", Łódź, 27–29 maja 1996, s.114.
9. Żarczyński A., Gorzka Z. and M.Kaźmierczak, Termiczne i termokatalityczne utlenianie 1,2-dichloropropanu, Ochr. Pow. Probl. Odp., 1997, *1*, 10.
10. Lou J.C. and S.S. Lee, Destruction of trichloromethane with catalytic oxidation, Appl. Catal. B: Environ. 1997, *12*, 111.
11. Gorzka Z. and M. Kaźmierczak, Thermocatalytic treatment of hazardous industrial wastewaters, J. Hazard. Mater., 1994, *37*, 127.
12. Anonim, Chemical review: propylene dichloride, Dangerous Prop. Ind. Mater. Rep., 1992, *3*, 296.
13. Polish Standard PN-81/Z-04029.01.
14. Polish Standard PN-76/C-04617.
15. Polish Standard PN-76/Z-04045.02.

18

THERMOCATALYTIC TREATMENT OF SULPHUR ORGANIC COMPOUNDS

Marek Kaźmierczak

Institute of General and Ecological Chemistry
Technical University of Łódź
90-924 Łódź, 36 Żwirki Str., Poland

ABSTRACT

Results of investigations on thermocatalytic oxidation process of some organic sulphur compounds (diphenyl sulphone, thiazole, thiourea), phenol in the presence of sulphur dioxide, and industrial wastewaters which contain sulphonated mineral oils, are presented. Platinum, copper-zinc and vanadic catalysts were used.

Intermediate products of the reaction and sulphur oxides gernerated in the process have a negative effect on catalysts in oxidation process.

The process of thermocatalytic treatment of exhausted cooling-oil emulsions with and without organic sulphur compounds was compared. The efficiency of oxidation of emulsion without sulphur was higher.

Key words: catalytic treatment, sulphur organic compounds, platinum, copper-zinc, vanadic catalysts.

1. INTRODUCTION

Industrial wastewaters are often treated using biological treatment methods. These wastewaters contain often various toxic substances able to decrease significantly the rate of biological treatment process. In such cases, it is necessary to apply pretreatment step involving selected physico-chemical methods. The thermocatalytic method is one of these methods suitable for treatment of liquid wastes with a high content of organic substances. The method involves the total oxidation of wastewater compounds by air on catalysts and

Chemistry for the Protection of the Environment 3, edited by Pawłowski *et al.* Plenum Press, New York, 1998

at elevated temperature. Platinum and copper-zinc catalysts are most frequently used in temperatures ranging from 250 to 800°C.

The thermocatalytic method used for treatment of industrial wastewaters with high content of toxic organic compounds is very effective. For example the thermocatalytic treatment of wastewaters from production of phenol-formaldehyde resins with an initial COD concentration of about 172 gO_2/L and a phenol concentration of 32 g/L produced at 500°C (on a copper-zinc catalyst) a colourless and transparent condensate with a COD concentration of about 10 mg/L and with a phenol concentration reduced to zero[1].

Liquid industrial wastes often contain organic and inorganic sulphur compounds, which are known to have adverse effect on catalytic processes. In a case of oxidation of organic compounds containing sulphur, catalysts like platinum or vanadium were used because their effectiveness do not deteriorate easily in a presence of these compounds. Intensive research work was recently carried out on preparation of very active catalysts containing TiO_2 or ZrO_2 with proper dopes of precious metals[3,4]. These catalysts can be used in a deep oxidation process of organic compounds and oxidation of SO_2 to SO_3.

A copper-zinc catalyst was found to be applicable at temperatures higher than 600°C to treatment of wastewaters with a low content of sulphur compounds[5].

The aim of this work was to determine an effect of sulphur organic compounds on catalysts, with special attention paid to the copper-zinc one. The aim of this work was also to determine an influence of these compounds on efficiency of the treatment process.

Diphenylsulphone (DPS) and thiazole were used in experiments. Both these aromatic substances are resistant to thermic decomposition. In thiazole, atom of sulphur is located in an aromatic ring. Thus, generation of sulphur dioxide needs an opening of the ring. Sulphonated mineral oil and thiourea were also investigated. Sulphonated oil is resistant to thermic decomposition, on the contrary to thiourea. Moreover, investigations on oxidation of phenol by air containing sulphur dioxide were carried out.

2. EXPERIMENTAL

Tests on thermocatalytic oxidation were carried out in an apparatus operated in a continuous mode. The apparatus consisted of an evaporator and a reactor. Both were 10 mm in diameter. The evaporator and the reactor were made of glass and quarz, respectively. Electrical heating was used to maintain their proper temperature. Raw solutions were injected with air to the evaporator. The mixture was subsequently passing from the evaporator to the reactor.

Unreacted compounds and intermediate products were analysed using chromatographic methods. Carbon dioxide, sulphur dioxide and nitrogen oxides were analysed using an automatic gas analyzer and colorimetric methods. Concentration of H_2SO_4 in condensates was analysed using a turbidity method. COD, TOC and pH values were also analysed using well known standard methods.

The following solutions of substrates were used: diphenylsulphone (DPS) in the form of an aqueous-ethanol solution at concentration of 0.125 to 1.25 g/L, thiazol in the aqueous solution at concentration of 1,2 g/L, thiourea and phenol in aqueous solution at concentration of 5 g/L. The used sulphonated oil was a component of spent cooling-oil emulsions.

The applied flow rate of substrate solutions was ranging from 5 to 15 ml/h. The flow rate of air ranged from 15 to 40 L/h. Temperature in evaporator was 210°C and in reactor ranged from 200 to 800°C.

Copper-zinc catalysts TMC-2 produced by Zakłady Azotowe in Tarnów, platinum catalyst KP-910 produced in Technical University in Wrocław and vanadic catalyst produced in Luboń near Poznań were used in our investigations.

3. RESULTS AND DISCUSSION

Results of experiments are presented in form of diagrams on Figures 1 - 10. Points marked in these figures present mean values obtained during experiments repeated 3 to 5 times.

Figure 1 shows the DPS conversion rate as a function of temperature and type of catalyst. High conversion rates were obtained at contact times of 0.01s for the platinum and copper-zinc catalyst, and 0.18 s for vanadium catalyst. It results from Figure 1 that platinum catalyst allows to achieve high efficiency of oxidation in the range of significantly lower temperatures (400 to 550°C) than other catalyst.

Figure 2 shows significantly more advantageous effect of platinum catalyst by presenting the content of sulphur dioxide in combustion gases from DPS oxidation. The lowest content of sulphur dioxide (0.007 mg/L) was observed when platinum catalyst was used. The highest content was obtained in the case of vanadic catalyst.

However, platinum catalyst lost its activity after about one hundred hour period of application at a temperature of 400°C (Figure 3). Activity of this catalyst was stable at a temperature of 500°C.

A possibility of using copper-zinc catalyst at temperature higher than 400°C. was also investigated. Figure 4 presents results of experiments during which high conversion rates of DPS were obtained at a contact time of 0.01 s and in a temperature range from 400 to 500°C. However, the activity of this catalyst decreased after experiment duration time of 5–10 hours under influence of sulphur compounds. It is possible to regenerate the catalyst by heating at a temperature higher than 670°C[6].

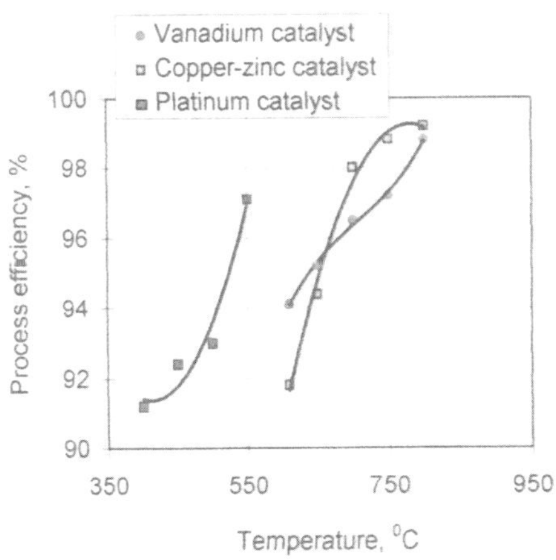

Figure 1. Effect of temperature and type of catalyst on treatment efficiency of DPS.

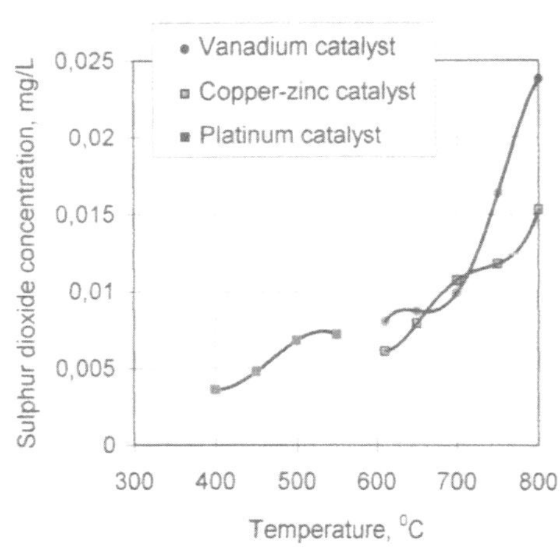

Figure 2. Thermocatalytic treatment of DPS. Sulphur dioxide concentration in combustion gases versus type of a catalyst and temperature.

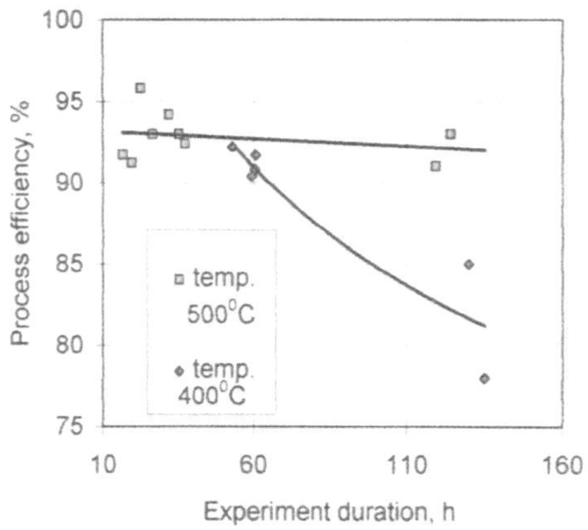

Figure 3. Thermocatalytic treatment of DPS. Effect of duration time of an experiment on activity of platinum catalyst KP-910.

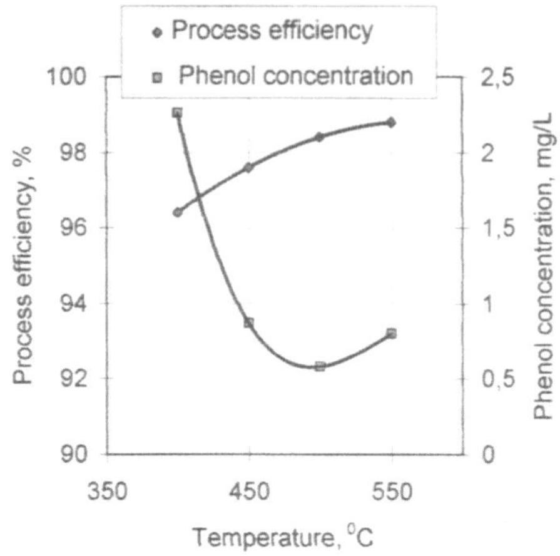

Figure 4. Thermocatalytic treatment of DPS. Effect of temperature on process efficiency and phenol formation on copper-zinc catalyst.

It is important to assure formation of not toxic products during treatment of toxic substances. However, data in Figure 4 shows that phenol is formed at low concentrations as an intermediate product of oxidation during treatment of DPS.

Test results of experiments on phenol oxidation using copper-zinc catalyst in air with addition of sulphur dioxide are showed in Figure 5. Concentrations of intermediate reaction products in a condensate were determined by comparison of general concentration of organic substances and concentration of phenol which did not react. Experiments were carried out at contact time of 0.002s.

p-benzoquinone was found among products of not complete phenol oxidation[6–8]. Figure 6 shows that its concentration increased when sulphur dioxide was present in a reaction mixture.

Organic acids are the other intermediate products of catalytic oxidation. If sulphur dioxide is present in reagents, the condensate contains except organic acids also sulphuric acid. Figure7 presents an effect of temperature and sulphur dioxide concentration in a reagent mixture on concentration of hydrogen ions in a condensate.

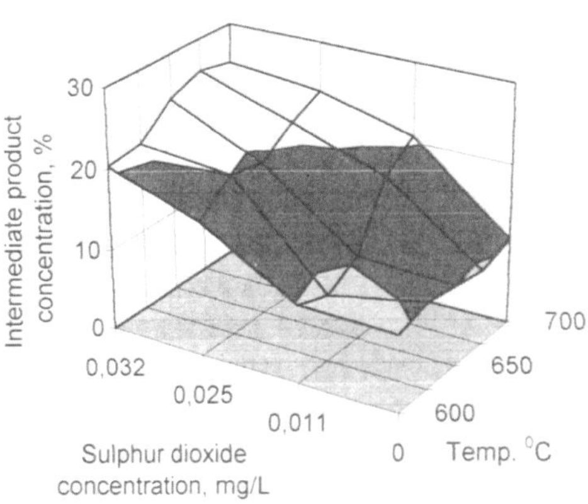

Figure 5. Thermocatalytic treatment of phenol at copper-zinc catalyst. Concentration of intermediate reaction products versus sulphur dioxide concentration and temperature in a reagent mixture.

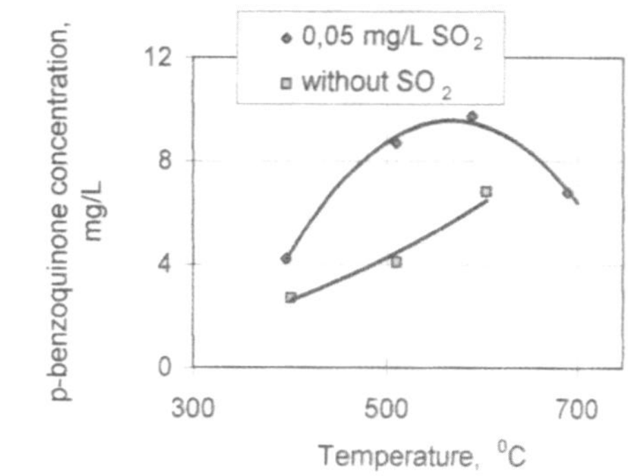

Figure 6. Thermocatalytic treatment of phenol at copper-zinc catalyst. Effect of temperature and sulphur dioxide on concentration of p-benzoquinone.

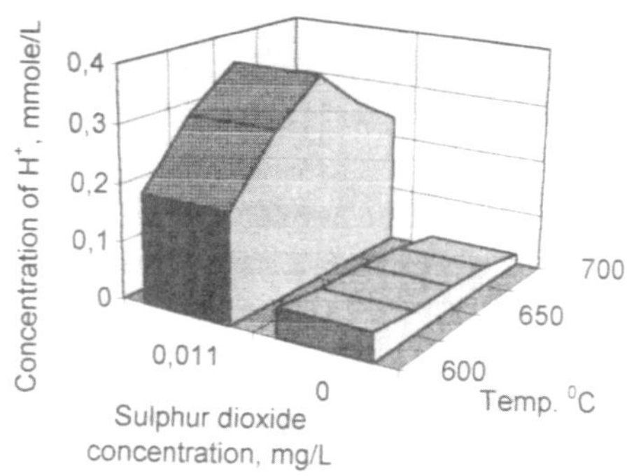

Figure 7. Thermocatalytic treatment of phenol on copper-zinc catalyst. Effect of temperature and content of SO_2 in a reagent mixture on hydrogen ion concentration in a condensate.

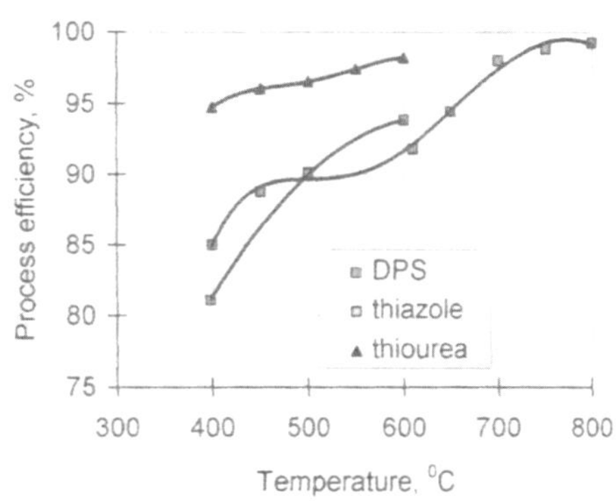

Figure 8. Effect of temperature of copper-zinc catalyst on thermocatalytic treatment efficiency of DPS, thiazole and thiourea.

Figure 8 shows effect of temperature of copper-zinc catalyst on conversion rate of DPS, thiazole and thiourea (contact time to 0.01s). Efficiency of oxidation of DPS and thiazole were similar. However, conversion rate of thiourea was higher.

An effect of presence of sulphur compounds on the oxidation of spent cooling-oil emulsions using copper-zinc catalyst is presented in Figure 9. The efficiency of oxidation of emulsions which contain mainly a product of ethylene-oxide addition to ricinus oil, are a little bit higher than treatment efficiences of emulsions made of sulphonated mineral oil. Experiments were carried out at contact time of 0.35 s.

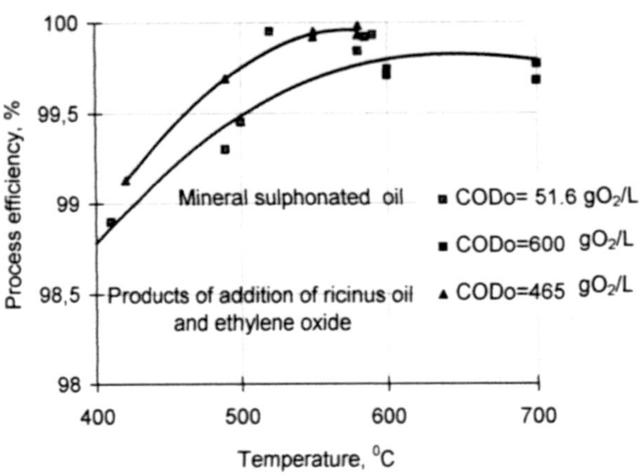

Figure 9. Effect of temperature of copper-zinc catalyst on thermocatalytic treatment efficiency of spent cooling-oil emulsions.

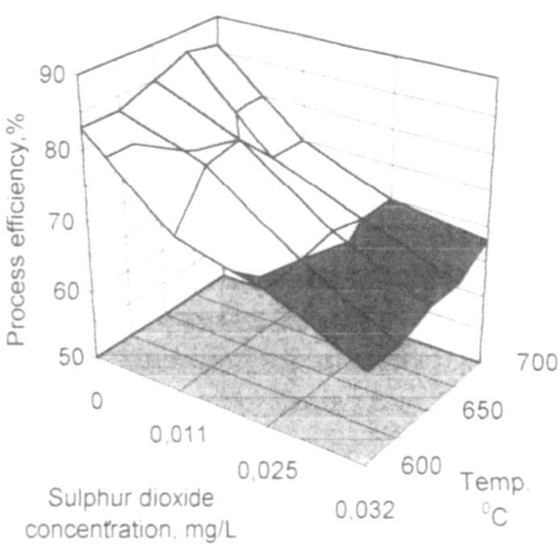

Figure 10. Effect of temperature of Cu-Zn catalyst and concentration of sulphur dioxide in a reagent mixture on oxidation efficiency of phenol.

Very clear effect of sulphur dioxide on efficiency of phenol oxidation is presented in Figure 10. The oxidation efficiency calculated from COD changes in a condensate depends much more on sulphur dioxide concentration in a reagent mixture than on temperature.

4. CONCLUSIONS

Results of experiments lead to the following conclusions:

1. Catalyst: platinum KP-910, copper-zinc TMC-2 and vanadic are very active in an oxidation process of organic sulphur compounds contained in industrial wastewaters. Conversion degrees of 95 to 99% were obtained at a contact time of 0.01s. An increase in contact time to 0.35s in experiments with exhausted cooling-oil emulsions caused an increase in the process efficiency to 99.8%.
2. Platinum catalyst was the most active among tested catalysts. Activity of this catalyst was stable at a temperature of 500°C.

3. Application of copper-zinc catalyst also allows to carry out oxidation process of organic sulphur compounds in the temperature of 500°C. However, periodical thermic regeneration of the catalyst is necessary.
4. Copper-zinc and vanadic catalysts were stable in the range of temperature from 600 to 800°C. The copper-zinc catalyst was found to be more active.
5. During thermocatalytic treatment of substrates tested, some intermediate products such as organic acids, phenol in DPS oxidation and p-benzoquinone in phenol oxidation, were formed.
6. Concentration of intermediate products increased when the content of sulphur dioxide in a reagent mixture increased.
7. Efficiency of the treatment process of both cooling-oil emulsions (one was a sulphur compound) was investigated. In the case of emulsion-sulphur compound, lower conversion degree was obtained.
8. Efficiencies of treatment of aromatic sulphur compounds, DPS and thiazole were similar. It can be important in explanation of a process mechanism and development of its mathematic model.
9. Efficiency of thiourea oxidation at copper-zinc catalyst was higher than in the case of other tested compounds.
10. 10. If a temperature of thermocatalytic treatment of organic sulphur compound is to be lower than 500°C, then catalyst more active from tested ones should be applied. Promising results were obtained during investigations on preparation of active catalysts which can simultaneously oxidize organic compounds and sulphur dioxide.

ACKNOWLEDGMENT

This work was partially supported by Polish State Committee for Scientific Research—grant No 3 P045 003 06

REFERENCES

1. Gorzka Z., Kaźmierczak M., Thermocatalytic Treatment of Hazardous Industrial Wastewaters, *Journal of Hazardous Materials* 1994, 37, 127–136.
2. Gorzka Z., Kaźmierczak M., Neutralization of Industrial Wastes Containing Detergents by Means of Catalytic Oxidation Method, *Physicochemical Methods for Water and Wastewater Treatment*, Pergamon Press, Oxford 1980, pp.175–183.
3. Babcock Deutsche Babcock Anlagen AG, Verfahren zur Verringerung der Emission von organischen Produkten unvollstaendiger Verbrennung, Eur. Patent Appl., 1991, No 447,537.
4. Lewicki A., Paryjczak T., Rynkowski J., Dwutlenek cyrkonu w katalizie. Właœciwosci i zastosowania, *Wiadomosci chemiczne*, 1996, 11–12, pp. 879–902.
5. Gorzka Z., Kaźmierczak M., Socha A., Michalska-Jednoralska A., Utilization and Treatment of Exhausted Cooling Oil-Emulsions, *Waste Management*, 1992, 12, 345–348.
6. Kaźmierczak M., The Effect of Sulphur Dioxide on Reaction of Total Phenol Oxidation at Copper-zinc Catalyst, *Chemistry for the Protection of the Environment 2*, Plenum Press, New York 1996, pp. 295–302.
7. Walsh M.A., Katzer J.R., Catalytic Oxidation of Phenol in Dilute Concentration in Air, *Ind. Eng. Chem. Proc. Des. Develop.*, 1973, 12, pp. 477–481.
8. Devlin H.R., Harris I.J., Mechanism of the Oxidation of Aqueous Phenol with Dissolved Oxygen,. *Eng. Chem. Fundam.*, 1984, 23, 387–392.

19

SIMULTANEOUS ELECTROOXIDATION OF CYANIDES AND RECOVERY OF COPPER ON CARBON FIBRE

A. Socha, E. Kuśmierek, and M. Kaźmierczak

Institute of General and Ecological Chemistry
Technical University of Łódź, Poland

ABSTRACT

An effect of medium pH on distribution of cyanide and copper ions in cyanide and hydroxide complexes with copper is presented. Total cyanide oxidation at carbon fibre and simultaneous total copper recovery was obtained at double electric charge theoretically necessary to complete cyanide oxidation. Electric charge necessary for total reduction of copper on carbon fibre caused simultaneous cyanide oxidation in 71.3%. Electric energy consumption in the case of these two processes was 8.6 kWh/kg CN^- and 4.3 kWh/kg Cu, respectively.

Key words: electrochemical treatment, cyanide complexes, carbon fibre.

1. INTRODUCTION

Wastewaters containing cyanide are usually treated with chlorine or sodium hypochloride after pH adjustment to alkaline levels. However, this method has several disadvantageous[1–3]. Other chemical treatment methods involve ozonation, hydrogen peroxide or air oxidation, sulphur-based technologies, etc.[4].

At present, the electrochemical treatment of cyanide is the most favoured method used for cyanide destruction. There are two techniques employed: direct oxidation at an anode[5–7] and indirect oxidation with hypochloride or chlorine generated electrochemically in cyanide wastes to which NaCl is added[3,6–8]. However, the direct oxidation eliminates the formation of toxic compounds containing chlorine which may occur during the indirect oxidation. The electrochemical treatment is more cost effective than the conventional alkaline chlorination[7]. It also offers the possibility of simultaneous recovery of dissolved

Chemistry for the Protection of the Environment 3, edited by Pawłowski *et al.* Plenum Press, New York, 1998

metals[7]. Electrolytic treatment of cyanide effluents is effective for higher cyanide concentrations (> 100 ppm). In the case of dilute cyanide solutions (< 50 ppm) such as rinsewater from plating operations, electrolytic recovery of metal and cyanide destruction are difficult[7] and inefficient[8, 9]. This is because of a mass transfer limitation and a low reaction rate per an unit of electrode area. Thus, the selection of a proper electrode material for treatment of dilute cyanide wastewaters is very important.

The reaction of cyanide oxidation was found to be slow on platinum and graphite anodes. Other anode materials such as stainless steel, magnetite or lead allow the rate of cyanide oxidation to increaserate but they are not resistant to corrosion[4]. A certain compromise between corrosion rate and efficiency was found in the case of lead dioxide on titanium[4, 5] or ferrite plates[9]. Recently, electrode materials with a large surface area e.g., conductive fibrous materials like graphite cloths, reticulated vitreous carbon, carbon fibre and porous electrodes were tested for a variety of applications in electrochemistry and for electrochemical reactors[10]. Porous and fibrous materials are often suggested for recovery of metals from solutions by electrochemical methods[10–12].

The aim of this work was to test the application of carbon fibre in the form of cloth in the process of electrochemical destruction of cyanides and copper recovery from a dilute solution of cyanide complexes with copper.

2. EXPERIMENTAL

The treatment of cyanide solution was carried out in an undivided electrochemical cell. Carbon fibre in the form of cloth was used as the anode. The real area of this electrode was about 0.1 m^2. A saturated calomel electrode was chosen as a reference electrode. Platinum wire or carbon fibre was the cathode. Argon gas was bubbled into the electrolyte during electrolyses.

Reagent grade potassium cyanide, cuprous cyanide and NaOH were dissolved in doubly distilled water. A stock solution of 2.5 mM $K_3[Cu(CN)_4]$ was prepared by mixing one part of CuCN with three parts of KCN and dissolving it in 0.5 M NaOH.

The concentration of cyanide ions was determined from a calibration curve[13] by recording square wave voltammograms at a hanging mercury drop electrode (HMDE). In order to determine the concentration of copper ions in solutions, small samples of 0.2 ml were added to 10 ml of 0.5 M H_2SO_4. Next, square wave voltammograms were recorded at the HMDE and the concentration of copper ions were determined from a calibration curve.

All voltammetric measurements were preceded by purging solutions with argon in order to remove dissolve oxygen. Current-potential (i–E) curves were recorded using an AUTOLAB system (EcoChemie, Holland).

3. RESULTS AND DISCUSSION

Plating rinsewater from copper plating with cyanide complexes contains copper ions at concentration of 78 to 310 mg/l and cyanide ions at concentration of 80 to 590 mg/l. Its pH is highly alkaline. Thus, investigations on treatment of plating rinsewater containing cyanide complexes with copper were carried out using a model solution of 2.5 mM $K_3[Cu(CN)_4]$ in 0.5 M NaOH (pH = 13.6). The content of cyanide ions in complexes with various numbers of cyanide ligands differs with pH (Fig. 1.).

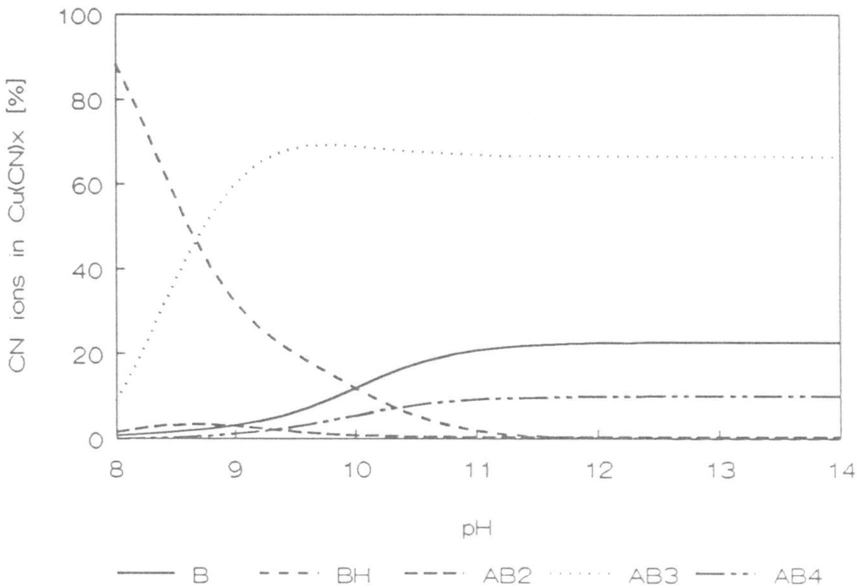

Figure 1. Distribution of cyanide ions in complexes with copper at different pH values; at constant c_{CN} = 10 mM, c_{Cu} = 2.5 mM; A: Cu, B: CN^-.

The content of cyanide complexes with copper at pH = 13.6 is as follows:

$K_3[Cu(CN)_4]$	=	10.12%
$K_2[Cu(CN)_3]$	=	66.64%
$K[Cu(CN)_2]$	=	0.49%
free cyanide ions CN^-	=	22.71%

The content of copper ions in cyanide complexes also differs with pH of the solution (Fig. 2.)

The content of copper in different complexes in the model solution at pH 13.6 is as follows:

$K_3[Cu(CN)_4]$	=	10.12%
$K_2[Cu(CN)_3]$	=	88.85%
$K[Cu(CN)_2]$	=	0.98%

Copper complexes with hydroxide ions do not occur in solutions with pH higher than 11.

Cyanide complexes with copper were found to be easier oxidized the carbon fibre than free cyanides. Complexes with a lower ligand number were oxidized at lower potential values[13]. This can be explained by values of instability constants and standard reduction potentials of cyanide complexes (Table 1).

Total destruction of cyanides and total copper recovery from a solution was observed at the electric charge which is two times higher than a charge (Q_{tCN}) theoretically necessary to complete oxidation of cyanides[14]. Then, electric energy consumption was about 8.6 kWh/kg CN^-. In the charge range from $0.5Q_{tCN}$ to $1.5Q_{tCN}$, formation of copper complexes with hydroxide ions was observed. Modification of the carbon fibre (used as an

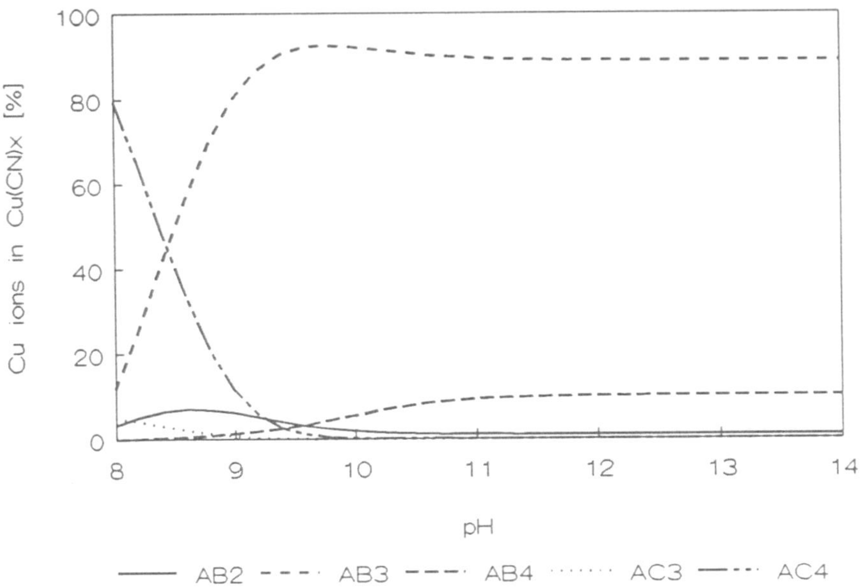

Figure 2. Distribution of copper ions in cyanide and hydroxide complexes at different pH values; at constant c_{CN} = 10 mM and c_{Cu} = 2.5 mM; A: Cu, B: CN^-, C: OH^-.

anode) with copper oxides caused an increase in a cyanide conversion by 20% if the electric charge corresponded to Q_{tCN} [15]. Simultaneously, electric energy consumption decreased by about 2 kWh/kg CN^-.

Modification of the carbon fibre also influenced copper recovery (Table 2).

As it results from Table 2, a degree of copper removal from solution was about 52% if a non modified anode was used. All copper ions removed from the solution were recovered at the cathode. The degree of copper removal increased by 21% if modified carbon fibre was used. However, only about 71% of copper removed from the solution was deposited at the cathode. The rest of the copper was deposited at the anode surface in the form of oxides. It means that anode was additionally modified.

In order to determine the potential range of copper reduction, differential pulse voltammograms were recorded in the solution of 2.5 mM $K_3[Cu(CN)_4]$ at the carbon fibre (Fig. 3).

Table 1. Instability constants and standard reduction potentials of copper complexes

Copper complex with CN^- ions	Instability constant pK	Standard reduction potential [V]
$[Cu(CN)_2]^-$	24.0	0.38*
$[Cu(CN)_3]^{2-}$	28.6	−0.37*
$[Cu(CN)_4]^{3-}$	30.3	−0.48*

*in 1 M NaOH

Table 2. Results of copper recovery during oxidation of 2.5 mM $K_3[Cu(CN)_4]$ at modified and non-modified carbon fibre; process parameters: potential of 0.6 V, electric charge of 20 C (Q_{tCN})

Copper amount at the anode in the form of oxides before oxidation [mg]	Initial copper ion amount in stock solution of $K_3[Cu(CN)_4]$ [mg]	Copper ion amount in the solution after oxidation [mg]	Copper amount deposited at the cathode [mg]	Copper amount at the anode in the form of oxides after oxidation [mg]
0*	1.59	0.76	0.83	0
1.05	1.58	0.43	0.83	1.37

*carbon fibre was not modified

Further, reduction of copper ions at the carbon fibre was carried out in the potential range from −1.3 to −1.7 V which was chosen on the basis of Fig. 3. Some results are presented in Table 3.

The highest amount of copper ions was deposited at the potential of −1.4 V and electric charge of 2.5 C—Q_{tCu} (Q_{tCu}—electric charge theoretically necessary for total reduction of copper ions). The degree of copper removal from the solution was about 53%. Simultaneously, some amounts of copper ions were deposited in the form of oxides at the anode surface during electrolysis in the investigated potential range. They caused modification of the anode surface. The conversion of cyanide ions was relatively low and varied from 22 to 33%. This was caused by the very low electric charge passing through the electrolyser. This charge (Q_{tCu}) theoretically corresponds to a conversion degree of about 13%. On the other hand, the conversion degrees of cyanides were higher from these calculated from Q_{tCu} which results from modification of the anode surface with copper oxides.

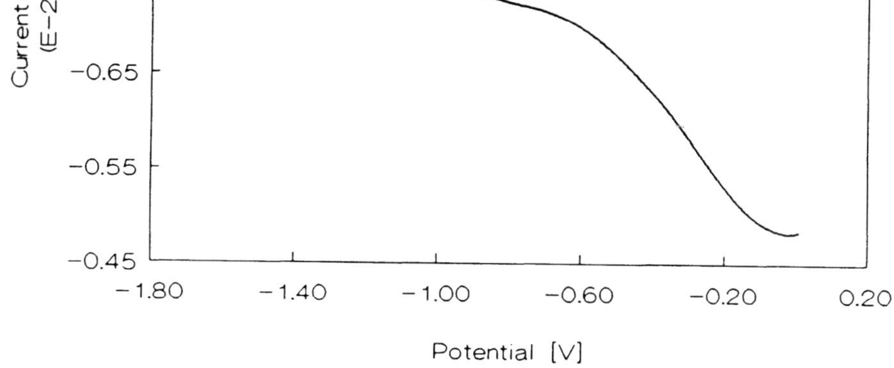

Figure 3. Differential pulse voltammogram recorded in the solution of 2.5 mM $K_3[Cu(CN)_4]$ at carbon fibre in the potential range from 0.1 to −2 V; scan rate: 0.1 V/s.

Table 3. Results of copper recovery obtained during its reduction at a carbon fibre anode at different potentials; process parameters; constant electric charge: 2.5 C (Q_{tCu}), initial copper ion amount in the stock solution: 1.61 mg

Electrolysis potential [V]	Copper ion amount in a stock solution [mg]	Copper ion amount deposited at the cathode [mg]
−1.30	1.11	0.10
−1.40	0.76	0.61
−1.45	1.01	0.33
−1.50	0.90	0.31
−1.60	0.92	0.25
−1.70	1.29	0.18

In order to determine the influence of the electric charge passing through the electrolyser, the reduction of copper ions at the cathode was carried out in the charge range from 2.5 (Q_{tCu}) to 15 C ($6Q_{tCu}$). Some results of investigations are presented in Table 4.

Total removal of copper from the solution was obtained at the electric charge of 12.5 C ($5Q_{tCu}$). An increase in electric charge caused increased in the amount of copper deposited at the cathode. However, in the investigated range of electric charge, small amounts of copper (0.30–0.40 mg) were deposited also at the anode in the form of oxides. The cyanide conversion obtained during copper reduction increased from 32 to 79% when electric charge was increased. This was also higher than theoretical. However, the higher electric charge, the lower the difference between theoretical and observed levels. At the charge of 15 C, the theoretical and obtained cyanide levels were almost the same. This effect on cyanide conversion degree can be explained by modification of the anode surface and its destruction. The higher electric charge caused deposition of higher amounts of copper at the cathode. The amount of free cyanide in solution increased and caused dissolution of the anode modification. It resulted in a decrease in the difference between theoretical and experimental cyanide conversions.

Electric energy consumption during electrolysis at the potential of −1.5 V and 12.5 Q ($5Q_{tCu}$) was about 4.27 kWh/kg Cu.

Table 4. Results of copper recovery during reduction at a carbon fibre anode at different electric charges; process parameters: initial copper ion amount in the solution: 1.61 mg, constant electrolysis potential: −1.5 V

Electric charge [Q]	Copper ion amount in the solution after electrolysis [mg]	Copper amount deposited at the cathode [mg]
2.5	0.98	0.30
5.0	0.47	0.71
7.5	0.38	0.85
10.0	0.15	1.00
12.5	0	1.17
15.0	0	1.27

4. CONCLUSIONS

Electrochemical oxidation of cyanide complexes with copper at a carbon fibre anode resulted in:

1. electric energy consumption during cuprocyanides oxidation at the carbon fibre at the electric charge ($2Q_{tCN}$) of twice that necessary for total cyanide destruction and copper removal was about 8.6 kWh/kg CN^-,
2. modification of the anode surface with copper oxides caused an increase in cyanide conversion (at $2Q_{tCN}$) by 20% and adecrease in electric energy consumption of 2 kWh/kg CN^-,
3. copper removal was about 52% (at $2Q_{tCN}$) and increased to 73% if a modified anode was used,
4. during cyanide oxidation, modification of anode surface with copper oxides was observed.

Electrochemical reduction of copper ions from cyanide complexes at the carbon fibre anode resulted in:

1. 53% copper ion removal at the potential of −1.5 V and electric charge (Q_{tCu}) theoretically necessary for complete removal,
2. simultaneously conversion of cyanides was 22 to 33% and was higher than resulted from electric charge (13%), due to modification of the anode surface,
3. electric energy consumption during reduction of copper at −1.5 V and at electric charge of 12.5 C ($5Q_{tCu}$) which caused total removal of copper ions, was about 4.27 kWh/kg Cu.

The results clearly show that carbon fibre in the form of cloth is applicable to the treatment of dilute cyanide wastewaters. Modification of the carbon fibre (used as an anode) with copper oxides enhances of cyanide oxidation and facilitates removal of copper at the cathode.

Electric energy consumption during theseexperiments seemed to be lower than in the case of other electrode materials and different types of reactors[7, 16], although they were calculated on the laboratory scale.

ACKNOWLEDGMENT

This work was supported by the Polish State Committee for Scientific Research—grant No. T09B 183 08.

REFERENCES

1. Tamura, H., Arikado, T., Yoneyama, H., Matsuda, Y., *Electrochim. Acta* **19**, 273 (1974)
2. Arikado, T., Iwakura, C., Yoneyama, H., Tamura, H., *Electrochim. Acta* **21**, 1021 (1976)
3. Shi-Chern, Y., Chih-Ta, W., Jin-Shen, W., *Chem. Eng. Comm.* **109**, 167 (1991)
4. Tissot, P., Fragniere, M., *J. Appl. Electrochem.* **24**, 509 (1994)
5. Hine, F., Yasuda, M/ Iida, T., Ogata, Y., *Electrochim. Acta.* **31**, 1389, (1986)
6. Jin-Shen, W., Chih-Ta, W., Shi-Chern, Y., *Proc. Natl. Sci. Counc. ROC(A)* **15**(6), 529 (1991)
7. Zhou, C.D., Chin, D.T., *Plat. Surf. Finish.*, **70** (1994)
8. Hwang, J.S., Wang, Y.Y., Wan, C.C., *Bull. Electrochem.* **11**(5), 241, (1995)

9. Balasubramanian, G., *Bull. Electrochem.* **6**(4), 446, (1990)
10. Oren, Y., Soffer, A., *Electrochim. Acta* **28**(11), 1649 (1983)
11. Abda, M., Gavra, Z., Oren, Y, *J. Appl. Electrochem.* **21**, 734 (1991)
12. Kirk, D.W., Foulkes, F.R., *J. Electrochem. Soc.* **131**(4), 760 (1984)
13. Socha, A., Kuśmierek, E., Proceedings of the 3rd International Symposium "Electrochemistry in Practice and Theory", Łódź University Press, Poland 1995, pp. 117–125
14. Socha, A., Kuśmierek, E., Chemistry for the Protection of the Environment 2, Plenum Press, New York and London, 1996 pp. 283–293
15. Socha, A. Kuśmierek E., Proceedings of the 4th International Symposium "Electrochemistry in Practice and Theory," Łódź University Press, Poland 1996
16. Lin, M.L., Wang, Y.Y., Wan, C.C., *J. Appl. Electrochem.* **22**, 1197 (1992).

20

NEUTRALIZATION OF HAZARDOUS WASTES COMBINED WITH CLINKER MANUFACTURING

Lucjan Pawłowski, Zdzisław Kozak, Ryszard Gierżatowicz, and Marzenna R. Dudzińska

Department of Environmental Protection Engineering
Technical University of Lublin
ul. Nadbystrzycka 40
20-618 Lublin, Poland

ABSTRACT

The Polish Government has financed a research project on utilization of a cement kiln for neutralization of hazardous wastes.

For research on technical scale, a cement plant using a wet process was selected. The results obtained are summarized in this paper.

Key words: hazardous wastes, clinker manufacturing, waste utilization, PCDF, PCDD.

1. INTRODUCTION

Clinker is manufactured by heating of a mixture of clay and limestone to over 1450°C in an oxidizing atmosphere with the combustion gases reaching 1900°C. Under such conditions, all organic substances passing through the high temperature zone in the cement kiln are decomposed and reduced to simple inorganic compounds.

Inorganic substances under conditions of clinker manufacturing are also oxidized and sintered into the clinker structure, and immobilized there.

These conditions are very advantageous for combination of clinker manufacturing process with utilization of both combustible and mineral wastes. Neutralization of wastes in cement plants is popular in developed countries (*see Table 1*). Recommendation of the European Union (EU) suggests that 40% of total energy used for cement manufacturing should come from waste combustion[1]. Currently, in Poland this method is not used on an industrial scale. Because the caloric values of some wastes are event higher than that of

Table 1. Characterization of an amount of energy produced from incineration of wastes in relation to total energy used for clinker manufacturing in 1994 according to Cembureau (in T.M.Loves[1])

Country	Part of energy produced from incineration of wastes [%]
France	52.4
Belgium	18.0
Switzerland	16.8
Republic Czech	9.7
Island	6.0
Italy	4.1
Sweden	2.0
Slovak Republic	1.6
Portugal	1.3
Spain	1.0
Hungary	0.1

coal (*see Table 2*), incineration of wastes in a cement kiln is an economically attractive and environmentally sound method of waste management.

The method of combining the clinker manufacturing and the wastes neutralization processes, strongly depends on the wastes properties. From that point of view, wastes can be classified into three following groups:

- flammable organic wastes, which can be used as substitutes for conventional fuels,
- non-flammable inorganic wastes, which can be used as substitutes for raw materials in clinker manufacturing,
- non-flammable inorganic wastes, which are not substitutes for raw materials but can be neutralized in the process of clinker manufacturing by thermal decomposition and immobilized in a clinker.

30% of the coal used for clinker manufacturing could be substituted by combustion of all wastes from plastics production in Poland.

Table 2. Characterization of caloric values of some polymers in comparison with calorific values of a wood and a hard coal

Polymer	Caloric values [MJ/kg]	Composition [% WW] C	H	Other
Polyethylene	43	85.6	14.4	—
Polypropylene	44	85.6	14.4	—
Polystyrene	40	92.3	7.7	—
Polyethylterephtalane	31	74.9	5.0	O 20.0
Polyvinyl chloride	18	38.4	4.8	Cl 56.8
Polyacetylene	45	92.3	7.7	—
Wood	about 18	50.0	6.0	O 44.0
Coal (standard)	32	80–85	4–6	O, S, N

Table 3. Typical chemical composition of sludge produced in a metal finishing plant

Content of water	61.00%
Contents of solid matter	39.00%
Components	
H_2O	61.00%
CaO	19.45%
MgO	0.30%
Fe_2O_3	5.71%
MnO_2	0.13%
Na	0.11%
K	0.02%
Zn	0.78%
Cd	< 0.0001%
Co	< 0.0001%
Cr	0.13%
Cu	0.078%
Pb	0.07%
Ni	0.03%
COD	185.6 mg O_2
Organic compounds and chemically bounded water	12.26%

As it results from the Table 2, the cement industry can be one of the attractive and environmental safe methods of hazardous wastes utilization. The Polish Government has financed the research project on combustion of wastes combined with clinker manufacturing.

2. NEUTRALIZATION OF MINERAL WASTES

Hazardous mineral wastes from metal finishing plants are a wide spread and difficult problem for waste disposal. The wastes are mainly composed of iron, calcium and aluminum oxides with addition of small amount of heavy metals oxides and cyanides.

An iron oxides, in the clinker manufacturing process, can be treated as substitutes for ferruginous materials (for example, iron ore) and calcium oxide as substitute for a limestone (calcium carbonate).

However, wastes also contain others impurities (*see Table 3*), mainly heavy metals. Our studies confirm that heavy metals are incorporated into the clinker structure thereby, they are immobilized. Therefore, they can be mixed in suitable proportion with raw materials used for clinker manufacturing. To avoid overdosing on some components, especially heavy metals, strict controls are necessary.

3. NEUTRALIZATION OF FLAMMABLE WASTES

As mentioned above, flammable wastes can be used as fuel substitute (in Poland a substitute for coal). Combustion of chlorine free wastes is not a problem because they are completely mineralized in the temperature of cement kiln.

Combustion of chlorine containing wastes may cause some problems due to formation of very toxic polychlorinated dibenzo-p-dioxins (PCDDs) and dibenzo-p-furans (PCDFs). Polychlorinated dibenzo-p-dioxins exist in 75 various forms while polychlorinated dibenzo-p-furans in 135 forms. In total, the group consists of 210 chemical compounds of different toxicity, called congeners. In such a mixture, the values of concentrations are calculated in relation to the most toxic form and determination of only 17 PCDFs and PCDDs, of the highest toxicity is required under the legal regulations in most countries. Such determination does not give the exact composition of products formed during the combustion of chlorine containing wastes. From a thermodynamic point of view, under conditions of a cement kiln, PCDF/PCDDs should not be formed if wastes in the suitable form, were introduced to either flame or calcination zone and held for a sufficient retention time.

The laboratory and technical studies we conducted, have confirmed the original assumptions.

It should be noted that emission of PCDF/Ds and dibenzo-p-furanes from waste dumps is as high as emission during combustion the same amount of wastes in a cement kiln[2].

Presence of chlorine in cement kiln leads to the formation of more volatile chlorides of heavy metals (especially of lead) and alkaline elements. Currently, this process is being carefully tested.

As mentioned in the case of flammable wastes, the procedure of wastes input is of great importance. Under this project, several systems of a waste input were tested.

4. CHARACTERISTIC OF WASTE INPUT SYSTEMS

The following three systems of waste input systems to a cement kiln have been developed:

- pneumatic ejector,
- sluice way
- primary wastes incinerator,

4.1. Pneumatic Ejector

Pneumatic systems for a load shooting in to the hot end of a rotary kiln are well known in the cement industry. Under pressure of about 5 MPa such installations can charge loads to about 30 kg between 35–40 m. In the installation developed in the project, the same results were achieved under much lower pressure and without strict fitting of load dimensions to the barrel bore. The developed installation has just been constructed and subjected to technical testing (*See Fig. 1*).

Despite the positive test results, the system is expensive and requires special manufactured containers at cost of about 1.5 USD.

However the above waste input system is very useful for hospital wastes treatment, handling and neutralization. The containers designed and constructed under this project are hermetically sealed without the possibility of opening and a guarantee of their safe transport.

4.2. Sluice Way

The installation developed under this project was called "sluice way", and gives the possibility of wastes input in any place in the kiln, similar to the solution introduced by

Figure 1. Pneumatic ejector—photo of installation in Rejowiec Cement Plant.

CADENCE[3]. The main difference between the two installations is the position of the pipe. In the *CADENCE* waste input system, wastes are loaded by a pipe situated radially to the kiln and projecting to a long distance inside the kiln, while in our project the connection pipe is tangent to circumference of the kiln with a charging hole directed in opposition to the kiln rotation. In the *CADENCE* system wastes loaded into the kiln are dropped on a charge surface, while in the sealed block developed under this project, the wastes are slide down along the inside surface of the kiln (*Fig. 2*).

4.3. Primary Wastes Incinerator

In the primary incinerator wastes are thermally pretreated. As a result, gases and solid residues are formed. Solid residues are grindable and can be ground, mixed with coal and then loaded into the rotary kiln.

Gases from the primary wastes incinerator are introduced into rotary kiln at the hot end. A rotary kiln is used to incinerate both gases and residues from the primary incinerator and allow to conduct process without residues. Flowsheet of the process is depicted in Fig. 3.

5. CONCLUSIONS

The studies performed, show that a cement kiln can be used for environmentally safe neutralization of hazardous wastes. The high temperature in the kiln guarantees complete

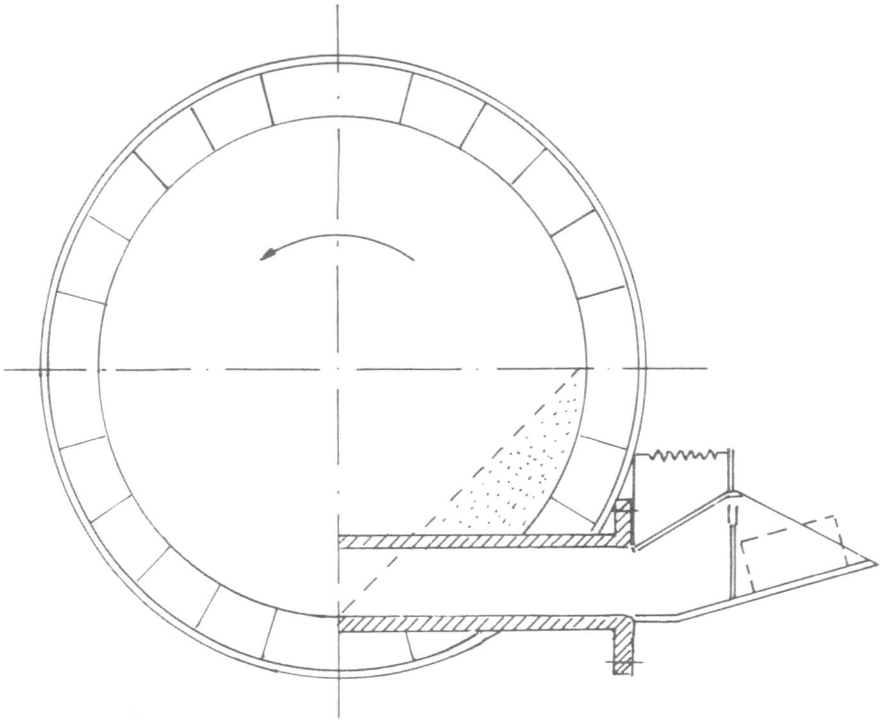

Figure 2. Schematic diagram of "sluice way."

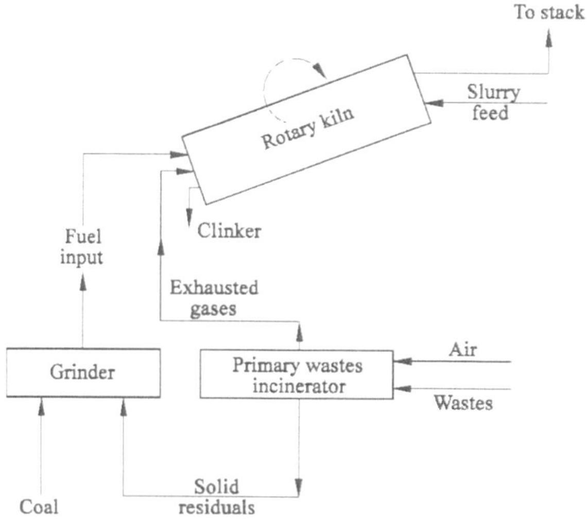

Figure 3. Schematic diagram of primary waste incinerator.

mineralization of all organic wastes. Additionally, acid components of combustion gases are neutralized by the alkaline dusts in electrofilters and mineral residues are easily sintered and immobilized into a clinker structure. Sintering in cement kilns is the only accessible method of neutralization of mineral wastes.

It can be concluded that the cement industry is safe and relatively cheap method of treatment and wastes neutralization. To benefit from all advantages of such neutralization, the sensible political and economic regulations on a national level are required.

REFERENCES

1. Lowes, T.M., Utilization of combustible wastes in cement manufacture, BCG Technical Center, 1988.
2. Pawlowski, L., ed. Utylizacja odpadów niebezpiecznych w piecach cementowych, Proc. of State Conference on Hazardous Waste Utilization in Cement Kilns, ..., Published: Politechnika Lubelska, Lublin, 1996.
3. All Fired Up. Burning Hazardous \Waste in Cement Kilns, Environmental Toxicology International (ETI), Inc., 1992.

21

AN ATTEMPT TO ESTIMATE THE PCDF/PCDD EMISSIONS FROM WASTE INCINERATED IN CEMENT KILNS

Marzenna R. Dudzińska, Zdzisław Kozak, and Lucjan Pawłowski

Department of Environmental Protection Engineering
Lublin University of Technology
40, Nadbystrzycka Str., 20-618 Lublin, Poland

ABSTRACT

This paper gathers the current state of knowledge on waste incineration in cement kilns with special respect to emissions of organic hazardous compounds such as PCDF/PCDDs, which tend be a crucial point, causing the strongest local opposition to the location of waste incinerator.

Emission from the process of conventional clinker production (with coal, as a fuel) were discussed based on a survey of the literature as well as actual measurements in a selected cement plant in the Lublin region. All data were compared with emissions from processes involving partial substitution of conventional fuel by solid and liquid wastes, inorganic and organic. Although, condition of clinker production are hardly suitable for "dioxins" formation, some mechanisms allowing the inhibition of the PCDDs formation were found and discussed.

Key words: solid waste utilization, incineration in cement kilns, PCDD emissions.

1. INTRODUCTION

Some say, that the earth's population by the end of the XX century might be called—"the throw away society" but persons knowledgeable about the environmental problems put "waste production" among the most serious issues of our time. The amount of waste generated is growing and all possibilities of pollution prevention, minimalization, utilization and destruction have to be taken into consideration.

Chemistry for the Protection of the Environment 3, edited by Pawłowski *et al.* Plenum Press, New York, 1998

2. WAYS OF WASTE UTILIZATION

Generally, the methods applied for waste utilization can be disposal on landfills, composting, incineration and recycling.

2.1. Disposal to Landfills

Disposal to landfills historically has been the simplest and cheapest method. It is still applicable for radioactive and hazardous wastes as well as for most municipal wastes. Although new technologies allow one to control the incidental problems of contaminants, leakage to ground water, emissions of gases that contribute to the "green house effect" are not yet completely solved. Shortage of land near urban areas makes such a solution only temporary. There are also problems associated with utilization of old, closed landfills and location of new ones.

2.2. Composting

The efficient method for organic and biodegradable wastes utilization appears to be composting. Unfortunately, it requires a careful selection of material to produce a compost, which may be used in food production. The latest tendency for separate wastes collection will help to improve the compost "quality" but still in many countries the market for composts made from such wastes has not been sufficiently developed. Another unexpected aspect of composting will be discussed later, together with consideration of hazardous emissions.

2.3. Incineration

In 1960s and '70s incineration seemed to be a way for total destruction and the solution for many problems, specially if the incinerators were used as sources of additional energy (called refuge derived fuel). This situation has changed as local communities have become more concerned about emissions and the incineration byproducts. It is generally believed that incineration plants are sources of strong secondary pollution. In such a situation, it is not easy to find a site for construction of a new facility. Construction of a modern incinerator, following all the environmental, health and safety restrictions and emission limits is very costly. Modernization of old plants is usually not economical and meeting the standards requires a careful selection of wastes input. Such selection usually reduces the caloric value of the wastes and limits the using of incinerators as energy sources.

2.4. Recycling

The latest tendency in waste utilization is a recovery and recycling, economically the most efficient method, and the most environmentally friendly. Developing the technologies, which help to reuse or recycle as much product as possible is the better way, but unfortunately not all materials can be recycled and due to sanitary demands recycled products cannot served all purposes.

2.5. Reduction

Limitation of waste at the place where they are produced seems to be the ideal solution. This generally means wasteless technologies, that is the latest tendency in industry, but the total elimination of waste seems to be a somewhat futuristic goal. Limitation and reduction of wastes may also mean changes in our ideas about wrapping and packaging, maybe even aesthetic and fashion changes.

Currently, one has to consider all known methods for pollution prevention and wastes utilization of primary importance and the total destruction, specially of hazardous wastes should also be a priority. In such a situation, incineration of solid wastes appears to be the preferred solution for the growing amount of solid hazardous wastes.

3. INCINERATION IN REGULAR AND ADOPTED FACILITIES

Wastes incineration could be carried out in specially designed and constructed incinerators as well as other technological processes which require heat. Combustion at high temperatures led to the decomposition of complex organic compounds ultimately to water and carbon dioxide. The amount of pollutants emitted in such a process depends on the composition of input, amount of wastes incinerated and the way of the process performing.

Although incineration is called "total destruction," it has to be strongly pointed out, that normally incineration in specially designed and build incinerators reduces the volume of wastes up to 80%. Incineration of wastes results in emission of fly ashes, acid gaseous and heavy metals as well as selected volatile organic compounds. Special emission reduction systems "produce" high amounts of liquid wastes[1]. Although all of these problems are serious, the public attentions is mostly attracted to the "dioxins," which means emissions of polychlorodibenzo-p-dioxins (PCDDs) and polychloro-dibenzo-p-furans (PCDFs), which are volatile, extremely stable and toxic substances.

Problems with location of any new, even very modern, low emitting waste incineration plant make all the alternative solutions very interesting and to the public a "hot issue." Separate problem is the very high costs of such "effective" incinerators. One of less expensive solutions appears to be the utilization of existing industrial installations for waste incineration.

The only requested condition to prevent formation of PCDDs and PCDFs is the high temperature (at least 1473 K—1200^{0}C) in which fumes residence times of at least 2 seconds.

Optimal conditions, recommended by the directive CEE/COM(93)296 for EU countries are presented in Table 1.

According to these data only three burning facilities are qualified to be applicable for waste incineration:

1. waste incinerator with a after burning system,
2. cement kiln for clinker production
3. facility for lime production.

Table 1. Optimal parameters for waste incineration, according to the EC directive CEE/COM(93)296

Temperature	
For 'organic' Cl content below 1%	1123 K
For 'organic' Cl content above 1%	1473 K
Oxygen demand	above 6%
Reaction time	more than 2 secs
CO content	
in 100% of day average	< 50 mg/Nm^3
in 95% of 10 min. average	< 150 mg/Nm^3

These recommendations focus the attention on the 'cement kiln' option, specially that adoption of cement rotary kilns for solid and liquid waste incineration does not meet bigger problems. Both organic and inorganic wastes can be easily combusted. Organic wastes can substitute fuel needed in the process and results in additional economic profits. Inorganic wastes can be built into the clinker structure, and yields the advantage of no solid residuals [2,3].

In spite of the above advantages there are still two important points for consideration:

1. are any leaking of inorganics (heavy metals) from the product possible?
2. are emissions from such a utility higher then from a cement plant run on conventional fuel?

Emissions considered refer to both inorganic (heavy metals) and organics (mostly PCDD and PCDFs) volatile products. The latter problem being the main issue of this paper.

4. PCDFs AND PCDDs FORMATION

PCDFs and PCDDs are considered by medical, environmental, and safety experts to be very toxic, hazardous substances emitted mostly during incineration of chlorine containing materials. Because of different number of chlorine atoms in the molecules, 75 polychlorodibenzo-p-dioxins and 135 polychlorodibenzo-p-furans can be formed. All of them have been found in the environmental matrices (air, soil, sediments and animals tissue)[4]. Among them, only 17 congeners are usually measured as they are considered to be of the highest toxicity. The PCDFs/Ds level is presented as TEF (toxicity equivalent factor), e.g. sum of 17 congeners levels multiplied by their toxicity factors, and only this value is regulated by law in most countries[5]. So far, there are no such regulations in Poland, but projection of The Clean Air Policy Act update proposes the value of TEF = 0,1 ng/ N m^3, that means the typical limit in EC countries.

PCDFs and PCDDs are formed in a number of chemical, photochemical and combustion reactions from their precursors like PCB, chlorophenols, chlorobenzenes and sometimes even aliphatic monomers[6]. They are easily accumulated in different organisms and show adsorptive properties, therefore, they can be found not only in gaseous emissions, but also adsorbed on dust and ashes. Although waste incineration is not considered to be a main source of PCDDs/PCDFs emission (according to the US EPA data, only up to 4% of all PCDDs and PCDFs emission come from incineration)[7], it is one of the main public concerns.

The presence of PCDDs and PCDFs has been detected in a number of combustion processes. Although the data on fossil fuel combustion are conflicting, in some investigations polychlorinated dioxins were found in ppb range in the emissions from coal burning installations[8]. Measurements made by Grochowalski and Wybraniec for selected Polish power plants operating on coal, showed the presence of PCDF/D (at low levels) in flue gas as well as in fly ash[9]. PCDD/PCDFs were also formed at low levels during burning of untreated wood (mostly detected on fly ashes, and only insignificant amounts in the soot)[10]. Another combustion process of considerably high emission of PCDD/PCDF is from automobile exhaust, the mechanism is not fully explained, as "dioxins" were found in systems where chlorinated additives to motor oils might be precursors[11] and in engines using unleaded petrol without chlorinated scavengers[12].

Among non-combustion sources of "dioxins," the paper pulp industry[13] and copper smelters[14] are also of concern. Latest research have identified 'dioxins' levels in domestic sewage and industrial wastewaters[15] as well as in the composting processes[16]. This latter source resulted in new laws and regulations, for example permission of composting restricted to organic material from separate collection and with limits below 17 ng TEF/ kg of dry mass. Usually the average content of PCDF/Ds in compost does not exceed 14 ng TEF/kg of dry mass. Polluted plants (by ashes), assimilation from the atmosphere (to less extend) and possible formation via enzymatic reactions from natural precursors are considered to be the main sources[16].

The public has to be made aware of all processes which could led to 'dioxins' formation, but one cannot deny, the presence of PCDDs and PCDFs in flue gases and fly ashes from the incineration processes is universal. It can be attributed to the following three possibilities[12]:

1. PCDDs/PCDFs are already present in trace amounts within the fuel and are not destroyed by the combustion process,
2. PCDDs/PCDFs are formed during the combustion process from precursors which are present within the fuel,
3. 'de novo' synthesis from other, non-chlorinated organic substances.

Additionally, dust particles from uncompleted combustion, are cited for catalyzing this process[17]. Cool soot present in fly ash from insufficient combustion process can react with HCl, O_2 and H_2O resulting in dioxin formation, according to the scheme:

$$C + HCl + O_2 + H_2O \rightarrow PCDF/PCDD$$

The most efficient inhibitor for PCDFs and PCDDs formation founded thus far, appears to be sulphur dioxide. Reports about its inhibiting properties[18] have been found so promising that in some studies the suphur additives to the municipal solid waste incinerators have been recomended[19].

Considering all know facts about "dioxins" formation mechanism it seems that the crucial points during combustion processes appear to be as follows:

- temperature occurring for sufficient time (more than 3 sec)
- presence of surface of fly ashes (catalytic effect)
- presence of traces of Cu,
- presence of SO_2, which was found to be an inhibitor of the process.

5. PCDF AND PCDD EMISSION FROM WASTE COMBUSTION IN CEMENT KILNS

The results from measurements made in Rejowiec Cement Plant, where experimental incineration of wastes have been held for one year (project PBZ 20-04) are presented in Table 2. EPA method 8280 was used with GC/MS from Finnigan Matt. Gas samples were taken under isokinetic conditions, based on EPA Method 23.

As the temperature in rotary cement kiln exceeds 1470°C for more then 3 seconds, the conditions are not preferable for dioxins formation. In Polish cement plants, were mostly high sulphur coal content is used as a fuel, natural occurrence of SO_2 in the process might suggest another disturbance in PCDF/D formation. Organic wastes incinerated in cement kilns for measurement purpose did not contain copper. Inorganic wastes contain

Table 2. PCDF and PCDD emissions from Rejowiec Cement Plant (Kiln No 4—.6% O_2)

	I-TEQ, ng /N m^3
Coal	0.0515
Coal + 10% inorganic wastes	0.0819
Coal + 10% organic wastes	0.0698

0.08% of copper, but slightly higher emission of PCDF/Ds from the process, where inorganic wastes were utilized might be rather related with higher content of chlorides in inorganic wastes from metal finishing processes. However, it has to be pointed out, that in examined cases, emissions were lower than EC limit, e.g. 0.1 ng /N m^3.

6. CONCLUSIONS

Measurements do not confirm exceeding the 'dioxin' emission from clinker production using organic wastes for fuel substitution, but, on the contrary in one case, reduction of emission was found, we can risk the statement that waste incineration in clinker rotary kilns do not cause any additional harassment to the environment connected with PCDF/Ds emissions. In some cases, for selected waste of low chlorine contend, even reduction of emission compare with process on raw coal combustion were notice. This may give another advantage of waste incineration in cement kilns—minimization of 'dioxin' problem.

REFERENCES

1. Steinmuller Report on "Flue Gas Cleaning," L.&C. Steinmuller GmbH, Environment Protection Division, Gummersbach, 1994.
2. Dudzinska, M.R. and L. Pawlowski, Application of the cement plants for wastes incineration, in: "Risk Management Strategies Applied to Environmental Cleanup in Central and Eastern Europe," Proc. of the International School of Innovative Technologies for Cleaning the Environment," ed. R.C.Ragaini, in: *The Science and Culture Series - Environmental Sciences*, series Editor: A.Zichini, World Scientific, New Jersey, 1997.
3. Herat, S., Protecting the environment from waste disposal: the cement kiln option," *Environmental Protection Engineering*, 1997, *23*, 25–34
4. Hagenmaier, H. Lindig, C. and J.She, Correlation of environmental occurrence of PCDDs and PCDFs with possible sources, *Chemosphere* 1994, *29*, 2163–2174.
5. Safe, S., Development, validation and limitations of toxic equivalency factors, *Chemosphere*, 1992, *25*, 61–64.
6. Heindl, A. and O. Hutzinger, *Chemosphere* 1986, *15*, 2001.
7. Travis, C.C. and H.A. Hatten-Frej, *Chemosphere* 1990, *20*, 729.
8. Kilble, B.J. and M.L. Gross, *Science* 1980, *207*, 59.
9. Grochowalski, A. and S.Wybraniec, Levels of PCDDs and PCDFs in flue gas and fly ash from coal combustion in power plant, *Chem.Anal.*, 1996, *41*, 27.
10. Clement, R.E., Tosine, H.M. and B.Ali, *Chemosphere*, 1986, *14*, 815.
11. Rappe, C. et all., Overview on fate of chlorinated dioxins and dibenzofurans. Sources, levels and isomeric pattern in various matrices. *Chemosphere* 1987, *16*, 1603–1618.
12. Hutzinger, O. and H.Fiedler, Sources and emissions of PCDD/PCDF, *Chemosphere*, 1989, *18*, 23.
13. Santl, H., Bichmaier, A., Gruber, L. and E.Stohrer, *Chemosphere* 1994, *28*, 1633.
14. Karasek, F.W., Viau, A.L. and M.F. Gonnard, *J.Chromatogrhaphy*, 1983, *270*, 227.
15. Fiedler, H., Sources of PCDD/PCDF and impact on the environment, *Chemosphere*, 1996, *32*, 55–64.

16. Krauss, T. Krauss, P. and H. Hegenmaier, Formation of PCDD/F during composting?, *Chemosphere*, 1994, *28*, 155–158.
17. Konduri, R. and E. Altwicker, Analysis of time scales pertinent to dioxin/furan formation on fly ash surfaces in municipal solid waste incinerators, *Chemosphere*, 1994, *28*, 23–45.
18. Jay, K. and L. Stieglitz, On the formation of polychlorinated aomatic compounds with copper(II) chloride, *Chemosphere* 1991, *22*, 987–995; Ogawa, H. et. all., Dioxin reduction by sulphur component addition, *Chemosphere*, 1996, *32*, 151–157.
19. Lindbauer R.L., Wurst, F. and T.Prey, Combustion dioxin suppression in municipal solid waste incineration with sulphur additives, *Chemosphere*, 1992, *25*, 1409–1114.

22

THE USE OF EDTA TO INCREASE THE LEACHABILITY OF HEAVY METALS FROM MUNICIPAL SOLID WASTE INCINERATOR FLY ASH

Peter Van Herck, Carlo Vandecasteele, and Dirk Wilms

Katholieke Universiteit Leuven
Department of chemical engineering
W. de Croylaan 46
3001 Heverlee Belgium
E-mail: Peter.VanHerck@cit.kuleuven.ac.be

ABSTRACT

Fly ash from municipal waste incinerators constitutes an environmental problem, as it is polluted with heavy metals. The heavy metals from fly ashes can partly be removed and possibly recovered by extracting with an acid solution. The remaining fly ash can be landfilled or used as construction material. The controlling factor of this extraction process is generally the solubility of the metal salts on the fly ash. By adding ethylene diamine tetraacetic acid (EDTA) to the extraction solution the solubility restrictions can be avoided and the available heavy metals will dissolve even at high pH.

Another possible application of the addition of EDTA to the leaching solution is as a test method for the evaluation of the total availability of the heavy metals on the fly ash. This is the portion of the metals that is not encapsulated in the matrix structure or in insoluble oxides. By using EDTA, solubility restrictions and the need for a low pH could be avoided.

The evolution of the leaching efficiency as a function of the pH was investigated with and without addition of EDTA and those leaching efficiencies were compared. The experiments show that the leaching efficiency for several heavy metals increases when EDTA is added.

Keywords: fly ash, leaching, EDTA.

1. INTRODUCTION

In Flanders, Belgium, around 2.8 million tonnes of municipal waste is generated yearly, of which 30% is selectively collected; 43% of the rest fraction is processed in incinerators and 57% is landfilled[1]. Incineration of one tonne of municipal waste leads to the formation of 10 to 50 kg of fly ash depending on the type of incinerator. Combustion residues in general, and fly ash in particular, form a major environmental problem. This fly ash is contaminated with heavy metals and polychlorinated dibenzodioxins and dibenzofurans. It must be considered as hazardous according to the Flemish environmental legislation[2].

Generally, dust particles are removed from the incinerator flue gas by means of an electrostatic precipitator. Wet scrubbers remove in a first stage HCl and HF from the flue gas and in a second stage SO_2. The first stage produces an acid waste water mainly containing HCl and to a lesser extent HF. This acid solution can be used e.g. to leach the fly ash, in order to remove part of the heavy metals, as in the 3R process developed in the Karlsruhe Nuclear Research Centre[3]. Of course, other (waste) acid solutions can in principle be used for the same purpose. The aim can be twofold: obtain a waste material containing a smaller amount of heavy metals so that it can be used as a construction material[4] or landfilled under less stringent conditions; and recover some heavy metals.

Solubility restrictions are a controlling factor in the leaching process. By adding ethylene diamine tetraacetic acid (EDTA) to the solution the solubility restriction is avoided and all the available metal will go into dissolution even at high pH.

Another possible application of the use of EDTA is the use as a test method for the evaluation of the total availability of the heavy metals on the fly ash. Usually the total concentration of the metals is taken as the available amount of metal for leaching. In this case it is asumed that e.g. on a landfill the metal release continues until it depleted completely. However there is always a portion of the metal that is not available for leaching e.g. the metals encapsulated in the matrix structure of the fly ash or in crystalline or amorphous oxides. Therefore it would be better to define a total availability of the metal.

2. TOTAL AVAILABILITY

Several definitions for the total availability were found in the literature. According to Garrabrants et al[5] total availability can be defined in two ways[5]:

1. the potential mobile content of an element;
2. the element content which serves as the thermodynamic source or driving force for release.

The NEN 7341 norm gives as definition:

The total availability is the (cumulative) emission (in mg/kg) that will occur over a very long period and under extreme conditions.[7]

In practice this means that if a certain mineral of a metal dissolves slightly under e.g. acidic circumstances, it will be totally dissolved when the extraction time is very long and the leaching solution is renewed. The total availability is then not the total concentration of the metal in the fly ash but the amount of minerals of that metal that can be dissolved under e.g. acidic circumstances.

Table 1. Total composition of the fly ash (g/kg)

Component	g/kg fly ash total dissolution	Component	g/kg fly ash total dissolution
Ag	0.029 ± 0.004	K	45 ± 1
Al	80 ± 3	Mg	12.6 ± 0.5
Ca	110 ± 4	Mn	0.87 ± 0.05
Cd	0.24 ± 0.01	Na	30 ± 1
Cl	54.0	Ni	0.078 ± 0.004
Co	0.0205 ± 0.0008	Pb	3.9 ± 0.1
CO_3	22.1	PO_4	27.0
Cr	0.50 ± 0.08	Sn	1.3 ± 0.1
Cu	0.78 ± 0.03	SO_4	58.0
F	0.6	Zn	12.9 ± 0.5
Fe	11.6 ± 0.4		

3. MATERIAL

Samples of MSWI fly ash were obtained from the Houthalen Waste Incineration Facility (Houthalen, Belgium), a municipal solid waste facility with an annual capacity of 98 000 tonnes. The fly ash was collected by a classical electrofilter. Table 1 gives the total composition of the fly ash.

4. ETHYLENE DIAMINE TETRAACETIC ACID (EDTA)

EDTA is a chelating agent and has a high affinity for a wide range of cationic metals. EDTA will further be presented as H_4L. The formation of the chelate involves two steps: the acid dissociation and the actual complexation. The consecutive dissociationconstants of H_4L are respectively 2.00, 2.67, 6.16, 10.26. Adding acid to the solution leads to a lower stability of the metal-EDTA complexes. The form H_4L has a low solubility in water. 1 gives the formula for the complexation constant.

$$M^{2+} + L^{4-} \leftrightarrow ML^{2-} \quad K_{ML} = \frac{[ML^{2-}]}{[M^{2+}][L^{4-}]} \tag{1}$$

Table 2 gives the most important K_{ML} values

Table 2. Stability constants of metalcomplexes, logarithmic value (Kennedy, J. H.)

	log K_{ML}		log K_{ML}
Na^+	2.61	Cd^{2+}	16.46
Mg^{2+}	8.69	Zn^{2+}	16.50
Ca^{2+}	10.70	Pb^{2+}	18.04
Mn^{2+}	13.79	Cu^{2+}	18.80
Fe^{2+}	14.33	Fe^{3+}	25.1
Al^{3+}	16.13		

5. DEFINITIONS

The leachability of the fly ash can be influenced by several parameters. Two of these parameters are acid dose (mol/kg) and L/S-ratio (l/kg).

The parameter AD, the acid dose, is defined as:

$$AD = \frac{\text{amount of } H^+ - \text{ions added}}{\text{mass of fly ash}}\left(\frac{\text{mol}}{\text{kg}}\right)$$

$$AD = \frac{\text{amount of } OH^- - \text{ions added}}{\text{mass of fly ash}}\left(\frac{\text{mol}}{\text{kg}}\right) \qquad (2)$$

In the former case AD is given a positive value, in the latter case a negative value. H^+-ions were added as HCl; OH^--ions as NaOH. The L/S-ratio is defined as:

$$\frac{L}{S} = \frac{\text{volume of leaching solution}}{\text{mass of fly ash}}\left(\frac{1}{\text{kg}}\right) \qquad (3)$$

The results are given as leaching efficiency:

$$\text{Leaching efficiency} = \frac{\text{leached amount of the element}}{\text{total amount of the element in the fly ash}}(\%) \qquad (4)$$

6. EXPERIMENTAL METHOD

The influence of the addition of EDTA to the leaching solution is investigated as a function of the final pH. The fly ash is leached with a L/S-ratio of 10 l/kg during 3 hours at room temperture. For 1 g fly ash 1g Na_2EDTA is added. After leaching the solution is filtrated over a glass microfiber filter and pH and metal concentrations were measured. By adding HCL or NaOH the final pH is varied. The results are compared with results of the same leach tests without the addition of EDTA.

7. RESULTS AND DISCUSSION

7.1. pH

Figure 1 gives the final pH after extraction. The addition of EDTA lowers the pH for acid doses around 0 mol/kg because during the complexation reaction H^+ is released. When water and EDTA is used to leach the fly ash a pH of 4.8 is established while without EDTA the pH would be 10.22. At an acid dose lower than -5 mol/kg and between 4 and 13 mol/kg the final pH is the same. At acid doses higher than 13 the pH of the leaching solution is higher when EDTA is added because the formation of H_3L^- and H_4L.

7.2. Metals

In the following figures the experimental data for the leaching without EDTA are presented along with the data of the experiment with EDTA. Besides the heavy metals also the

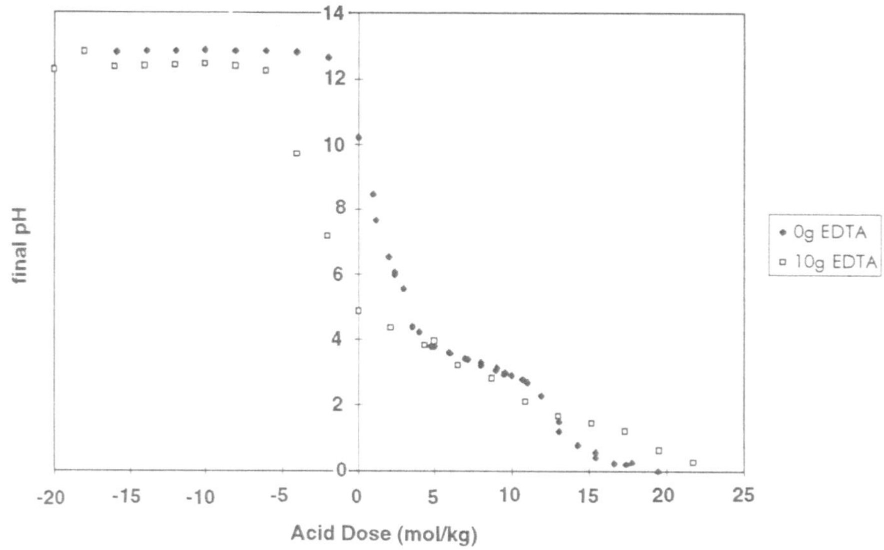

Figure 1. Final pH after extraction with and without the addition of EDTA.

matrix elements are discussed. Because during the complexation reaction the pH of the solution will be lowered the leaching efficiencies of both experiments are plotted as a function of the final pH of the solution in order to see the effect of the addition of EDTA.

From Table 2 the affinity of EDTA for several metals can be deduced. The metals will be discussed in order increasing affinity of EDTA for the metals. EDTA has a low affinity for Na and K. Therefore the figures will show almost no influence of the addition of EDTA. Because Na_2EDTA was added the measurement of Na was difficult. Only the figure for K is therefor given here. Figure 2 shows that there is no influence of the addition

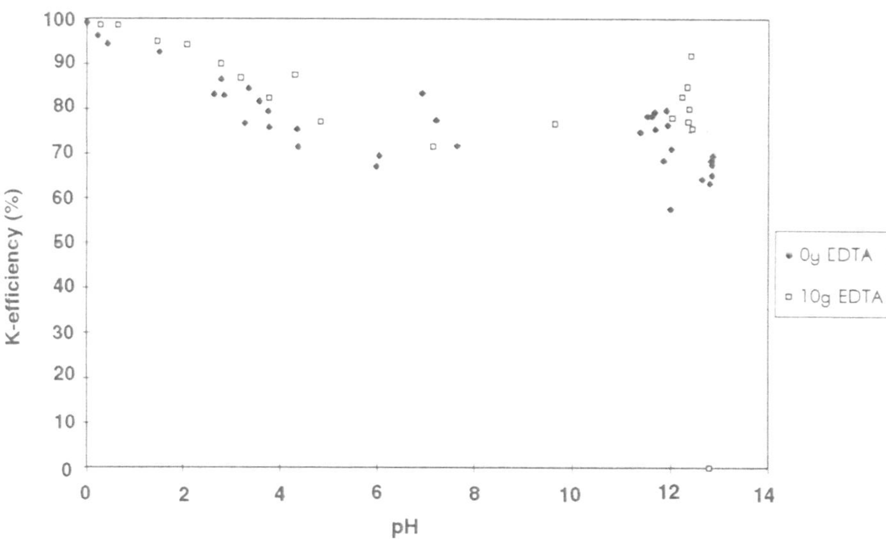

Figure 2. Leaching efficiency of K (%) as a function of the final pH; with and without the addition of EDTA.

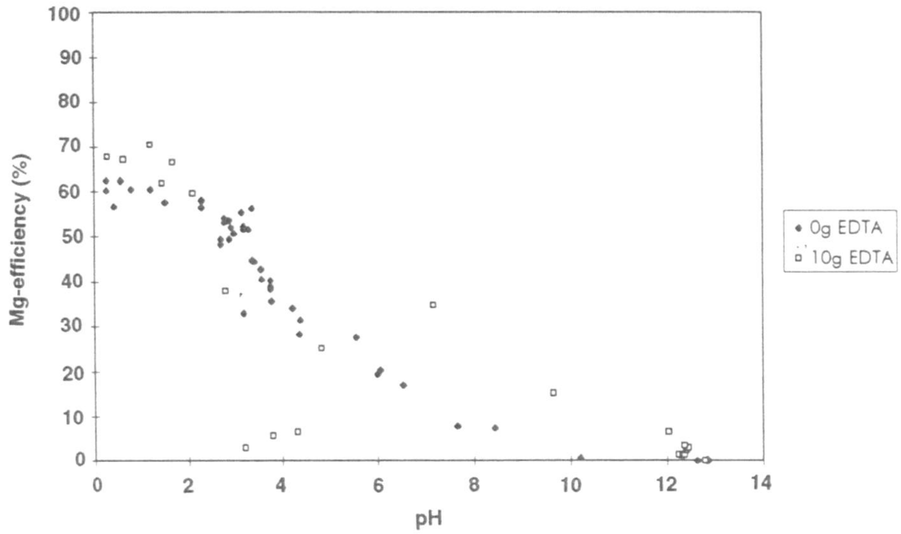

Figure 3. Leaching efficiency of Mg (%) as a function of the final pH; with and without the addition of EDTA.

of EDTA on the leaching efficiency of K^+. The total availibility of K and Na can be measured without EDTA.

According to Figure 3 also for Mg there is no influence of the addition of EDTA to the solution although the affinity of EDTA for Mg is higher than for Na and K. At high pH the complexation should be better but here the complexation constant is to low to compete with the precipitation reactions of the Mg minerals. To determine the total available amount the NEN 7341 test uses 2 leach experiments (pH 7 and 4) at a L/S-ratio of 50A l/kg and an extraction time of 3 hrs[7]. Figure 3 shows that at a pH of 4 the leaching process is not yet finished. Even with a L/S-ratio of 50 instead of 10 l/kg it is not certain that the available Mg would be leached. It is clear that about 60% of the total amount of Mg is available for leaching (under acidic circumstances). This amount is reached when a pH of 1 is used during the leaching experiment. At this pH it could be possible that the matrix structure is partially deteriorated leading to leaching of metals normally encapsulated in the matrix structure. When EDTA is used the leach experiment could be carried out at a higher pH and with only 1 test. In the case of Mg the leaching of the fly ash with EDTA will give no better indication of the total availibility of Mg than leaching without.

Figure 4 gives the leaching efficiency for Ca. It is clear that the addition of EDTA has an influence on the leaching of Ca. When the NEN 7341 test is used it is clear that at a pH of 4 the leaching process is not finished. When EDTA is used in water to determine the total availability the final pH would be 4.8. At that pH the maximum leachability for Ca (90%) is almost reached. So the addition of EDTA to the solution gives a good approximation for the total availibility of Ca and only one experiment must be performed at a moderate pH.

Figure 5 gives the leaching efficiency of Mn as a function of the final pH. EDTA has a higher affinity for Mn than for Ca but it is clear that the leaching of Mn does not improve very much by addition of EDTA. Between pH 12 and 3 there is a slight improvement by adding EDTA. When the fly ash is leached with water and EDTA (pH=4.8) the

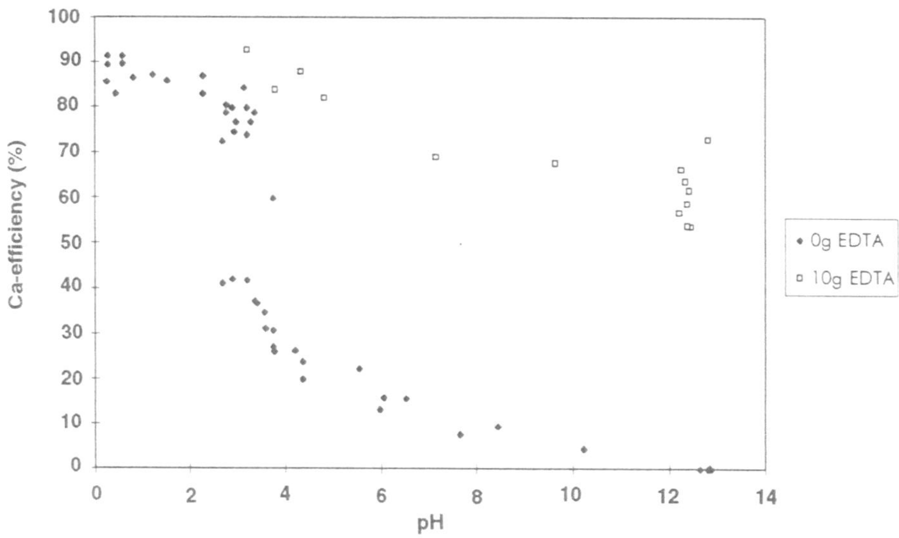

Figure 4. Leaching efficency of Ca (%) as a function of the final pH; with and without the addition of EDTA.

leaching efficiency is about 30%. So leaching of the fly ash with water and EDTA does not give a good view of the total availibility of Mn. When the pH of the solution is below 3 there is no influence of the addition of EDTA. At this pH EDTA starts to occur as H_4L. This form will not form complexes with the metals and has also a very low solubility leading to precipitation. Therefore there is no influence of EDTA anymore because it is removed from the solution.

Figure 6 gives the leaching efficiency of Fe as a function of the final pH. The leaching efficiency is based upon the total amount of Fe. But it is possible that Fe occurs both

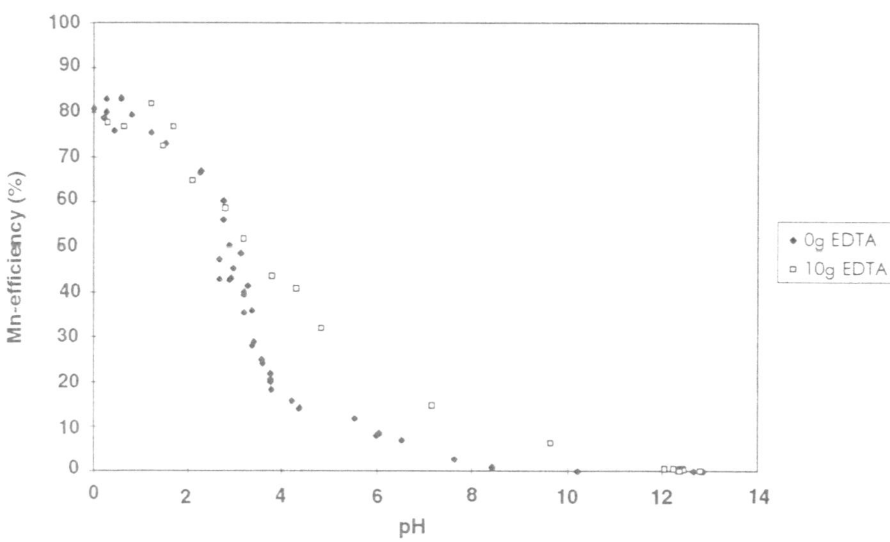

Figure 5. Leaching efficiency for Mn (%) as a function of the final pH; with and without EDTA.

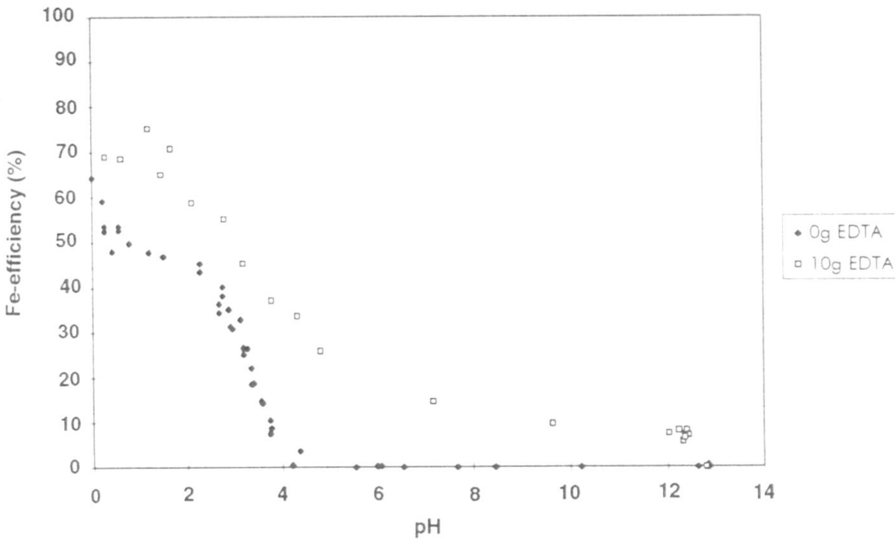

Figure 6. Leaching efficiency of Fe (%) as a function of the final pH, with and without EDTA.

as Fe^{2+} and Fe^{3+}on the fly ash. Normally Fe^{3+} would stay precipitated until pH 2–1.5 but the affinity of EDTA for Fe^{3+} is very high. So when EDTA is added to the solution all of the available Fe^{3+} will form an EDTA-complex even at very low pH. Figure 6 shows an evolution of the leaching efficiencies that is similar to the behaviour of Mn. The only difference is the shift of about 10% of the leaching efficiency with EDTA in comparison to the leaching efficiency without EDTA. This shift exists over the whole pH range indicating that this is the amount of Fe^{3+} available on the fly ash. The EDTA-complex constant is high enough to compete with the precipitation reactions. At a pH of 10 the leaching efficiency of Fe starts to increase just like the leaching efficiency of Mn. Here the effect of the addition of EDTA on the leaching of Fe^{2+} starts. At pH of 3 the effect of EDTA on the leaching of Mn stops. At this point the leaching efficiency of Fe with EDTA is 45% and without EDTA between 30 and 40%. The difference is about 10% just like at pH 12. The influence of EDTA on the leaching of Fe^{2+} stops. At pH lower than 3 the difference between the leaching efficiency with and without EDTA is between 10 and 15%. The affinity of EDTA for Fe^{3+} is so high that nevertheless EDTA exists as H_4L and is even precipitated the complexation still occurs. At very low pH Fe^{3+} starts to dissolve without the addition of EDTA and the difference between the leaching efficiencie diminish.

When the fly ash is leached with water and EDTA (pH=4.8) the leaching efficiency is about 20%. So this experiment gives no good value for the total availibility of Fe.

Figure 7 gives the leaching efficiency of Al as a function of the final pH. Just like Mn the addition of EDTA does not have much influence and when a pH 3 there is no more the influence is gone. At pH of 4.8 the leaching efficiency with EDTA is about 10%. So the test with water and EDTA does not give a good view on the total availability of Al.

Figure 8 gives the leaching efficiency of Cd as a function of the final pH. Cd leaches easily, also without the addition of EDTA. Nevertheless there is an influence for pH higher than 7. There the leaching of Cd with addition of EDTA is better than without EDTA. At pH 4.8 the leaching of Cd is already completed so the benefit of the addition of EDTA to the leaching solution is the lowering of the pH.

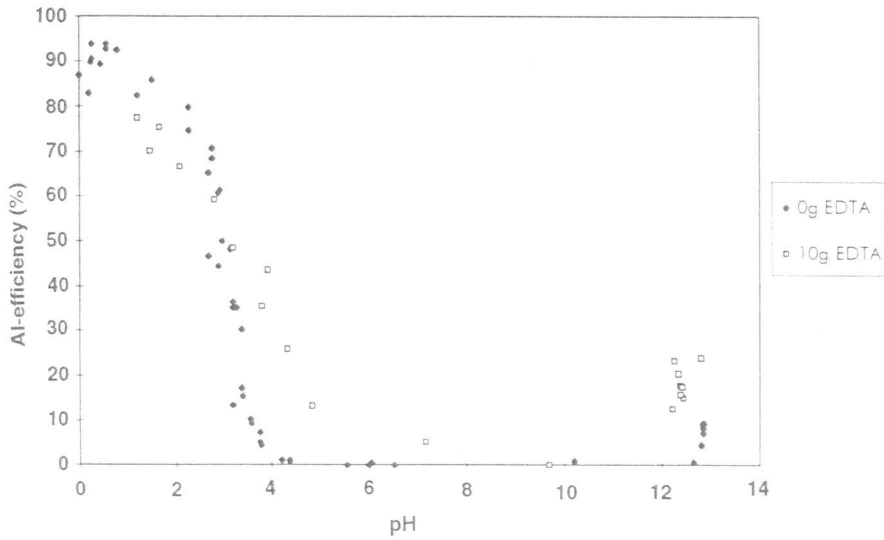

Figure 7. Leaching efficiency of Al (%) as a function of the final pH; with and without EDTA.

Figure 9 gives the leaching efficiency of Zn as a function of the final pH. The influence of the addition of EDTA is more obvious. Below pH 3 the complexation of Zn with EDTA gives a high increase in leaching efficiency. At pH 10 the efficiency increases from a few percentages to about 45%. For pH lower than 3 there is no influence anymore. The leaching of Zn is almost complete and EDTA doesn't give an additional benefit. At a pH of 4.8 the leaching efficiency of Zn with EDTA is 68% so this test doesn't give a reliable result for the total availibility.

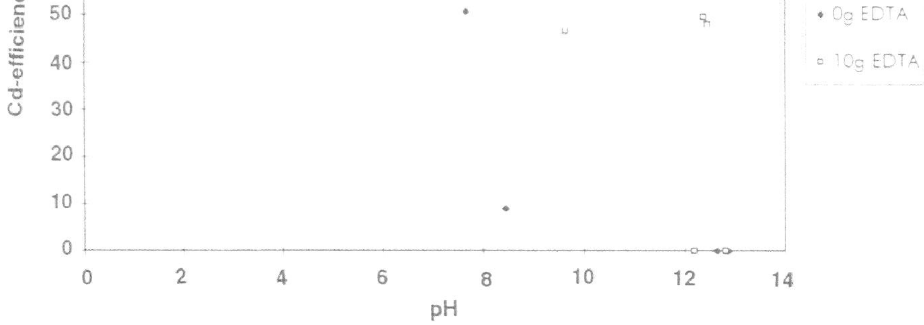

Figure 8. Leaching efficiency of Cd (%) as a function of the final pH; with and without EDTA.

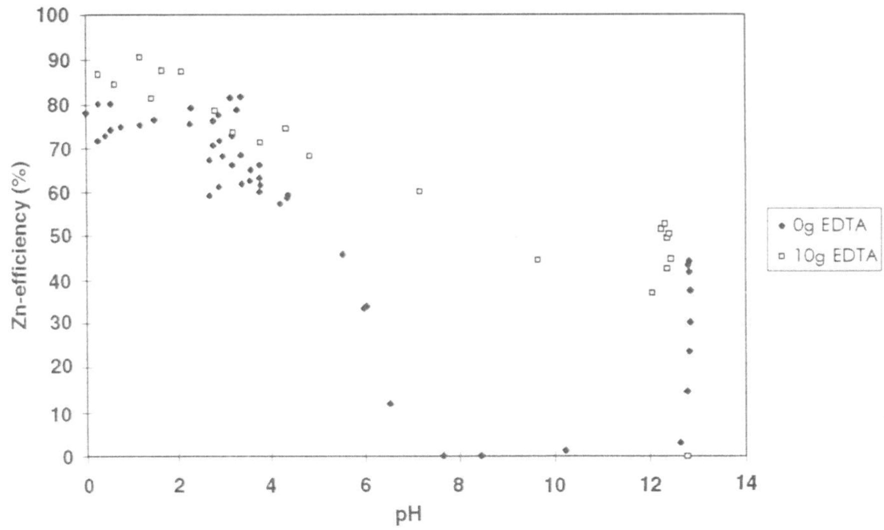

Figure 9. Leaching efficiency of Zn (%) as a function of the final pH; with and without EDTA.

Figure 10 gives the leaching efficiency of Pb as a function of the final pH. The influence of the addition of EDTA is now very clear. When leaching solution without EDTA is used Pb leaches only at very low pH while the addition of EDTA increases the leaching efficiency until the value of the total availibility for acid extraction. So the test of leaching the fly ash with water and EDTA is useful to determine the total availibility of Pb. At a pH lower than 3 there is still an influence of the addition of EDTA.

Figure 11 gives the leaching efficiency of Cu as a function of the final pH. Here also addition of EDTA has a large influence. At pH 4.8 the leaching efficiency it gives a rather good view of the total availibility of Cu.

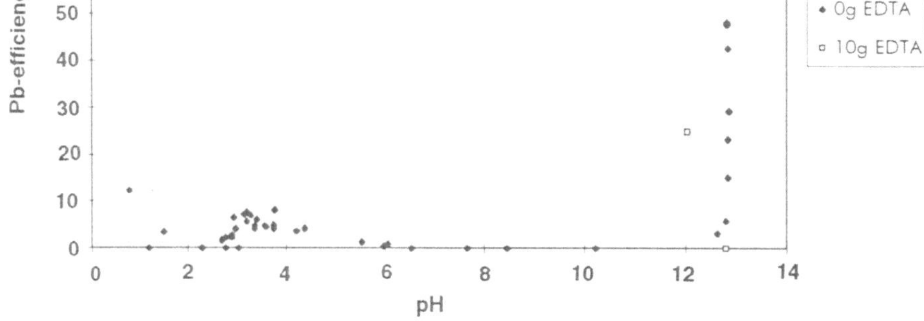

Figure 10. Leaching efficiency of Pb (%) as a function of the final pH; with and without EDTA.

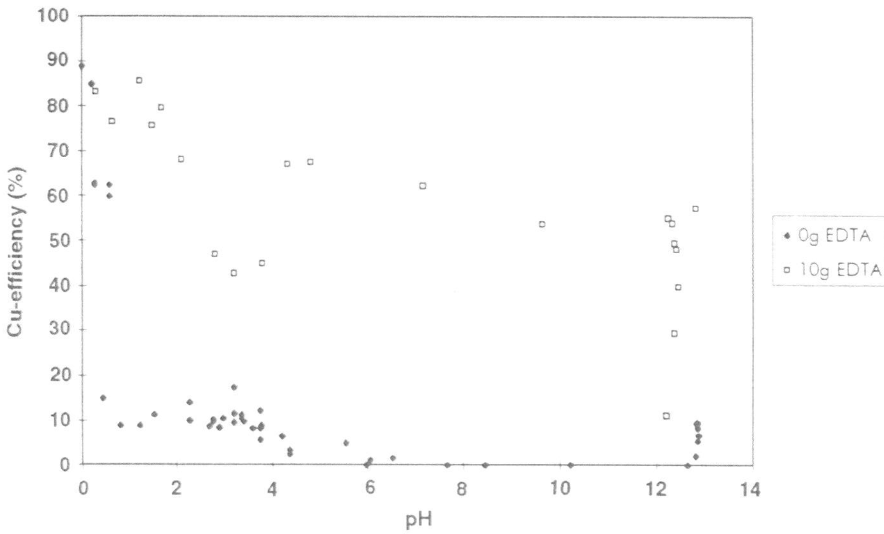

Figure 11. Leaching efficiency of Cu (%) as a function of the final pH; with and without EDTA.

8. CONCLUSION

According to the stability of their EDTA-complex the metals show a different behaviour when EDTA is added. The addition of EDTA doesn't influence metals like Na an K because the stability of the complex is too low. Cu and Pb on the contrary have an increased leaching efficiency over the total pH range when EDTA is added. The other metals are in between these two behaviours according to the stability of their EDTA-complexes. It is obvious that the pH is a major parameter for the extraction of heavy metals from the fly ash because it determines the solubility of the metal salts.

In the NEN 7341 test the result is influenced by solubility restrictions. By adding EDTA it is possible to avoid these solubility restrictions. But it must be pointed out that it is important to investigate if the elevated value is the true total availibility. Not all the metal-complexes have a high stability and this difference can influence the results. The results for the heavy metals Cd, Zn, Pb and Cu however are good and can be used as the total availability.

REFERENCES

1. Wille D. and De Boeck G., *Inventarisatie Huishoudelijke Afvalstoffen in Vlaanderen in 1994, Productie, Inzameling en Verwerking,* Openbare Afvalstoffenmaatschappij voor het Vlaamse Gewest, Publicatie nr. D/1996/5024/4, April **1996**.
2. Senelle R., Dujardin J. and van Damme M., VLAREM II, *Die Keure La Charte*, Brugge, **1995**.
3. Vehlow J., Brown H., Horch K., Merz A., Schneider J., Stieglitz L. and Vogg H., *Semi-Technical Demonstration of the 3R-Process*, Waste Management and Research, **1990**, vol. 8, 461–672.
4. Mulder, E, *Pre-Treatment of MSWI-Fly Ash for useful Application*, Waste Managment, **1996**, vol. 16, nr 1–3, 181–184.
5. Garrabrants A.C., Kosson D.S., *Use of Chelating Agent to Determine the Metal Availibility for Leaching from Soils and Wastes*, Proceedings of The international Conference for the Environmental and Technical

Implications of Construction with Alternative Materials (WASCON '97), June 4 1997, Houthem, The Netherlands, 1997.
6. Van der Bruggen, G. Vogels, P. Van Herck, C. Vandecasteele; *Simulation of Acid Washing of Municipal Solid Waste Incineration Fly Ashes in order to Remove Heavy Metals*. Accepted for publication in Journal of Hazardous Materials.
7. NEN 7341, *Determination of the availability of inorganic components for leaching*, NNI Delft, 1992

23

ECOLOGIC AND ECONOMIC ASPECTS OF UTILIZATION OF FLY ASHES FOR ROAD CONSTRUCTION

Jan Kukiełka

Technical University of Lublin
40 Nadbystrzycka Str., 20-650 Lublin, Poland

ABSTRACT

About 20 million Mg of ashes and slags are generated annually by electric power plants in Poland but only 50% has been utilized. Furthermore, due to significant decrease in road construction in 1990, highway engineering in Poland used only about 0,5 million Mg of ashes. The first practical use of ashes from waste gas desulfurization from the Lublin-Wrotków thermal-electric power plant was stabilization of existing slag pavements before covering it with a bituminous mat. Active ashes were also used for stabilization of sands with addition of small amount of cement at the surface of the upper layer of road foundation) and for stabilization of dusty clay in the foundation layer (Świdnik, Partyzantów Street, Poland).

Key words: volatile ashes, ecologic aspects, road making, stabilization.

1. INTRODUCTION

The ashes of brown coal originating from the Konin region (Poland), containing more than 15% free CaO and greater than 3% SO_3 and called calcium-sulfate ashes of the third group (C-form), were used partly as cement or lime substitutes for road making. They were used either as a binding agent or, after mixing with cement, to improve subgrade and low layers of bituminous road foundations.

More than 100 km of mostly local roads were built using the ashes, especially in Rzeszów and the Lublin region in 1970. It was found that fly ashes of brown coal originating from the Konin region show different physical and chemical properties, even when ashes come from the same electric power station. This is due to collection of ashes from

Chemistry for the Protection of the Environment 3, edited by Pawłowski *et al.* Plenum Press, New York, 1998

different places in the ash removal system. Only fine and very fine fractions of ashes containing suitable amount of CaO are useful as binder for subgrade stabilization.[1]

The above ashes are suggested to be used for stabilization of loesses only after mixing with cement (6% of cement and 6% of ash in relation to loess weight).[2,3,4]

According to specialists, waste produced utilizing a dry method of desulfurization may be used for cement and building industries.

The dry method of gas desulfurization is used in Opole and Rybnik thermal-electric power stations. Composition of ashes formed in heat and power generating plant in Opole is as follows [6]:

SiO_2	—	42,7% (by weight)
Fe_2O_3	—	7,8%
Al_2O_3	—	21,8%
CaO	—	16,0%
CaO (free)	—	8,8%
MgO	—	3,0%
SO_3	—	2,7%
N_2O	—	0,6%
K_2O	—	2,3%
ignition loss	—	1,5%

Despite the high content of free calcium oxide, the above ashes are not very useful as the binding agent for road construction. Ashes from Rybnik plants are used as fillings in mines, where due to high temperature, they show sufficient binding properties.

The ashes from Rybnik and Opole plants show rather low hydraulic activity.

Ashes formed during the desulfurization process of waste gas in Lublin-Wrotków thermal-electric power station showed the following content SiO_2—39%, CaO (free)—17,7%, SO_3—4,51% and were tested by calorimetric analysis. After adding 10% HCl solution to 36 samples of ashes, temperature rises were measured. Obtained temperature rises in the range from 35°C to 64°C were the result of high and very high activity in the above ashes.

A certificate of radiation monitoring (JBT No 049/A10/94) of ashes confirms their usefulness in building materials production. According to instruction JBT No 234, the content of ashes in finished products must be limited to 70% of the mass of all dry components.

Ashes from thermal-electric power station in Lublin-Wrotków contain 39% of SiO_2, 17,7% of CaO (free) and 4,51% of SO_3.

Chalk and then hydrated lime used as sorbents appeared to be effective in desulfurization of waste gas (average—38%, max. 60%) and formed in the process hydraulically active ashes were experimentally used for cement production and as binding material for road making.

2. GENERAL EVALUATION OF ECONOMIC EFFECTS AND POSSIBILITIES OF ASHES USE

Ashes and slags formed in amount of about 20 million Mg per year in electric power plants in Poland are used only in about 50%. Due to significant decrease of road construction, in 1990 highway engineering in Poland used only about 0,5 million Mg of ashes.

Potentially the ashes may be used for embankment making, "Cegran" aggregate production, subgrade improvement and foundation layers for expressways [5], as well as for filling material for bituminous mixes for ring, fast and local road construction and for hardening of unsurfaced farming roads.

The increase in demand for ashes formed in desulfurization processes in highway engineering and its decrease in demand in cement and building industry may be expected. Hydraulically active ashes may also be very useful material for forest road hardening where acid soils limit use of cement.

The "green programme" realized in West Germany in 1950 and 1960 assumed hardening of 10.000 km of farming roads per year. In Poland, hardened farming roads, except public roads do not exist and local roads in countries and towns have sometimes only slag pavements.

Waste material such as hydraulically active ashes may be a source of good materials for road construction and hardening of farming roads, as well as for modernization of slag pavements, which are troublesome due to dust.

The first practical use of ashes from waste gas desulfurization in Lublin-Wrotków thermal-electric power plant was stabilization of existing slag pavements before covering it with a bituminous mat. Active ashes were also used for stabilization of sands (with addition of small amount of cement at the surface of the upper layer of road foundation) and for stabilization of dusty clay in the foundation layer (Œwidnik, Partyzantów Street, Poland).

Laboratory studies[7] lead to the following conclusions :

- hydraulically active ashes harden very slowly after humidification and thickening reaching great increase of strength between 1st and 3rd month (depending on temperature),
- calcium hydroxide formed in result of water addition to active ashes produce about 20% increase of volume what may be cause destruction of hardening material,
- addition of cement may be indicated, especially at the surface of the layer stabilized with active ashes,
- active ashes should be transported after wetting with water.

It should be noticed that use of hydraulically active ashes for road construction needs special attention with regard to organization of work, as well as suitable storage especially in winter.

Hardened ash is now useless waste material because there is no experience in the use of bounded ashes for embankment construction or subsurface improvements.

Because of its low permeability, the usefulness of hardened calcium-sulfate ashes, even for antiflood embankment building and sealing of waste dumps, should be taken into consideration[8]. Highly alkalic dust from fluid boilers have not been tested to determine the feasibility of utilizing it in highway construction.

3. ECOLOGIC ASPECTS OF ASHES USE FOR ROAD CONSTRUCTION

Road and railway embankments construction utilizing ashes and ash-slag mixtures aroused anxiety in 1970 because of their potential influence on ground waters and their potential adverse reactions with concrete and steel. It was stated that water samples collected several meters from storage wards or embankments made of the ashes did not show any influence of fly ash on changes in total hardness and pH, as well as significant increases in the concentration of water soluble chemical compounds characteristic of the ashes[1].

It should be noted that inactive ashes subjected to long-lasting washing with water show decreasing aggressiveness due to gradual washing out of soluble compounds from them. The samples taken from electro-filter showed a decrease of pH from 12,2 to 8,2 and a reduction of SO_4^{-2} and Ca^{+2}, with concentrations from 1091 to 209 mg/l and from 766 to 40 mg/l, respectively.

Analysis of ashes interacting with steel and ferroconcrete structural elements (culverts, bridges etc.) showed that even with being in contact with ash-slag mixture, anticorrosive protection in the case of tight pavements and sufficient protection of road crown and road slope against penetration of rainfalls to embankments, is not necessary. Components of hydraulically active volatile ashes are not so easily washed out from hardened layer of road foundation as from unbounded ashes.

The road foundation layer stabilized by hydraulically active ashes is made after wetting with water up to optimal humidity (according to Proctor's method). Changes of humidity are caused by evaporation or by rainfall when the road foundation layer is not covered by bituminous pavement.

Humidity increase in winter depends on capillary pulling up of water from, for example a high level of ground water, or may be caused by a temperature difference between the layer and soil. Drying of the layer after spring defrosting may cause washing of its components to lower layers of subgrade.

In laboratory studies, extreme conditions of washing out were assumed.[7] Water was pumped under pressure inside samples (No. 1 and 2) through performated metal tube. Another sample of ash was placed in water and extract tested for comparison. Table 1 shows chemical composition of extracts.

Sample No. 1 consisted of sand stabilized with 15% of ash after 6 months of hardening. The volume of 1000 cm^3 of water was filtered five times through the sample and the content of washed out components was estimated in 1 dm^3 of extracts. The concentration of sulfates after the first cycle of washing was equal to 231 mg/dm^3 and increased to 568 mg/dm^3 after the fifth cycle.

Table 1. Chemical composition of extracts

		Amount of washed out components per 1 kg of sample [mg]						
		Sample 1			Sample 2			
Parameter unit		After 1st cycle	After 3rd cycle	After 5th cycle	From 1st cycle	From 4th cycle	From 6th cycle	Sample 3
1	Chlorides	12,59	11,59	22,47	114,32	26,70	4,36	331,80
2	Sulphurates	285,19	550,62	701,23	1917,05	1101,14	311,36	2812,00
3	Calcium	93,58	161,73	222,22	939,77	927,27	173,86	1250,00
4	Magnesium	3,47	3,83	3,77	0,05	0,09	0,05	0,15
5	Sodium	13,21	11,81	12,96	98,75	7,28	2,75	130,00
6	Potassium	308,64	71,73	31,48	460,23	24,09	17,27	754,00
7	Cadmium	< 0,005	< 0,005	< 0,005	< 0,005	< 0,005	< 0,005	< 0,005
8	Copper	0,02	0,04	0,07	0,03	< 0,006	< 0,006	0,05
9	Lead	< 0,05	< 0,05	< 0,05	< 0,05	< 0,05	< 0,05	< 0,05
10	Zinc	< 0,006	0,05	0,01	0,03	< 0,006	< 0,006	0,05
11	Nickel	0,43	0,35	0,37	0,05	0,01	< 0,01	0,15
12	Chromium	0,04	0,05	0,04	0,22	0,14	0,06	0,47
13	Iron	< 0,012	0,012	< 0,012	< 0,012	< 0,012	< 0,012	< 0,012

Sample No. 2 consisted only of hydraulically active ashes. In that case, in every cycle a new portion of 0,5 dm^3 of water was added. After the first cycle, the concentration of sulfates was equal to 1687 mg/dm^3 and after the sixth one—274 mg/dm^3.

The amount of washed out components from sample No. 3 was relatively high in the case of sulfates (2812 mg/kg), calcium (1250 mg/kg) and chlorides (323 mg/kg).

The following chemical analyse of extracts were carried out :

- measurement of cations concentration by means of plasmatic spectrometer IPC Hilger D 824,
- measurement of anions concentrations by means of liquid ion-exchange chromatography,
- measurement of pH.

Since the addition of 15% of hydraulically active ashes is usually sufficient for subgrade stabilization, results of sample 1 studies are the most similar to real conditions existing in road foundation layer.

It should be noted that the permeability of cohesive soil stabilized with binding materials is very small, but under the layer of stabilized sand there is very often the sand layer of relatively small humidity, but usually greater than humidity of road foundation.

Results of the laboratory studies are rather promising from the environmental point of view; however further studies of soils situated under road foundation layers made of ashes, as well as penetration of washed out substances, should be undertaken.

4. CONCLUSIONS

Because of their limited use in both cement and building industries, hydraulically active ashes may be very useful for road construction.

Both laboratory studies, as well as practice in road construction confirmed the lack of environmental hazard because of ashes radioactivity and dusting problems.

The probability of ground water pollution by compounds washed out from foundation layer is also low. Studies concerning the presence of substances washed out from ashes to the subgrade situated under the foundation layer have not been conduct up to now.

REFERENCES

1. Pachowski J., "Volatile ashes and their use in road making", WKŁ, Warszawa, 1976.
2. Kukiełka J., "Stabiliesierung von Löss mit Zement und Braunkohlenflugaschen aus Konin", Berlin, 1980.
3. Kukiełka J., "Road foundations from loesses stabilized with cement and ashes of brown coal origin", Drogownictwo, 1, 1985.
4. Kukiełka J., "Cement-ash-loess mixes for road foundations of road pavements", Drogownictwo, 3, 1985.
5. Kukiełka J., "Use of active ashes for subgrades improvement and expressway pavements making", Przegląd Budowlany, 9, 1996.
6. Giergiczny Z., "Use of puzzolanic and hydraulic activity of volatile ashes in binding material industry", symp. mat., Poznań, 1996.
7. Kukiełka J., Wasąg H., Kawa M., "Evaluation of influence of humidity changes in subgrade and soil layer stabilized with ash on soil and ground water under road pavement designed in Świdnik, Poland", Technical University of Lublin, 1996.
8. Sztomajer S., Sztomajer Z., Książek L., Przedecki T., "Testing of some industrial waste materials of Łódź region towards their use as sealing materials", symp. mat., Łódź, 1996.

24

SOLIDIFICATION/STABILISATION OF HAZARDOUS WASTE CONTAINING ARSENIC

Effect of Waste Form Size on the Leachability

Veronika Dutré and Carlo Vandecasteele

Katholieke Universiteit Leuven
Department of Chemical Engineering
W. de Croylaan 46—3001 Heverlee, Belgium
E-mail: veroniek.dutre@cit.kuleuven.ac.be

1. ABSTRACT

Industrial waste, containing arsenic (approximately 32 wt%), antimony and lead was studied. The waste was solidified with inorganic materials such as cement and pozzolanic materials (lime, slags) in order to reduce the leachability of the contaminants from the waste. The solidification process was optimised by measurement of the influence of all the additives used in the process on the concentration of arsenic in the leachate of the solidified waste, since arsenic is the most toxic component in the waste material.[1]

Semi-dynamic leach tests were used to determine the mobility of the waste components through the monolithic matrix, by measuring the rate of leaching. The results from semi-dynamic leach tests can be presented as the cumulative fractions released as a function of the square root of the leach time. A linear relationship indicates that diffusion is the release mechanism. Leaching of the contaminants can then be described with an effective diffusion coefficient. Studies reported in the literature indicate that effective diffusivities can depend on the size of the Solidification/Stabilization (S/S) waste sample.[2] Therefore, in this paper, the validity of testing small-scale specimens to predict the leaching off full-scale waste forms was studied. Three sizes of cylindrical samples were prepared and subjected to the semi-dynamic leach test. For each sample, the effective diffusion coefficient was then calculated. The results found for the different sized samples were in good agreement with each other. It could be concluded that data obtained from small-scale waste forms can be used to predict satisfactorily the leaching of contaminants from larger-scale waste samples and thus from monolithic waste blocks to be placed in landfill.

Keywords: Solidification/stabilisation, semi-dynamic leaching, diffusion coefficient.

Chemistry for the Protection of the Environment 3, edited by Pawłowski *et al.* Plenum Press, New York, 1998

2. INTRODUCTION

Solidification/stabilisation (S/S) technology is used to transform potentially hazardous liquid or solid waste into less hazardous or nonhazardous solids before disposal in a landfill. In solidification/stabilisation the concentration of the considered hazardous component in the leachate can be reduced by: (a) formation of (an) insoluble compound(s); and (b) reduction of the mobility of the contaminants by encapsulation in the resulting monolithic matrix. Other objectives of S/S processes are, of course, the production of a monolithic solid mass and the improvement of the handling and physical characteristics of the waste.

Leaching of contaminants out of the cement-based waste form is mostly a diffusion controlled process. With the assumption of a constant diffusion coefficient, it can be shown that the cumulative fraction of a substance that has been released at time t (CFR) is given by:

$$CFR = \frac{2}{\sqrt{\pi}} \frac{SA}{V} \sqrt{D_e t} \tag{1}$$

with D_e = effective diffusion coefficient (cm^2/s), t = leach time (s), SA = surface area of specimen (cm^2), and V = volume of specimen (cm^3).

The effective diffusion coefficient can thus be calculated from the slope of the plot of CFR as a function of the square root of time. To evaluate the leachability of a diffusing species, a leachability index L may be defined as:

$$L = \log \frac{b}{D_e} \tag{2}$$

with b = 1 cm^2/s. This index can be used to compare the relative mobility of different contaminants on a uniform scale that varies from 5 (very mobile) to 15 (immobile).[3]

3. EXPERIMENTAL

3.1. Preparation of Solidified Waste Samples

Three sizes of cylindrical samples (small, medium and large) were prepared, using the same ratios of binder-to-waste mass for each sample. The solidification recipe used was as follows: per g of waste material, 1 g of lime and 1.1 g of cement (Cem II/A-M 32.5 R) are added together with 3 ml of water.[1] Sample 'small' was prepared using this recipe with 10-fold amounts. Sample 'medium' used 20-fold amounts, and sample 'large' 100-fold. Some information regarding the samples is given in Table 1.

3.2. Semi-dynamic Leach Test

Semi-dynamic leach tests are tests whereby the leachant is replaced periodically after intervals of static leaching. The leachate is analysed using ICP-MS. In this study, the timing of the leachant renewal is based on the equation[4]:

$$t_n = n^2 t_1 \text{ with } t_1 = 1 \text{ hour and } n = 1, 2, 3, 4, 5, \ldots \tag{3}$$

where t_n = duration of nth leaching interval.

Table 1. Physical parameters of the S/S waste

Sample	Surface area (cm^2)	Volume (cm^3)	V/SA (cm)	Volume of leachant (ml)
Small	65	36	0.56	650
Medium	103	79	0.77	1000
Large	322	437	1.36	3500

Distilled water is used as the leachant. The samples were subjected to the N^2 semi-dynamic leach test for a period of 196 hours.

4. RESULTS AND DISCUSSION

The results are presented for antimony in Figure 1 as the CFR (cumulative fraction released) and CAR (cumulative amount released) values. As can be seen from the CFR plots, the cumulative fraction of antimony that is released decreases as the S/S sample size increases. From the slope of these plots and the volume-to-surface area ratio (V/SA), the effective diffusion coefficients can be calculated. These are for the samples 'small,' 'me-

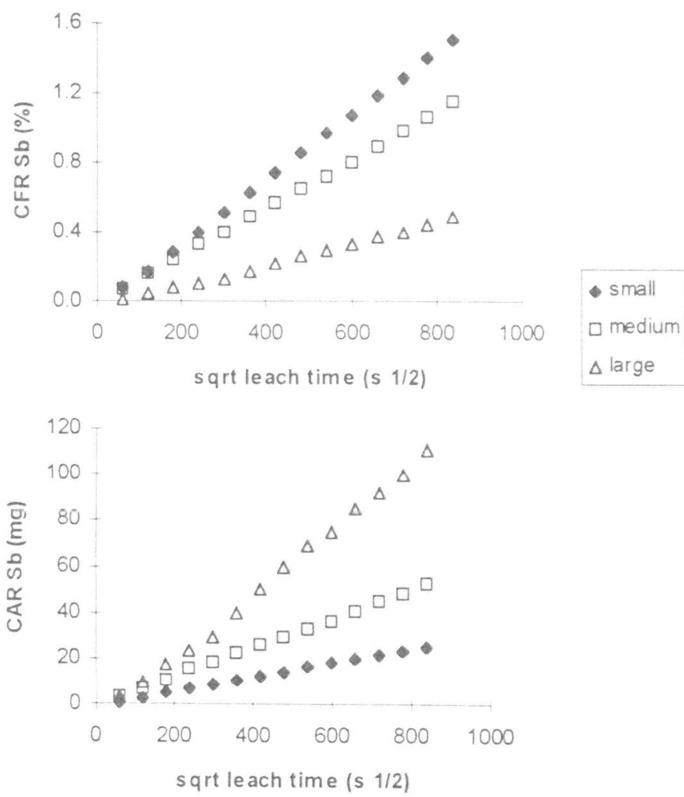

Figure 1. CFR and CAR plots for Sb for three sizes of solidified waste samples.

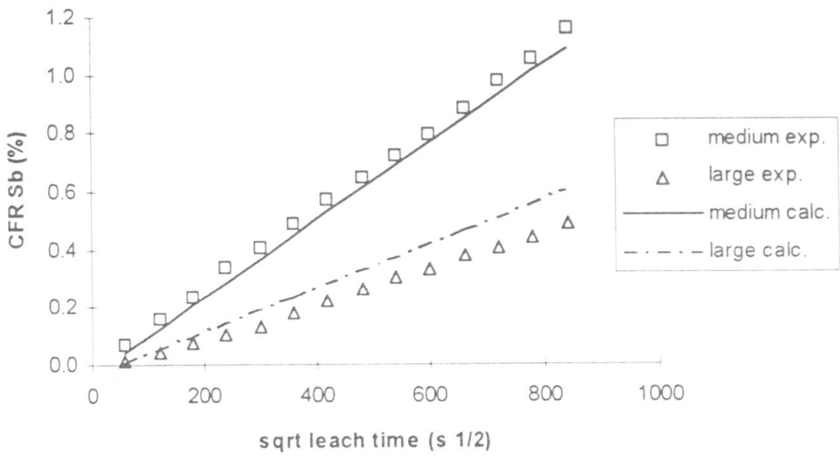

Figure 2. CFR plots (lines) for the 'medium' and 'large' waste samples deduced from the results obtained from the small waste sample, compared to the experimental results.

dium' and 'large' respectively: 8.48 10^{-11}, 8.73 10^{-11} and 5.53 10^{-11}, cm^2/s. The results found for the different sized samples are thus in good agreement with each other. The leachability indices can be calculated from these diffusion coefficients and are: 10.07, 10.06 and 10.26 for the samples 'small,' 'medium' and 'large.' The contaminant Sb thus has a low mobility from the S/S samples.

In order to evaluate the validity of using leaching results from small-scale samples to predict the leaching behaviour of contaminants from large-scale waste forms, CFR values for the 'medium' and 'large' waste samples were calculated using the effective diffusion coefficient from the small sample (equation 4).

$$CFR_{medium} = \frac{slope_{small}(V/SA)_{small}}{(V/SA)_{medium}}\sqrt{t} + intercept_{small}$$

$$\text{with } slope_{small} = 2\left(\frac{SA}{V}\right)_{small}\sqrt{\frac{D_{e,small}}{\pi}} \qquad (4)$$

These results are presented in Figure 2.

As can be seen from the figure, using the results from the 'small' waste sample, gives a good prediction of the behaviour of the medium size waste sample. The results obtained from the 'small' sample overestimate somewhat the leaching from the 'large' waste sample, but the agreement is still satisfactory.

5. CONCLUSIONS

Leach tests were conducted on specimens of different size. When the effective diffusion coefficient for the small sample was used to predict the release for the larger samples, the predicted CFR value for the medium sample was slightly lower, and for the large sample somewhat higher. When the release after one year of leaching is calculated, the following results are obtained. For the medium sample, 7.54% will have leached after one year, based on the data obtained from the small sample, whereas a value of 7.68% is calculated using the diffusion coefficient calculated from the experimental results for the medium sample. Thus, the amount leached is underestimated by a factor of 0.98. For the large sample, the predicted release (CFR) after one year is a factor of 1.24 higher than the actual release (4.25%/3.43%).

It can thus be concluded that data obtained from small-scale waste forms can be used to predict satisfactorily the leaching of contaminants from larger-scale waste samples.

From this, it can be inferred that the diffusion coefficients determined on small-scale S/S waste samples can be used for the calculation of the CFR values of the considered elements from large monolithic waste blocks on a landfill.

REFERENCES

1. Dutré, V. and C. Vandecasteele, Solidification/stabilisation of hazardous arsenic containing waste from a copper refining process. *Journal of Hazardous Materials*, 1995, *40*, 55–62.
2. McIsaac, C.V. and S.T. Croney, Effect of waste form size on the leachability of radionuclides contained in cement-solidified evaporator concentrates generated at nuclear power stations. *Waste Management*, 1991, *11*, 271–282.
3. Andrés, A., Ortiz, I., Viguri, J.R. and A. Irabien, Long-term behaviour of toxic metals in stabilized steel foundry dusts. *Journal of Hazardous Materials*, 1995, *40*, 31–42.
4. Côté, P.L., Constable, T.W. and A. Moreira, An evaluation of cement-based waste forms using the results of approximately two years of dynamic leaching. *Nuclear and Chemical Waste Management*, 1987, *7*, 129–139.

A NEW METHOD FOR TREATMENT OF CHROMIUM CONTAINING WASTES

Z. Kowalski[1] and A. Kozak[2]

[1]"Alwernia" Chemical Works
ul.K.Olszewskiego 8, 32-066 Alwernia, Poland;.
[2]Institute of Chemistry and Inorganic Technology
Cracow University of Technology ul. Warszawska
24, 31-155 Cracow, Poland

A new method of treatment for chromic tannery wastes containing chrome and organic compounds has been investigated. It was found that fluidized bed coal combustion ashes could be used for treatment of chrome waste containing high concentration of organic compounds. Produced sediment contain chrome ettringite as the main crystal phase. The method is very effective and Cr(III) level in treated wastes could be below 0.01 ppm.

Key words: chromic waste, tanneries, treatment.

1. INTRODUCTION

Leather tanning processes that use chromic agents produce large amounts of waste water. These contain, among other pollutants, Cr(III) and organic substances. Poland produces 0.5–0.7 million cubic meters of such wastes per year.

The concentration of chrome and organic pollutants varies considerably from one chromic tannery waste to another. The content of Cr(III) ranges from 0.5 to 2.0g/dm^3. The content of organic compounds depends mostly on the type of leather and the parameters of the tanning process. Differences in Chemical Oxygen Demand (COD) and Biochemical Oxygen Demand (BOD_5) respectively range from 2000 to 5000 g/m^3 and from 700 to 2000 g/m^3.[8,9,11]

Conventional method for treatment of chromium containing waste[1] is based on suitable precipitation of $Cr(OH)_3$ and filtering of the precipitate. Precipitated sludge containing mostly a chromium hydroxide is difficult to be separated because of the high concnetration of colloidal organic compounds in tannery wastes. Therefore, neutralisation of such wastes with $Ca(OH)_2$ or NaOH in Polish tanneries is only partially effective, and

Chemistry for the Protection of the Environment 3, edited by Pawłowski *et al.* Plenum Press, New York, 1998

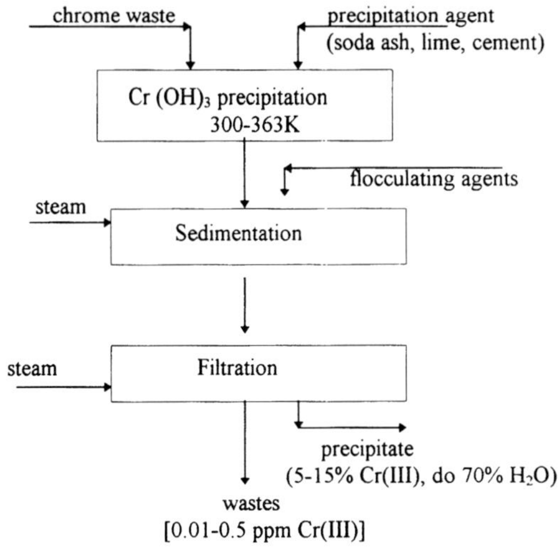

Figure 1. Flowsheet of the conventional method for treatment of chromium containing wastewaters.

only if the organic pollutant concentration is low. Nevertheless, it appears that the currently employed direct precipitation of $Cr(OH)_3$ from waste waters is the simplest and cheapest method of treatment.

It was found that an use of cement for precipitation of Cr(III) leads to rapid sedimentation of the slurry[8,11] and makes filtration easy. The filtrate after dilution (with e.g. municipal sewage) can be treated in conventional biological treatment plants. Over the last seven years, this very simple and effective method has been implemented in six Polish tanneries.

Research work on the utilisation of the precipitate[7] resulted in developed methods of using it as a component of concrete.

Rising prices of cement and energy encouraged us to undertake further research on finding new and less costly precipitation agents that could bring about sedimentation at lower temperatures. Different materials were examined and promising results were obtained with some types of coal ash, especially these from fluidized bed combustion (FBC) technologies. It would be a novel and attractive to employ a waste material, such as ashes, to treat another waste.

Ashes produced from a conventional pulverised fuel in FBC installations have different chemical and phase compositions[2,6,12]. These differences are due to lower combustion temperatures in FBC technologies (about 850° C, both for bubbling, BFBC, and circulating, CFBC, beds), and the use of calcium based sorbents to combine the sulphur present in the fuel. The waste ash produced in that process can contain as much as 50% of $CaSO_4$ and CaO. In the presence of water, free CaO is converted into $Ca(OH)_2$, and anhydrite to gypsum. Additionally in the highly alkaline environment, aluminium oxide reacts with calcium sulphate and forms calcium sulpho-aluminate (ettringite, $3CaO{\cdot}Al_2O_3{\cdot}3CaSO_4{\cdot}32H_2O$).

There are different methods of preparation of the ettringite. According to Benstedt and Prakash Varma [3] basic reaction is

$$3CaO(AlO_4)_2 + 3CaSO_4 \cdot 2H_2O + 26H_2O = Ca_6 \cdot [Al(OH)_6]_2 \cdot (SO_4)_3 \cdot 26H_2O \qquad (1)$$

and according to Thiel [14]

$$3CaO \cdot Al_2O_3 + 3CaSO_4 \cdot 2H_2O + 26H_2O = 6CaO \cdot Al_2O_3 \cdot (SO_3)_3 \cdot 32H_2O \quad (2)$$

$$4CaO \cdot 3Al_2O_3 \cdot SO_3 + 6CaO + 5CaSO_4 + 96H_2O = 3[6CaO \cdot Al_2O_3 \cdot 3SO_3] \cdot 32H_2O \quad (3)$$

In water solutions for ettringite preparation pH should be rather high [10–13].

The Al^{3+} ions (having ionic radius - $0,51 \cdot 10^{-16}$m) in the ettringite structure can be replaced with Cr^{3+} ions which have a similar ionic radius ($0.69 \cdot 10^{-16}$m)[13]. Previous studies[10] have shown that the pure chrome-etringitte can also be synthesised under similar conditions to those for typical ettringite precipitation.

2. EXPERIMENTAL

The FBC CII ashes used in tests on precipitation of Cr(III) came from Canadian experimental CFBC installations [4,5]. The fly ash fraction major components (% by weight) are SiO_2—13.0; Al_2O_3—9.5; Fe_2O_3—9.4; CaO—41.8; SO_3—14.5 and ignition loss as 14.2%.

First, crude experiments were made using aqueous $Cr_2(SO_4)_3$, to assess how solutions containing Cr(III) would behave in contact with FBC ashes. Different quantities of the CII ash were added and mixed. After settling of the sludge, the Cr(III) content in the filtrate was determinated using the colorimetric method, and the pH of the filtrate was recorded.

The phase composition of the drained solids was determinated using the X-ray diffraction method. The results are presented in Table 1.

It was shown that the FBC ashes are an attractive precipitation agent for removal of Cr(III).

In further experiments, typical chromic tannery wastes from the "Niepolomice" tannery, with the composition Cr(III)—0.9 g/dm^3, BOD—1500 g/m^3 and COD—4500 g/m^3, were used. In 1993, this tannery started to use a cement as the flocculating agent. The consumption value of the cement is 15 kg/m^3 of treated wastes.

In the tests, 12.5 kg and 20 kg of FBC ash CII were added to 1m^3 of the chromium wastes. The average mixing time after ash addition was 10 minutes to allow any free CaO to be converted to $Ca(OH)_2$. Sedimentation tests were made at temperatures of 293, 313 and 333K (the optimum temperature for the cement method is 363K). The best results were obtained with the addition of 12.5 kg of the FBC ash per 1m^3 of the wastes at 333K and were practically the same as with the cement addition.

Table 1. Treatment of aqueous solutions containing Cr(III) ions using FBC ashes

Test no	[CR(III)] solution [mg/dm^3]	Ash/Cr mass ratio	Filtrate [Cr(III)] [ppm]	Filtrate pH	Main crystal phases component of solid residues
1	31.4	158	<0.01	11.3	not analyzed
2	19.0	60.6	<0.01	11.6	ettringite prominent, calcite, quartz
3	175	72.2	<0.01	11.6	as in sample 2
4	2080	19.2	<0.01	11.5	ettringite prominent, gypsum, calcite, quartz
5	3930	10.2	0.01	10.4	as in sample 4

The Cr(III) content in all samples of the thickened slurry was 1–1.5%, and the H_2O content 66–80%. Slurry samples were filtered at 323–333K, under a pressure of 0.03–0.04 MPa, using a laboratory frame press with BT-17 filter cloth. The filtrate contained less than 0.01 ppm of Cr(III). After dilution with municipal sewage (typical proportions could be 1:5), the filtrate could be treated in a conventional biological plant.

The mean water content in filter cake (for 4 samples) was 50%. After 2h drying at 378K the filter cake contained 6.4% Cr_2O_3, 30.5% CaO, 9.1% Fe_2O_3, 8.8% Al_2O_3, and 12.7% SiO_2. Its ignition loss after calcination at 873K for 3h was 24.2%. The precipitate could be used as a raw material for sodium chromate production [9].

The fact that good results of sedimentation were obtained in tests at pH 9 and at pH 11 indicates that the effect of the addition of the ashes might be different in the two cases. At pH 9, the presence of the ash causes the aggregation of solid particles and the formation of large agglomerates with an apparently higher sedimentation rate than the initial particles. At pH 11 either some substitution of Cr(III) ions for Al(III) ones in the ettringite structure takes place, or the Cr compound is precipitated alongside ettringite.

Synthesis of the chrome ettringite was another part of our works. It could be made according to reaction:

$$Cr_2(SO_4)_3\ aq + CaO = 3\ CaO \cdot Cr_2O_3 \cdot 3CaSO_4 \cdot 32H_2O \qquad (4)$$

Another method, (according Murdie method[13]) was also used. In this method, $Al_2(SO_4)_3$ solution was added to saturated $Ca(OH)_2$ solution.

Reaction times for both methods were 3 to 18 days. The produced slurries were filtrated. Results of the chrome ettringite synthesis, filtrate and filter cake analysis are presented in Table 2 (positions 4–7 for the Murdie method).

X-ray phase analysis of five samples of solids after sedimentation at pH 8.8 to 9.1 showed that they contained gypsum, quartz and calcite. At pH 11 to 11.5, however, the solids also contained ettringite, and there was evidence of the presence of $CaSO_4$ (hemihydrate).

The goal of our research was the comparison of different type precipitation agents. In experiments presented in Table 3, we used FBC ashes, cement and calcium oxide. Using different ratios of Ca contained in the used precipitation agents and the Cr(III) contained the treated chromium solution, we wanted to achieve equilibrium system. The experiment lasted from 7 to 9 days at temperature 293 K. During this period, the samples were occasionally mixed, then filtrated. Filtrate and filter cake analysis are presented in Table 3.

Table 2. Results of the chrome ettringite synthesis

Sample no	Content in filtrate [mol/dm³]			Filtrate pH	Reaction time [d]	XRF analysis- main phases in filter cake
	Ca^{2+}	SO_4^{2-}	Cr^{3+}			
1	0.0305	0.0127	< 0.01 ppm	12.15	18	ettringite-Cr, $CaSO_4 \cdot 2H_2O$
2	0.0084	0.0043	< 0.01 ppm	11.29	18	ettringite-Cr, amorphic phase
3	—	—	< 0.01 ppm	12.20	5	ettringite-Cr, $CaSO_4 \cdot 2H_2O$, $Ca(OH)_2$
4	0.0211	0.0124	—	11.09	9	ettringite-Cr, $CaSO_4 \cdot 2H_2O$
5	0.0234	0.0128	0.1 ppm	10.85	13	ettringite-Cr, $CaSO_4 \cdot 2H_2O$
6	0.0254	0.0114	0.1 ppm	11.14	4	ettringite-Cr, $CaSO_4 \cdot 2H_2O$
7	0.0121	0.0093	—	9.57	3	ettringite-Cr

Table 3. Treatment of aqueous solutions contained Cr^{3+} ions using FBC ashes (CII), calcium oxide and cement

Test no	Precipitation agent	Starting $Cr_2(SO_4)_3$ solution: Cr^{3+} content [g/dm³]	Starting $Cr_2(SO_4)_3$ solution: Precipitation agent [g/250cm³]	Mole ratio: CaO / Cr^{3+}	Mole ratio: CaO / Al^{3+}	pH	Filtrate after Cr treatment—composition: Cr [mg/l]	Ca^{2+} [mg/l]	SO_4^- [mg/l]	OH^- [mg/l]	pH	Main crystal phases of solid residue
1	FBC ashes	1.5788	2.1216	1.6	0	2.4	2.15	664	1651	—	5.3	$CaSO_4 \cdot 2H_2O$, ettringite traces, SiO_2
2	contained 22.9% of	1.5788	2.6250	2.0	0	2.4	<0.01	660	1584	—	6.8	as in sample 1
3	CaO, 7.1% of Al_2O_3	1.5788	4.0530	3.06	0	2.4	<0.01	660	1546	—	8.6	as in sample 1
4		1.5788	3.4508	1.6	1.6	2.4	<0.01	660	1584	—	7.2	$CaSO_4 \cdot 2H_2O$, ettringite prominent, SiO_2, Fe_2O_3
5		1.5788	4.3211	2.0	1.6	2.4	<0.01	660	1546	—	8.4	as in sample 4
6		1.5788	14.4000	3.0	3.0	2.4	<0.01	1120	1238	510	12.1	as in sample 4
7	Calcium oxide	1.5788	0.4000	1.01	—	2.4	600	92	1325	—	3.3	amorphic
8		1.5788	0.6810	1.6	—	2.4	<0.01	632	1507	—	6.6	$CaSO_4 \cdot 2H_2O$, ettringite traces
9		1.5788	0.8513	2.0	—	2.4	<0.01	612	1516	—	8.3	as in sample 8
10		1.5788	1.0640	2.5	—	2.4	<0.01	1120	1363	467	11.4	as in sample 8
11		1.5788	1.2982	3.05	—	2.4	<0.01	1100	1248	493	12.1	$CaSO_4 \cdot 2H_2O$, ettringite
12		1.5788	1.4900	3.5	—	2.4	<0.01	1120	1200	518	12.2	as in sample 11
13		1.5788	1.9147	4.5	—	2.4	<0.01	1128	1200	530	12.3	$CaSO_4 \cdot 2H_2O$, ettringite, $Ca(OH)_2$
14	Cement contained	1.5788	1.0640	1.1	—	2.4	405	572	2131	—	4.3	as in sample 12
15	44.3% of CaO	1.5788	1.2144	1.26	—	2.4	228	520	2112	—	4.6	$CaSO_4 \cdot 2H_2O$ prominent, ettringite traces, $Cr(OH)_3$
16		1.5788	1.4187	1.48	—	2.4	<0.01	876	2112	—	6.7	as in sample 14
17		1.5788	2.1244	2.2	—	2.4	<0.01	668	1603	—	8.6	as in sample 14

Results presented in Table 3 to conclude that all the CaO from FBC ashes was utilised to precipitate chromium (no reaction between Ca and Al). The Ca/Cr mole ratio on the level 2 would be sufficient for treatment of the solution from chromium. The CaO from cement was utilised not only to precipitate chromium, but to set the cement. This is important for possibly reducing of the quantity of FBC ashes required for chrome waste treatment and should result in a much lower quantity of sediment compare to the method that require use of cement as a precipitation agent.

3. SUMMARY

It was found that the ashes from fluidized bed coal combustion could be used for treatment of chrome wastes containing high concentrations of organic compounds.

The addition of the ashes promotes the sedimentation of the precipitated solids from wastewater containing organic compounds, and the obtained slurry enhances filtration. Chrome ettringite is main crystal phase of the sediment. Experiments on ettringite synthesis and the use of calcium compounds from different agents to precipitate chrome compounds should decrease quantity of the sediment and increase effectiveness of the proposed method.

Further examination could help in developing of the technology of the chrome waste treatment with using fluid bed combustion ashes as precipitation agent.

4. REFERENCES

1. Awierbuch,T.; Pawlow P.; *Tiechnologija sojedinienij chroma*, Izd.Chimija: Leningrad 1969.
2. Behrandres, C.B.; Hutzler, N.J.; *J. Environmental Engineering-ASCE.* 1994, 120, 6, 1288
3. Bensted, J.; Prakash Varma S.; Cement Technology May/June 1973.
4. Bulewicz, E.M.; Dudek, K.; Mazurkiewicz, M.; Piotrowski, Z.; Application of Ash Water Dense Suspension (AWDS) Technology for CFBC Ashes (PERD); Final Report for CANMET, DSS Contract No.23440-1-9166/01-SQ; 1992.
5. CANMET. Investigation of Advanced Ash Management Technologies of CFBC and LIFAC Residues. CEA Report 9141 G 890, Ottawa 1996.
6. Carr, C.E.; Colclough, P.; In *Pressurized Fluidized Bed Combustion.* Alvarez Cuenca M.,Anthony E.J., Eds.; Blackie Academic and Professional: London, 1995, 318.
7. Kowalski, Z.; *Z. Nauk.Polit. Warsz.,Chemia*, 1990, 51.
8. Kowalski, Z.; *J. Hazard. Mater.* 1994, 37, 137.
9. Kowalski, Z.; Walawska, B. Utilization of chrome tannery wastes into sodium chromate production process. *Polish J. Appl. Chem.* 1997 in press.
10. Kozak, A., Cracow University of Technology, personal communication, 1995.
11. Polish Patent. 141 964, 1988.
12. Mulder, E.; van Duin, P.J.; Grootenboer, G.J.; In *Atmospheric Fluidized Bed Coal Combustion*; Valk M., Ed.; Elsevier: Amsterdam, 1995.
13. Murdiel, M.; *Powder Diffraction Journal* 1986, 1
14. Thiel, A.; Doctoral Thesis, AGH Kraków 1982.

26

AGRICULTURAL USE OF SLUDGE IN CHINA

Cao Zhihong

Institute of Soil Science
Academia Sinica
Nanjing, 210008 P. R. China

ABSTRACT

Research on agricultural utilization of sewage sludge in China began in the early of 80's and criteria has been set for agricultural application of sewage sludge. Usually sludge is composted, it is used as fertilizer in China. During composting, straw, wood chips, bark and tree leaves are added as amending and/or bulking agent to improve fertility of composted sludge. Sometime potassium fertilizer and lime are also added to increase the K content and to reduce the bio-activity of heavy metals.

Compared to developed countries, the percentage of treated wastewater, especially domestic wastewater, is still very low in China. However, as environmental protection laws are implemented and domestic wastewater treatment becomes more usual. the total amount of sludge will increase dramatic ally. Therefore research on proper use of sludge should be given high priority.

Key Words: China, fertilizer, sludge.

INTRODUCTION

Sewage sludge is a by-product of wastewater treatment. The total amount of industrial wastewater in China has rapidly increased from 2,340 million tones in 1992 to 5,029 million tones in 1996 (Cao,1997). About 60% of the discharged industrial wastewater is pretreated, however, for domestic wastewater the fraction is only 20%. Nearly 10 million tonnes of sludge are generated annually from about 80 main wastewater treatment plants and about 27,000 medium and small size wastewater treatment plants in China.

With more wastewater treatment plants established, the amount of sludge will increase rapidly. On the other hand, the amount of sludge deposited into sea or used as land fill is expected to decrease sharply. So the amount of sludge applied as fertilizer in agri-

Chemistry for the Protection of the Environment 3, edited by Pawłowski *et al.* Plenum Press, New York, 1998

Table 1. The nutrients content (%) of sludge in China

Element	Range of content
organic matter	9.2–47.8
Total N	0.25–5.7
Total P	0.22–2.1
Total K	0.34–1.6

*The data come from Xue et al(1997) and Ge et al (1995).

culture should be increased. However, agricultural use of sludge in China is still limited to small scale field experiments and many problems remain to be solved.

The purpose of present paper is to review the agricultural utilization of sewage sludge in China. Problems related to application are also discussed.

COMPOSITION OF SLUDGE (BENEFITS AND HAZARDS)

The sewage sludge not only contains appreciable amounts of nutrients (organic matter, nitrogen and phosphate), but also many hazardous compounds (such as heavy metals, hazardous organic compounds and pathogens). These hazardous compounds are the main factors affected the recycling of sludge, and also represent a potential source of secondary pollution. The contents of organic matter, total N and total P in most sludge are high, but total K content is low (Table 1). Some heavy metal concentrations in sludge may be over the maximum permitted values recommended by the state (Table 2). The compositions of sludge depend on source of wastewater. If it comes from domestic wastewater, the content of heavy metals is low.

Sludge also contains many kinds of organic compounds as a result of industrial and household activities. The organic compound in sludge are synthetic compounds such as monocylic aromatics, polynuclear aromatic hydrocarbons (PAHs) and polychlorinated biphenyls (PCBs). Although their concentrations are low, the potential hazard is large if they accumulate over a long period. They can enter the food chain by uptake from the soil solution and accumulate in plants and animals.

Table 2. The heavy-metal contents of sludge in China

Heavy metal	Range of content (mg/kg)
B	51.2–60
Cu	28.4–3068
Zn	168.6–6912
Pb	42.3–492.1
As	5.6–20
Hg	0.61–6.96
Cr	42.6–1411.8
Ni	27.3–297.1
Cd	0.05–15

*data from Xue et al(1997) and Ge et al (1995)

In domestic sewage sludge, pathogenic organisms such as bacteria, protoza, parasitic worms and viruses are always found. So before utilization, sludge should be treated to kill these pathogenic organisms.

APPLICATION

Research on agricultural utilization of sewage sludge began in the early of 1980's and criteria for agricultural application of sludge have been made. However, up to now, the amount of sewage sludge applied as fertilizer is very low in China. Most of the sewage sludge was concentrated and used as land-fill or deposed in the sea.

COMPOSTING OF SLUDGE

Before used as fertilizer, sludge must be treated to kill pathogens, decompose toxic organic compounds and reduce activities of heavy metals. Composting usually is used to treat sludge in China. During composting, straw, wood chips, bark and tree leaves are added as amending and bulking agents. Sometime manure is also added to accelerate the composting process. Corn stalk is a good bulking agent in composting (Xue et al, 1997).

During composting, the temperature in the sludge may be as high as 70°C up to two weeks. It can be as high of 56°C even in fall and winter (Xue et al,1997; Zhao et al, 1993). Such high temperatures are enough to kill most of the pathogens. In a typical experiment after composting for one month, the bad odors disappeared and the concentrations of available N, P and K were significantly increased while the total content of N, P and K decreased (Table 3). The results also showed that composting substantially reduced the number of E.Coli. and roundworm ovum (Table 3). The composting process has no significant effect on content of heavy metal in sludge, but has some effect on the form of heavy metal in sludge (or its activity). The contents of heavy metals extracted by water were lower than 2% (Guo et al, 1995).

Table 3. Effect of composting process on composition of sludge*

	Before composting	After composting
Odor	strong	none
Total N (%)	2.81	2.46
Available N (%)	0.038	0.18
Total P (%)	0.39	0.34
Available P (%)	0.006	0.015
Total K (%)	1.38	0.68
Available K (%)	0.16	0.27
pH values	6.8	7.4
E. Coli (MPN/g)**	7.0×10^7	2.4×10^4
Roundworm ovum (n/kg)	11555.6	8809
The percent of survive ovum (%)	72	0–5

*Data from Ge et al (1997) and Zhao et al (1993)
**MPN: Maximum possible numbers.

Table 4. Effect of sludge on the content of nutrient elements in soil*

	The amount of applied sludge (t/ha)		
	0	75	150
Organic matter (%)	1.2	1.6	2.5
Available N (mg/kg)	52	147.4	293.6
Available P (mg/kg)	15.3	28.2	35.7
Available K (mg/kg)	184	190	264.3

*Data from Cao et al (1997), Lin et al (1994) and Zhang et al (1996b).
**The total N, P and K in sludge applied were 5.7, 1.7 and 1.6%, respectively.

EFFECT OF APPLIED SLUDGE ON SOIL FERTILITY

The physical and chemical properties of soils are improved after applying sewage sludge for some years. The results are consistence with that of other countries, such as lowering volume weight of soil, increasing CEC etc (Ge,1995). The contents of available N, P and K in soils increased markedly with the amount of sludge applied (Table 4). However it has also been reported that there is no significant increase of nutrient elements in soil when sludge has been applied only for one year (Guo et al, 1993). In general, the effect of sludge on soil fertility is very similar to that of farm yard manure.

EFFECT OF APPLIED SLUDGE ON YIELD AND QUALITY OF PLANTS

Sludge compost contains appreciable amounts of nutrients (Table 4) and is an ideal soil amendment. The results of sludge application showed that the yield increased (Table 5). Concerning crop yields, there was no difference observed between application of composted sludge and manure (Guo et al,1993; Yu et al, 1996). The effect of sewage sludge applied on ornamental plants has also been investigated. The number of the flowers and the blooming period of rose of Sharon (Hibiscus Syriacus L.), Chinese rose (R. Chinensis Jacq.), Nasturtium (Tropaeolum Majus L) and Canna (C. Indica L) were increased and prolonged with application of sludge (Zhang et al, 1996a).

ENVIRONMENTAL AND HEALTH ASSESSMENT

Sludge contains many toxic trace elements, organic compounds and pathogens. Its potential environmental and health problems should be considered in recycling of sludge to agricultural land. In China, sludge is generated mainly from industrial wastewater, so

Table 5. Effect of sludge used on yield of crops (Lin et al, 1993)

Sludge used (tone/ha)	Rice (kg/ha)	Wheat (kg/ha)	Maize (kg/ha)
Control	4358	3073	2784
37.5	5316	5454	5206
75.0	/	5944	6088

trace metal pollution has been the main concern in agricultural use of sludge. Many studies have demonstrated that the level of heavy metals in amended soils increased slightly. Accumulation of trace metals in plants has also been observed, but the content in edible part of crop was usually in the range of background values (Lin et al. 1994, Guo et al, 1994, 1995; Cao et al,1997, Xue et al,1997;Zhou et al, 1994). The heavy metals in soils amended with sludge mainly exist in acid-solved form and residual form. The contents of bioavailable heavy metals in soils increased with the amount of sludge applied. The fractions of bioavailable Cd and Zn in soil amended with 240 tonne/ha were as high as 17.8% and 20.1% of total Cd and Zn content, while both fractions in the control were below 2% of the total (Guo et al, 1995). Ouyang et al (1994)reported that the contents of Cd and Hg in soils in which sludge was applied for thirty years were over the limit values, and contamination of the products was detected: the contents of Cd in corn and wheat were 230 and 440 ug/kg respectively.

Because sludge contains high contents of nitrates and phosphate, the effect of sludge on groundwater should be of concern. Cao et al (1997) reported that the content of nitrates in the surface runoff water or infiltration water from greens area had increased significantly after sludge application, but the content of phosphate only changed slightly. They suggested that application up to 50 tonne/ha is safe.

The effects of organic compounds on soil and plants are not well understood. Some studies have show no significant changes in contents of toxic organic compounds in soils and plants after sludge application.

PROBLEMS RELATED TO AGRICULTURAL USE OF SLUDGE

Research on agricultural application of sludge in China is in an early stage. If the large amounts of sludge generated are concentrated and used as land fill, it becomes a potential pollution source. The amount of sludge and urban solid waste concentrated and used to land fill was as high as 646 million tonnes and took up 557 million m^2 of land of which were 3.8 million m^2. The amount deposited in sea will decrease in rapidly in recent years because of implementation of sea environment protection laws. On the other hand, the percentage of wastewater treatment is increasing, especially for domestic wastewater, and total sludge amount until certainly increase dramatically. Agricultural use of sludge is an economic way to consume sludge. Therefore the amount of sludge used in agriculture should increase greatly. But there are some important problems to be solved in agricultural use of sludge in China.

First, the government should make regulation for recycling of sludge. It is not only the demand for environment protection, but also the need for social and economic sustainable development. Agricultural use of sludge can feed back the nutrients in sludge to farm land, and these reduce the demand of inorganic fertilizers. On the other hand it is also environmental friendly to eliminate the possible hazards of sludge use.

Second, effects of sludge application on soil environment and crop quality are not until known in China. The standards for applications of sludge on arable land suffer from lack of experimental data and modifications are needed. For example, criticism of sludge application has been made based on the USA standard in 1984. At that time the research on agricultural use of sludge n China just began, few data were available. Some results have indicated that the limit value of Zn in sludge may be increased from 1000 mg/kg to 2500 mg/kg (Zhang, 1996a). The Tianjing municipal government has made a local law for agricultural use of sludge (Table 6). Its limit values for Zn, Cu and Ni are higher than the

Table 6. Comparison of limit values for the state and Tianjing (mg/kg)

Element	State	Tianjing
Cu	500	750
Zn	1000	2500
Ni	200	350
Cd	20	12.5
Hg	15	15
Pb	1000	1000
Cr	1000	1000
As	75	50
B	150	20

state's while As and Cd standards are lower. The fate of pollutant in soil depends on the characteristics of soil. In China there are many type of soils, therefore more data are needed to make more reliable standards.

Finally, we need to develop a better, economic and easy way to compost sludge in order to meet the demand for large scale use of sludge. Up to now, there is no special factories to produce cleaner sludge and/or sludge compost. Some new sludge-treatment techniques have been developed in England and USA since 1990, and the sludge compost product for agricultural use has been commercialized. Introduction of advanced techniques to accelerate the agricultural use of sludge in China should be considered.

CONCLUSIONS

In China, the percentage of sludge applied to farm land is low, but it is expected to increase rapidly in the near future. For this reason, effects of sludge application on the properties of soil, yield and quality of crops, environmental quality of ecosystems (plant, soil and groundwater) should be investigated in detail. Based on this research, more scientific and reliable standards for sludge compost for agriculture use should be made. Finally a more economic, cleaner and efficient treatment technology should be developed in order to accelerate the large scale agricultural use of sludge products.

REFERENCES

Cao Renlin Study on the environmental impact of sludge compost utilization on green area. *Research of Environment Sciences* (in Chinese). 1997, 10 (3):46–50

Ge Naifen, Ma Shuofang, Qin Huaiying and Zhou Lixiang The components of dewatered sewage sludge and its values used as manure. *Agro-environmental Protection* (in Chinese). 1995, 14:202–206

Guo Meilan, Wang Kui, Zhang Qingxi and Zhang Yongping The research of agricultural utilization of sewage sludge in Taiyuan. *Agro-environmental Protection* (in Chinese). 1993, 12:258–262

Guo Meilan, Tian Lutao, Wang Yanqin and Zhang Qingxi Effect of city sewage sludge and wastewater compost on the accumulation of heavy metals in Crops. *Agro-environmetal Protection* (in Chinese). 1995, 14:67–71

Guo Meilan, Mi Erfang, Tiang Luotao and Xi Mingqi Effects of city sewage sludge and sludge waste compost in the soil as a fertilizer resource. *Agr0-environmental Protection* (in Chinese). 1994, 13:204–209

Lin Chunye Effect of sewage sludge application to agricultural land on soils and crops. *Agro-environmental Protection* (in Chinese). 1994, 13:23–25

Ouyang Xihui, Cai Jin and Dong Qing Effects of long -term application of waste sludge on soils and crops in agriculture. *Agro-environmental Protection* (in Chinese). 1994, 13:271–274

Xue Chengze, Ma Yun, Zhang Zhangqiang and Gao Kunlui The production of compound organic fertilizer using municipal sewage sludge. *Agro-environmental Protection* (in Chinese). 1997, 16:11–15

Yu Caihong Effects of city garbage and sewage compost on the yield and quality of plant. *Agro-environmental Protection* (in Chinese). 1996, 15:264–266

Zhang Zhengqiang Study on the response of some ornamental plant to the sewage sludge compost. *Environmental Pollution & Control* (in Chinese). 1996, 18,1–4, 1996a

Zhang Zhengqiang The application of sewage sludge compost to city gardens and turfgrass lands. *Agro-environmental Protect* (in Chinese). 1996, 15:36–40, 1996b

Zhou Lixiang, Hu Aitang and Ge Neifeng Effect of municipal sewage sludge applied to agricultural land on crop and soil. *Agro-environmental Protection* (in Chinese). 1994, 13: 158–162

Zhao Ziding and Chang Yuehai Study on composting of chemical fibers sludge. *Agro-environmental Protection* (in Chinese). 1993, 12:11–13

27

A MODEL STUDY OF SOIL ACIDIFICATION IN A SMALL CATCHMENT NEAR GUIYANG, SOUTHWESTERN CHINA

Liao Bohan,[1] Hans Martin Seip,[1] Thorjørn Larssen,[1] and Xiong Jiling[2]

[1]Department of Chemistry, University of Oslo
P.O. Box 1033, Blindern, 0315 Oslo, Norway
[2]Guizhou Institute of Environmental Sciences
148 Xinhua Road, Guiyang 550002, P.R.China

ABSTRACT

Soil acidification has been reported in southwestern China and may become a serious problem in large areas. In order to understand the effect of acid deposition on soils in this region, detailed studies of deposition, throughfall, soils, soil waters and streamwater have been carried out in a small catchment (about 7 ha) near Guiyang since 1992. In the present paper, the MAGIC model with two soil reservoirs was selected to investigate long-term changes in soils and soil waters due to acid deposition for four different plots in the catchment, and to estimate likely responses to assumed future deposition scenarios. Data from the previous field and laboratory studies were used as far as possible to determine model parameters and to check results. The model simulations indicate that long-term acid deposition has resulted in soil acidification in the catchment, especially in the upper horizons. However, partly due to different topography and vegetation, soil parameters (e.g. cation exchange capacity, pH and base saturation) vary substantially resulting in large variations in sensitivity to acidification even within this small catchment. The model outputs depend heavily on values chosen for dry deposition factors for major ions, the solubility constant for $Al(OH)_3$, weathering rates of soil minerals, and to some extent on soil cation exchange capacity, sulfate absorption capacity, and ion concentrations in deposition in the background year (1895). The model results predict that soil acidification will accelerate if deposition of basic metal cations decreases or deposition of SO_4^{2-} increases by 30%; on the other hand, soil conditions will improve if a 30% reduction in SO_4^{2-} deposition is achieved in the future. Probably soil acidification will continue slowly, similar to the present situation in the catchment, if acid deposition continues at the present level.

Key words: MAGIC model, China, acid deposition, soil water, prediction.

Chemistry for the Protection of the Environment 3, edited by Pawłowski *et al.* Plenum Press, New York, 1998

1. INTRODUCTION

Soil acidification due to acid deposition has been reported in southwestern China by both field observation[1,2] and modelling work[3]. There are concerns that this may become a serious problem in large areas. On the other hand, some studies indicate that surface water in this region is not likely to be acidified in the near future[3,4].

China is a fast developing country in the process of expanding its economy. In the coming decades, the energy demand will increase dramatically and acid deposition will inevitably become a widespread threat to the environment[1,5]. It seems unrealistic to expect that significant reductions in SO_2 emission will be achieved in the near future[3]. Therefore, it is an urgent task to estimate present ecosystem impacts of acid deposition, to predict possible future trends and to offer these data to environmental managers and decision makers.

To achieve this goal, field and laboratory studies should be combined with modelling work. As a part of a research program, detailed studies of precipitation, throughfall, soils, soil waters and streamwater were initiated in a small catchment near Guiyang in April 1992. We have also carried out laboratory experiments with soils from this catchment[6,7,8]. Guiyang (26°34′N, 106°38′E), the capital of the Guizhou Province, is one of the cities suffering most serious acid deposition in southwestern China. The annual average SO_2 concentration was 200–500 $\mu g/m^3$ in the late 80s and early 90s in Guiyang city[9].

To estimate ecosystem trends for various assumptions about future deposition, a model is needed. Acidification of soils and waters is a result of a number of complicated processes and all models today have some shortcomings. In spite of this, modelling work may be useful[10] and numerical simulation models for effects of acid deposition have found increased usage in the past decade[11].

We decided to use the MAGIC model because it seems to have been more extensively tested at manipulation sites than any other acid-base chemistry model[11]. MAGIC was applied by Zhao and Seip[3] to predict soil acidification and has also been used to estimate critical loads of acidity in some other Chinese soils[1,12]. This model was originally developed for assessing the long-term acidification of surface waters[13,14] and it has been widely used throughout North America and Europe[11,15,16,17,18]. The model includes key chemical processes in soil, such as sulfate adsorption and desorption, weathering of soil minerals, exchange of base cations, and dissolution of soil aluminium-containing minerals. Major weaknesses seem to be in the aluminium submodel and in the neglect of mineralization of organic sulfur compounds.

In this paper we report model estimates of long-term changes in soils and soil waters in the small catchment near Guiyang, both hindcasts and forecasts for assumed deposition scenarios. The usefulness of the results in view of model deficiencies and uncertainties in input parameters is discussed.

2. SITE DESCRIPTION

The catchment is located about 5 km northeast of Guiyang city. The area is about 7 ha and the elevation ranges from 1320 to 1400 m above sea level. The major vegetation is Chinese fir (*Cunninghemia lanceolota*), pine (*Pinus massoniana*) and wild rose bushes. The soil type is generally *haplic alisol* corresponding to yellow soil in the Chinese classification system[19,20], but there are quite large variations within the catchment, both with respect to topography and soils. The climate is subtropical, monsoonal with high humidity.

Annual precipitation is 1175 mm with an average discharge of 630 mm, hence the evapotranspiration in the catchment is about 46%.

This catchment was selected because it (a) is relatively undisturbed by land-use activities, (b) is exposed to acid, SO_4^{2-} containing deposition, and (c) has a soil type sensitive to acidification which is representative for a large area in southern China. Since April 1992, 24 ceramic cup lysimeters (P80 cell from KPM, Berlin, Germany) and 20 throughfall samplers have been installed in seven plots within the catchment (Fig. 1). Wet precipitation, throughfall, soil waters and streamwater have been sampled and analysed regularly. The results of detailed studies in the catchment have been reported[9,21].

In the catchment, median pH in precipitation is 4.4 (quartiles: 4.19 and 4.77) and median sulfate concentration 228 μeq/L (quartiles: 147 and 339 μeq/L). Dry deposition of both SO_2 and alkaline dust is considerable. The sum of wet deposition of SO_4^{2-} and dry deposition of SO_2 has been estimated to be about 8.5 $gSm^{-2}yr^{-1}$.[21] Soil pH (in water) is generally below 4 in the upper (O, A) horizons. Base saturation (BS) decreases from the O-horizons down to the C-horizons, while aluminium saturation (AlS) shows the opposite trend. Generally AlS is high; even in the A-horizons it may be about 90%[9]. At concentrations corresponding to or somewhat higher than ambient levels in soil waters in this catchment, soil sulfate adsorption was found to be low, typically 2–4 meq SO_4^{2-}/kg.[6] Weathering rates determined by short-term laboratory studies for the soil at plot A were in the range 46–103 $meqm^{-2}yr^{-1}$ for total base cations[8].

3. MODEL DATA INPUTS

In this work, the MAGIC version with two soil reservoirs[22] was applied for four different plots: plots A, C, D, and E (see Fig. 1). Since there are more than two soil horizons at these plots, we decided to use average values weighted according to depths of the horizons. In most cases the A- and B-horizons in the catchment were included in the upper soils (soil 1 in the model) and the C-horizons in the deeper soils (soil 2 in the model). The model simulations included three parts: simulation of historical changes in the catchment (hindcasts), sensitivity of the results to changes in the model parameters, and soil responses to assumed future scenarios of deposition (forecasts).

To carry out the hindcast from the background year (1895) to the reference year (1995), estimates are needed for wet and dry deposition of major ions, weathering rates and vegetation uptake rates. Median values for compositions of wet deposition, based on field studies during 1992–1995, are given in Table 1. Corresponding values for the background year (1895) were estimated from observations at a remote site in a mountainous area (unpublished data, Guizhou Institute of Environmental Sciences). Assumed deposition trends are shown in Fig. 2. Present dry deposition factors listed in Table 1 are based on the observed concentration ratios of throughfall to wet deposition at the different plots with some adjustments.

In the model simulations we tried to use the soil parameters obtained from field observation and laboratory work as far as possible. From our previous laboratory studies of the soils in this catchment, we have values for sulfate adsorption for plots A and E and weathering rates of soil cations for plot A. However, it is difficult to transfer weathering rates from laboratory studies to field conditions, so adjustments were made by trial and error until the model outputs for soil-water chemistry matched values obtained from the field observation. Vegetation uptake and initial conditions were also obtained in this way. We do not have measurements for vegetation uptake in the catchment, which are given as

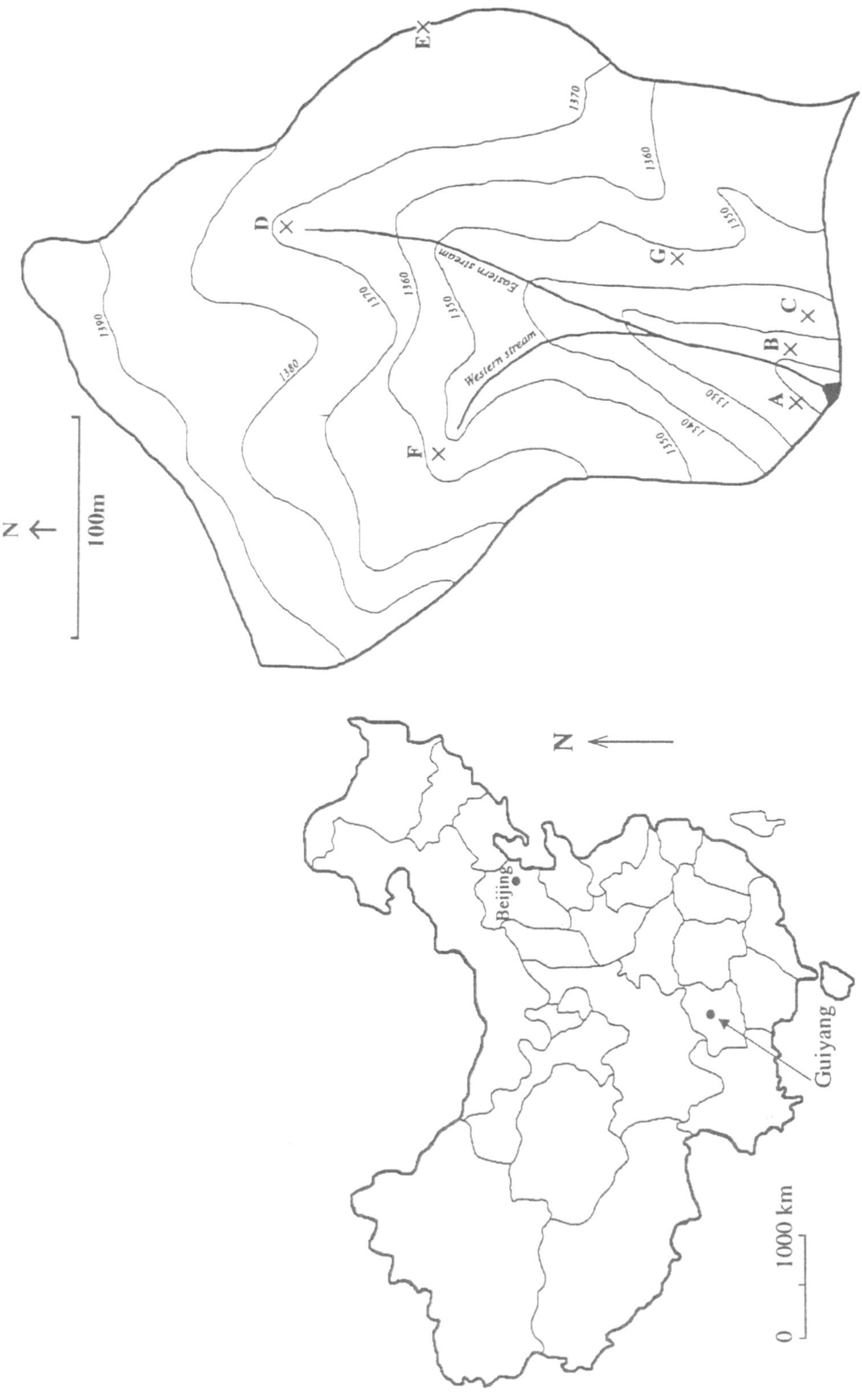

Figure 1. Map of China and the Guiyang catchment. Soil and soil water samples were collected from the points marked with X and letters (A-G).

Table 1. Ion concentrations (μeq/L) in precipitation and dry deposition factors for various plots in

	Ion conc. In precip., 1995 catchment	Dry dep. factor, 1995				Ion conc. In precip., 1985 catchment	Dry dep. factor, 1985 all plots
		Plot A	Plot C	Plot D	Plot E		
Ca^{2+}	133.4	2.5	1.5	3.0	2.0	20.0	1.3
Mg^{2+}	30.0	3.0	1.5	3.2	2.0	6.0	1.3
Na^{+}	8.3	2.0	1.2	1.5	1.5	7.0	1.0
K^{+}	6.9	2.0	1.2	2.8	1.5	1.1	1.0
NH_4^{+}	28.5	1.8	1.2	1.3	1.3	8.0	1.0
SO_4^{2-}	227.7	3.4	2.5	2.8	2.4	25.0	1.3
Cl^{-}	5.6	3.0	1.8	1.6	1.6	4.2	1.0
NO_3^{-}	19.0	1.9	1.5	2.6	1.6	9.0	1.0
F^{-}	4.5	2.0	1.5	1.5	1.6	1.0	1.0
pH	4.3					5.9	

percentages of external sources (i.e. total deposition and weathering). As a starting point we considered values in the range usually observed in forested catchments in Europe and the US (59–122 $meqm^{-2}yr^{-1}$; 21–46 $meqm^{-2}yr^{-1}$).[23,24] The deep soil at plot A has very high concentrations of SO_4^{2-} and Ca^{2+} (see Table 2) and it was necessary to use high weathering rates for both species. This horizon seems to be strongly influenced by sub-lateral flow affecting the soil water composition. In some cases it was necessary to introduce production of nitrate (e.g., in plot C). Since there is no option for nitrification, nitrogen must be produced by what is denoted 'weathering' in the model. The actual process is likely to be mineralization of organic material, ammonium oxidation or N-fixation[21]. From our previous laboratory studies for the soils in this catchment, we have present values for exchangeable cations and therefore also cation exchange capacity (CEC) and base saturation (BS), see Table 2. The final parameter values are given in Table 3.

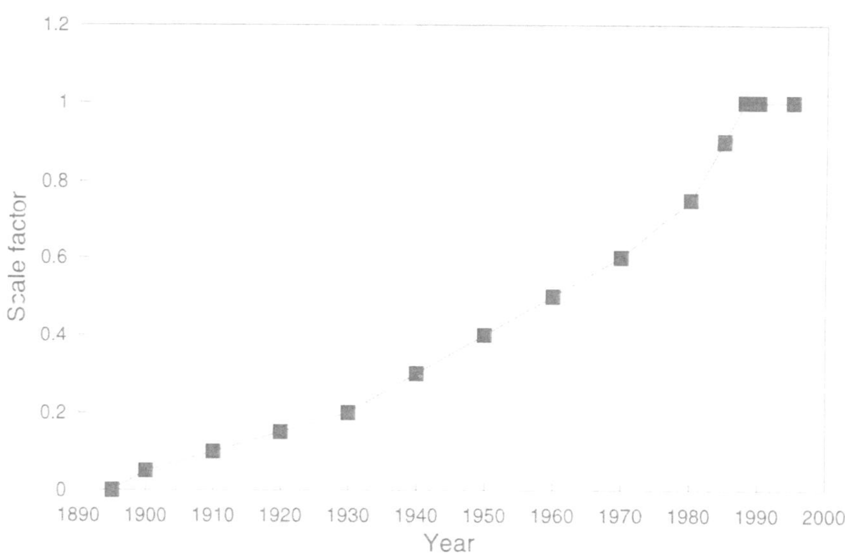

Figure 2. Assumed changes in wet and dry deposition from 1895 to 1995. Scale factors of zero and one correspond to deposition values in the years 1895 and 1995, respectively (see Table 1).

Table 2. Comparison of field values and model outputs for soil waters and soils in the Guiyang catchment in the reference year 1995

	Plot A				Plot C				Plot D				Plot E			
	Upper		Deeper		Upper		Deeper		Upper		Deeper		Upper		Deeper	
	Field	Model	Field	Model	Field	Model	Field	Model	Field	Model	Field	Model	Field	Model	Field	Model
pH	4.3	4.4	4.3	4.4	4.3	4.4	4.5	4.5	4.7	4.9	5.1	5.9	4.3	4.4	4.7	4.9
Ca^{2+}, μeq/L	685	684	1233	1231	311	310	198	196	827	829	913	912	538	534	599	594
Mg^{2+}, μeq/L	208	207	87	87	105	104	58	57	260	257	276	276	88	86	80	79
Na^{+}, μeq/L	22	22	26	26	14	14	11	11	14	14	31	30	21	21	22	21
K^{+}, μeq/L	59	59	29	29	19	18	6.6	6.8	78	77	25	25	9.1	8.8	3.4	3.4
NH_4^{+}, μeq/L	7.0	7.1	9.8	9.9	9.8	9.8	7.0	6.9	13	13	8.7	8.8	14	14	4.4	4.5
Al^{3+}, μeq/L	274	452	777	637	236	259	189	239	96	64	12	0.2	318	305	139	75
total SO_4^{2-}, μeq/L	1582	1581	2074	2074	615	614	425	424	1206	1204	1020	1020	1131	1131	764	765
Cl^{-}, μeq/L	141	141	249	249	88	88	27	28	17	17	56	56	4.2	4.3	51	51
NO_3^{-}, μeq/L	1.9	1.8	74	74	116	115	152	152	119	118	6.3	6.5	3.9	3.9	1.0	1.0
total F^{-}, μeq/L	37	37	20	20	2.7	2.7	2.2	2.2	30	30	1.7	1.7	15	15	2.0	2.0
ECa, %	22.4	22.4	7.2	7.2	4.4	4.3	2.3	2.3	11.7	11.7	25.1	25.1	13.0	13.0	4.1	4.1
EMg, %	3.5	3.5	1.9	1.9	1.2	1.2	0.8	0.8	3.0	3.0	4.2	4.2	1.5	1.5	0.6	0.6
ENa, %	0.1	0.1	0.1	0.1	0.1	0.1	0.2	0.2	0.1	0.1	0.2	0.2	0.1	0.1	0.2	0.2
EK, %	2.2	2.2	0.9	1.0	1.3	1.3	0.7	0.7	2.1	2.1	1.6	1.6	1.6	1.6	1.4	1.4
BS, %	28.2	28.2	10.1	10.1	7.0	7.0	4.0	4.1	16.9	16.9	31.1	31.1	16.3	16.3	6.3	6.2

Table 3. Parameter values used in the model simulations in the Guiyang catchment

	Plot A		Plot C		Plot D		Plot E	
Parameter	Upper soil	Deeper soil	Upper soil	Deeper soil	Upper soil	Deeper soil	Upper soil	Deeper soil
Soil depth (m)	0.11	0.19	0.33	0.57	0.22	0.38	0.30	0.20
Density (kg/m^3)	1000	1400	1000	1400	1000	1400	1000	1400
CEC (meq/kg)	117.5	77.1	68.5	25.4	90.3	82.5	64.5	72.4
SO_4^{2-} halfsat (meq/m^3)	27586	23797	14865	10000	20000	20000	14865	12804
SO_4^{2-} maxcap (meq/kg)	19.6	20.7	13.0	14.0	16.0	18.0	13.0	16.9·
log10 K[Al(OH)$_3$]	9.14	9.36	9.05	9.40	9.81	10.37	9.12	9.84
log10 (SAlCa)	−1.09	1.05	0.72	1.04	1.82	5.76	−0.23	2.73
log10 (SAlMg)	−0.23	−0.68	0.93	0.78	2.08	6.54	0.17	2.66
log10 (SAlNa)	−1.49	−1.37	−1.25	−2.65	−1.28	1.33	−1.16	−1.11
log10 (SAlK)	−4.07	−3.77	−4.53	−5.04	−2.73	−1.48	−5.89	−6.29
Weathering rates (meq/m^2/yr)								
Ca^{2+}	17.0	285.0	54.0	8.0	3.0	110.0	12.0	16.0
Mg^{2+}	17.0	0.0	17.0	4.0	24.0	20.0	8.0	8.0
Na^+	0.0	2.0	0.0	0.0	0.0	8.0	0.0	0.0
K^+	19.0	0.0	3.4	1.0	18.5	0.0	2.0	0.0
NH_4^+	0.0	1.5	0.0	0.0	0.0	0.0	0.0	0.0
SO_4^{2-}	20.0	260.0	0.0	0.0	0.0	0.0	23.0	0.0
Cl^-	63.0	57.0	40.0	0.0	0.0	20.4	0.0	24.7
NO_3^-	0.0	38.0	34.0	19.5	11.7	0.0	0.0	0.0
F^-	11.3	0.0	0.0	0.0	9.5	0.0	0.5	0.0
total BC	53.0	287.0	74.4	13.0	45.5	138.0	22.0	24.0
Uptake (% of external sources)								
Ca^{2+}	2.0	0.0	39.0	42.0	7.0	5.0	8.0	3.0
Mg^{2+}	1.0	58.0	15.0	52.0	0.0	14.0	38.0	29.0
Na^+	34.0	0.0	29.0	22.0	45.0	0.0	17.0	2.0
K^+	4.0	51.0	11.0	62.0	0.0	64.0	58.0	45.0
NH_4^+	93.1	0.0	85.7	30.0	83.0	30.0	81.0	68.0
SO_4^{2-}	0.0	0.0	46.0	30.8	5.5	15.3	0.0	32.4
Cl^-	0.0	0.0	0.0	68.5	3.0	0.0	76.0	0.0
NO_3^-	97.5	0.0	0.0	0.0	0.0	94.5	93.5	75.0
F^-	0.0	45.0	80.0	19.0	0.0	94.2	0.0	87.0
Ca^{2+}, meq/m^2/yr	8.2	0.0	112.6	72.0	33.1	27.4	26.0	8.9
Mg^{2+}, meq/m^2/yr	1.2	63.5	10.5	30.7	0.0	21.8	29.8	15.6
K^+, meq/m^2/yr	1.4	15.8	1.4	6.6	0.0	26.1	8.2	2.1
total BC, meq/m^2/yr	10.8	79.3	124.5	109.3	33.1	75.3	64.0	26.6
Initial exchangeable base cations (%)								
ECa	25.59	9.98	5.38	2.05	13.12	21.12	20.02	3.23
EMg	7.32	1.13	1.97	0.92	11.45	3.79	3.98	0.86
ENa	0.15	0.10	0.12	0.23	0.29	0.31	0.20	0.20
ENa	3.98	0.79	1.06	0.64	3.89	1.42	1.53	1.21
BS	37.04	12.00	8.53	3.84	28.75	26.64	25.73	5.50

Observed values for concentrations of major ions in soil waters were used for adjusting input parameters. Chemical compositions of soil waters for different plots are the median values of soil waters collected from lysimeters installed in different soil layers at corresponding plots. Where soil properties used in the model are averages for two horizons, soil-water concentrations were averaged in the same way.

The model also needs concentrations in water for the background year. These values may be calculated by the model from an assumption of steady state (input equals output) or they may be given. Most often the steady state assumption was applied, but in some cases, to be mentioned later, it was necessary to use the other option.

4. RESULTS

Present Characteristics of Soils and Soil Waters

Observed values for soil-water composition and soil properties are given in Table 2. Compared to the other plots, plot D has higher soil-water concentrations of Ca^{2+} and Mg^{2+} and lower Al^{3+} concentrations, as well as higher soil pH and BS in both soil layers. The conditions for plot A are fairly similar to those at plot D, except for lower pH and correspondingly higher Al^{3+} concentrations. Compared to plots A and D, concentrations of Ca^{2+}, Mg^{2+}, and SO_4^{2-} for plot E, and particularly plot C, are low. Low BS-values show that these two plots are in 'bad' conditions, i.e. further acidification may be critical for the vegetation. Although this catchment is quite small, large variations are found in soil water chemistry as well as soil properties between different plots, partly due to different topography and vegetation[21]. Thus soil sensitivity to acidification is also very different for these plots.

Simulated Historical Changes in the Catchment

Historical simulations in this catchment were carried out for the period of 1895–1995. The model outputs for ion concentrations in soil waters and for exchangeable base cations in soils in the reference year 1995 are compared to the field values in Table 2. Except for some of the Al^{3+} concentrations (this will be discussed later) and pH for the deeper soil at plot D, a very satisfactory agreement between the model outputs and the field values has been obtained, indicating that important parameters used for this model study are reasonable. The trends in concentrations of (Ca^{2+} + Mg^{2+}), SO_4^{2-}, Al^{3+}, pH, molar ratio of Al/(Ca+Mg) (R_{CL}) in soil waters and base saturation (BS) in soils during this period are given in Fig. 3.

Concentrations of SO_4^{2-} in soil waters increased considerably for all plots in both soil horizons in accordance with the increased deposition. Most of this increase is balanced by increase in concentrations of Ca^{2+} and Mg^{2+}. In the upper soils this loss of soil base cations is not fully compensated by increased deposition of these ions and weathering, resulting in decreasing base saturation. From the 1970s, concentrations of Al^{3+} also increased greatly, because of soil acidification and increased ionic strength in soil waters. The lowest Al^{3+} concentrations in soil waters are found for plot D where soil pH is relatively high.

According to these modelling results, decreases in pH (about 0.3–0.6 pH units) and BS (about 1.6–11.8 %) have occurred in the upper soils for all plots, indicating that acid deposition has resulted in soil acidification in the Guiyang catchment. The R_{CL} ratio in the upper soil waters increased, although not much for plot D. However, for all plots these ratios are at present far below 1.0, which has been suggested as a threshold for harmful effects[25,26]. In the deeper soils, both pH and BS were quite stable for all plots during the past

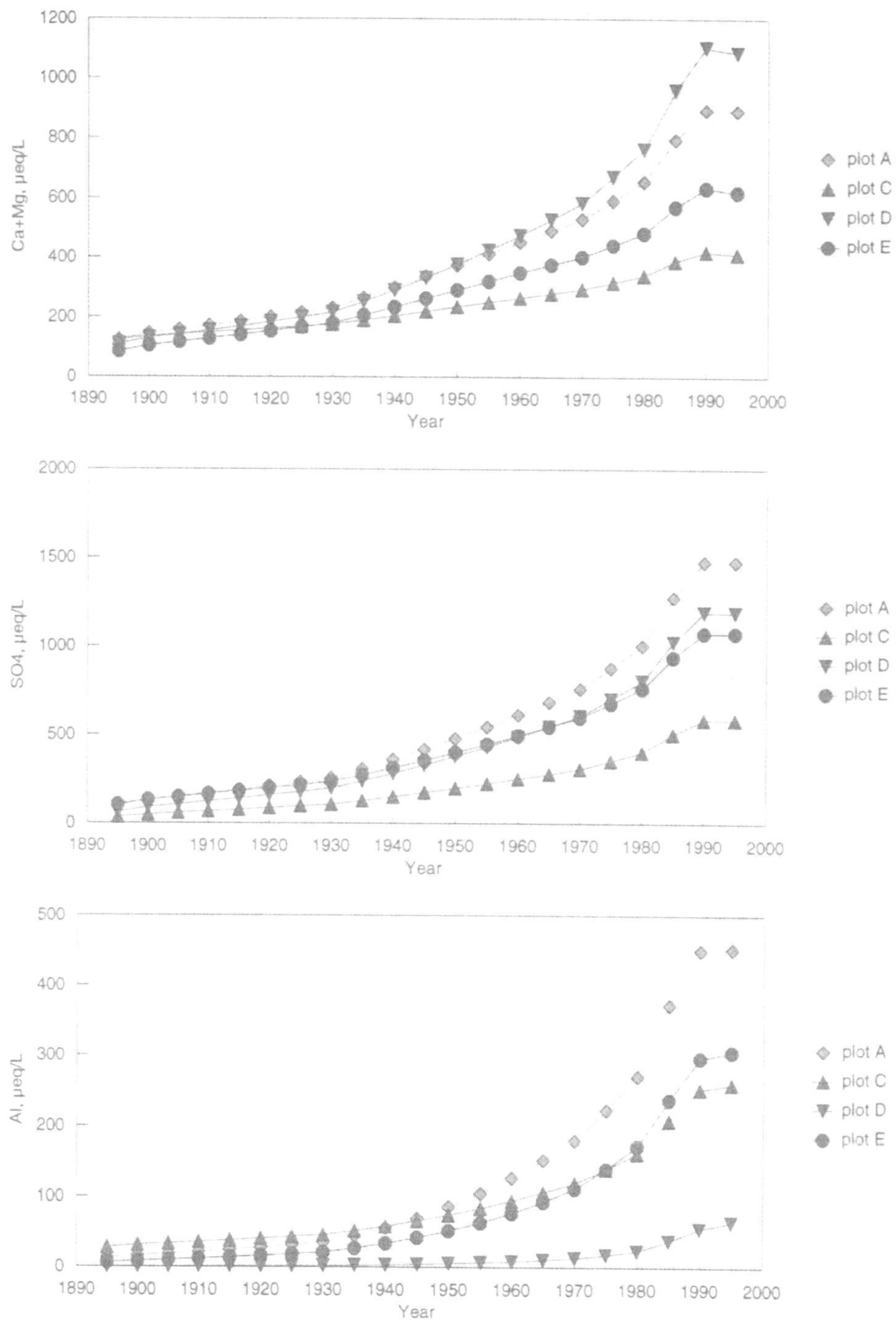

Figure 3. Simulation results for the upper (pp. 227–228) and the deeper (pp. 229–230) soils and soil waters in the Guiyang catchment for the period of 1895–1995.

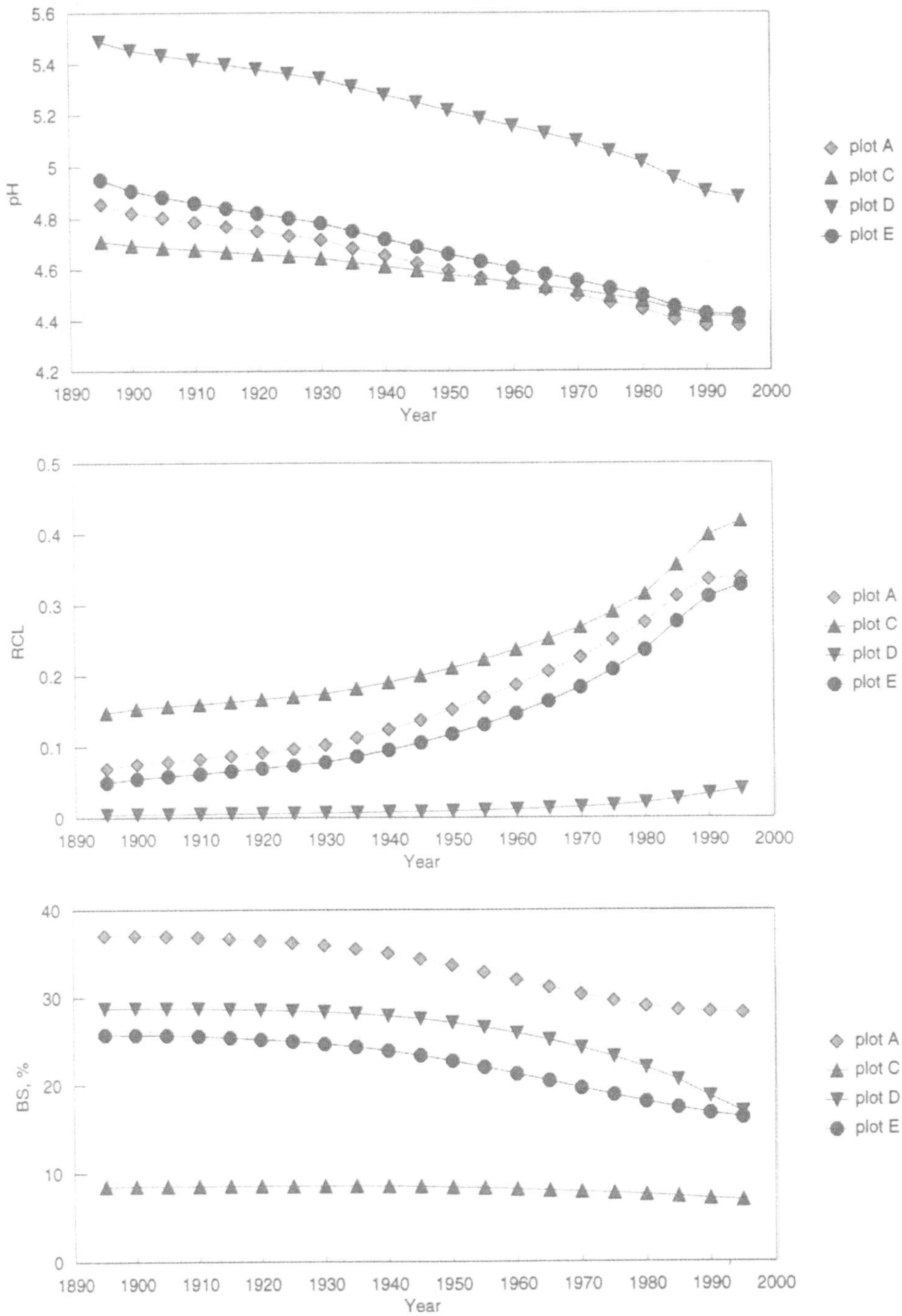

Figure 3. (*Continued*)

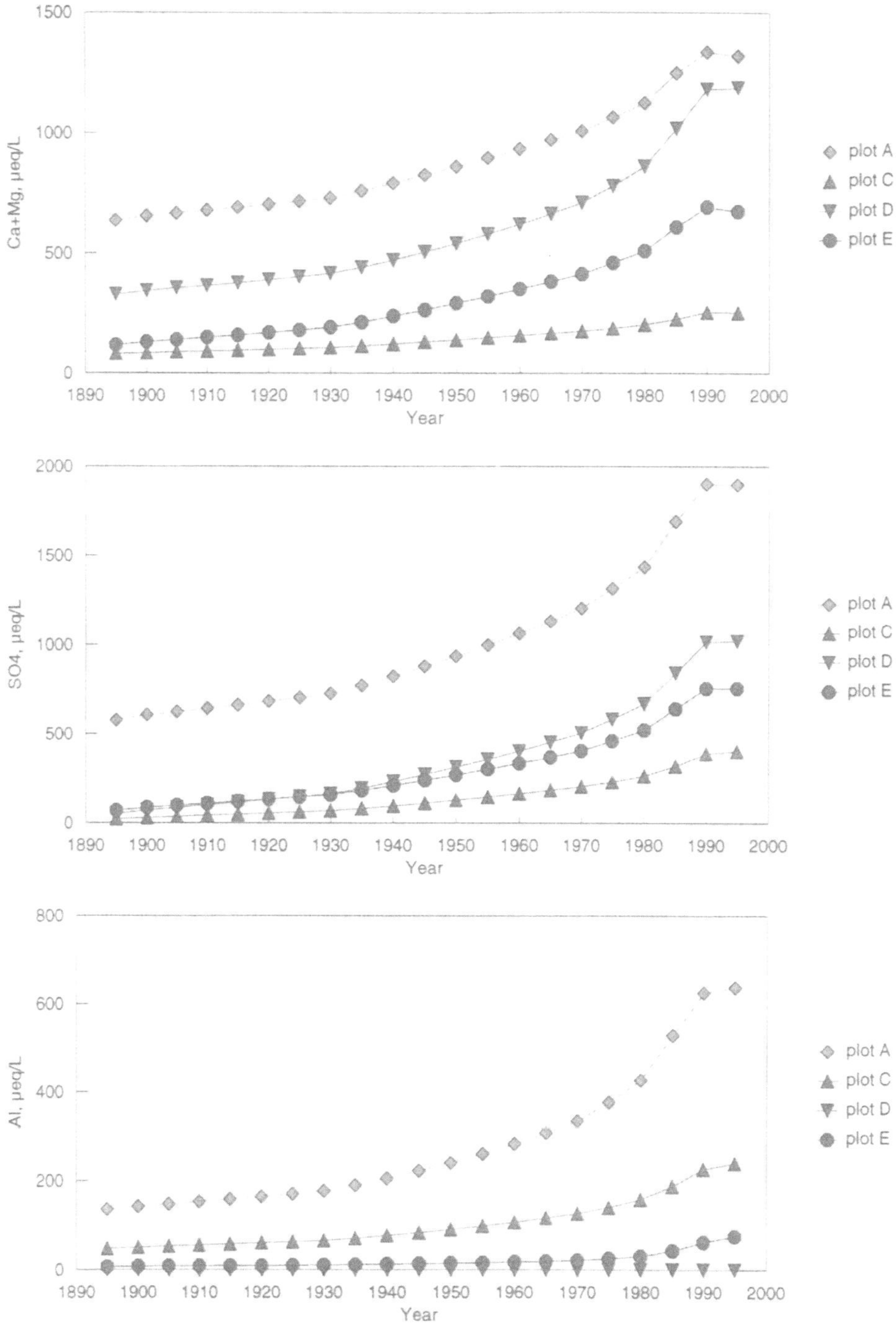

Figure 3. *(Continued)*

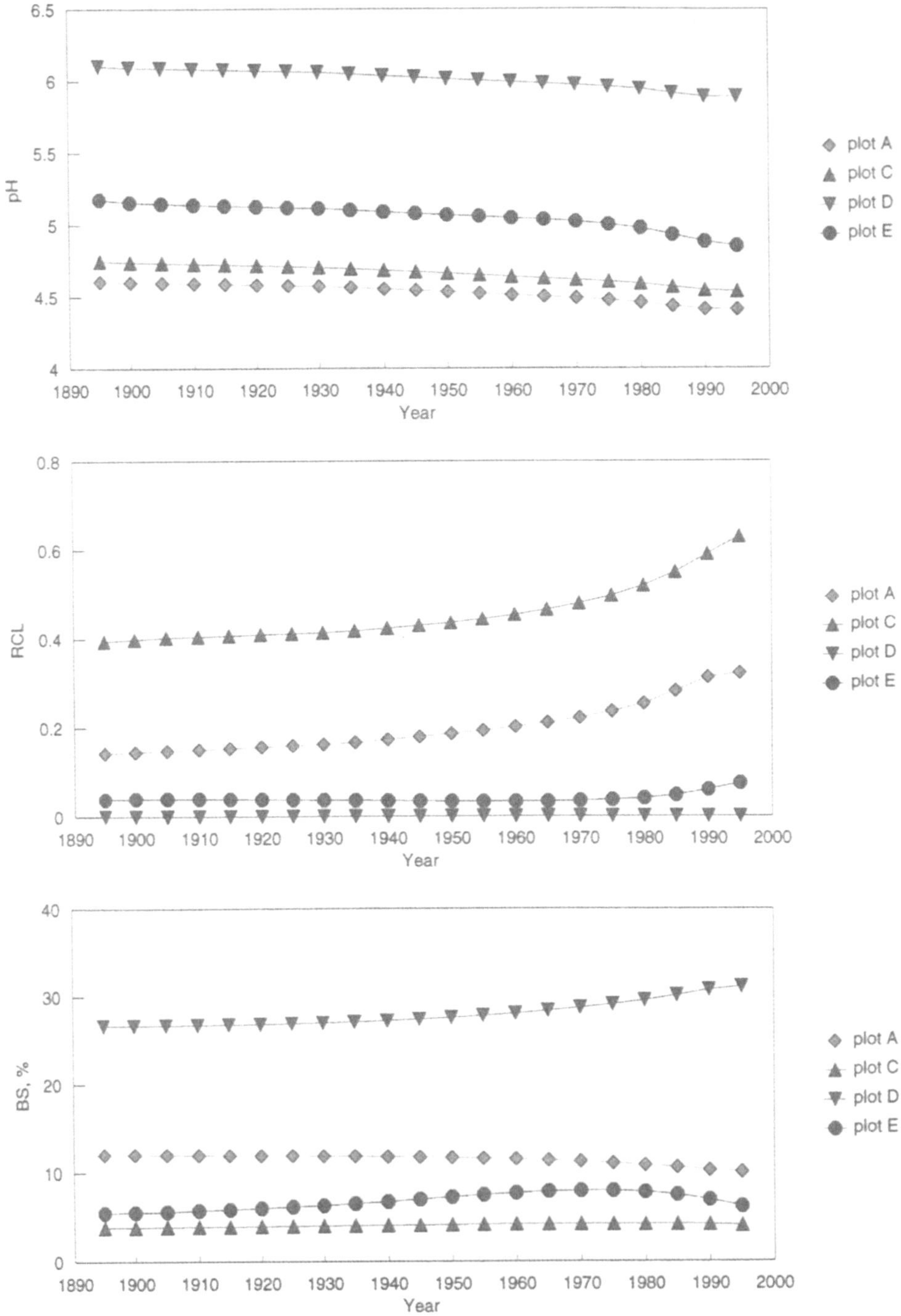

Figure 3. *(Continued)*

century, demonstrating that these soils have a large acid neutralizing capacity[3]. The changes that have occurred in soil waters in deeper horizons (mainly increased concentrations of SO_4^{2-}, Ca^{2+} and Mg^{2+}) are not likely to be harmful. The increase in Al^{3+} at plot A may be of some concern, although R_{CL} is less than 0.3. These results are similar to the conclusions of our previous laboratory work[7,8].

Sensitivity to Changes in the Model Parameters

There are many factors influencing the results of the MAGIC model, such as dry deposition factors, ion concentrations in deposition, the solubility constant for $Al(OH)_3$, weathering rates, soil sulfate adsorption capacity, sulfate half saturation, the selectivity coefficients for exchange of Al^{3+} with base cations, as well as soil CEC, depth, porosity, density, and temperature. Table 4 shows parameter values for 40 simulations carried out to test sensitivity. Generally, changes in the input parameters result in quite similar trends in output values for all four plots. However, there are differences in the magnitude of the changes. Results for plot E for concentrations of base cations, SO_4^{2-}, and Al^{3+}, pH, R_{CL}= Al/(Ca + Mg), and BS for 1995 are shown in Fig. 4.

In general dry deposition factors (s1–s10) and base-cation weathering rates (s36–s40) are critical parameters. Note for instance the dramatic changes in Al^{3+}, R_{CL} and BS when the dry deposition factors are changed. However, it should be noted that the

Table 4. Changes in model parameters used in sensitivity analyses

Part 1: dry deposition factors in 1995

	Ca^{2+}	Mg^{2+}
s1	4.0	4.0
s2	3.0	3.0
s3	2.5	2.5
s4	2.0	2.0
s5	1.0	1.0
	SO_4^{2-}	
s6	4.0	
s7	3.0	
s8	2.5	
s9	2.0	
s10	1.0	

Part 2: deposition concentrations in 1895 (μeq/L)

	Ca^{2+}
s11	40.0
s12	30.0
s13	20.0
s14	15.0
s15	10.0
	SO_4^{2-}
s16	40.0
s17	30.0
s18	25.0
s19	15.0
s20	10.0

Part 3: constants

	logK[$Al(OH)_3$]
s21	11
s22	10
s23	9
s24	8
s25	7
	SO_4^{2-} maxcap (meq/kg)
s26	30
s27	20
s28	15
s29	10
s30	5
	CEC (meq/kg)
s31	120
s32	90
s33	60
s34	30
s35	10

Part 4: weathering rates (meq/m²/yr)

	Ca^{2+}	Mg^{2+}
s36	100	50
s37	70	35
s38	40	20
s39	20	10
s40	10	5

Note: s0 is the base case in Figs. 4 and 6; part 3 and part 4 are used for both soil layers.

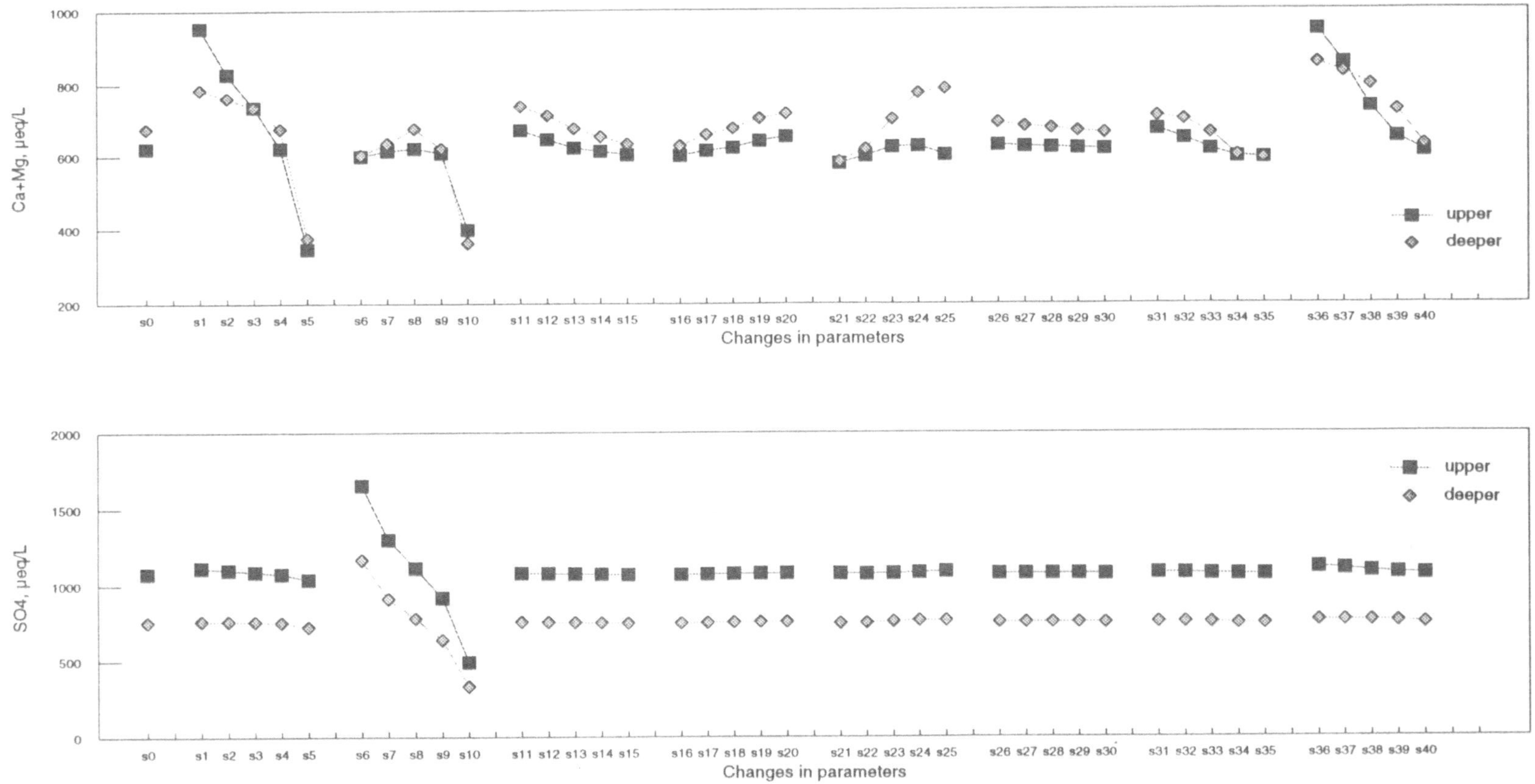

Figure 4. Results of soils and soil waters for plot E in 1995 corresponding to different model parameters. s0 is the base case. s1–s5 refer to dry deposition factors in 1995 for Ca^{2+} and Mg^{2+}, s6–s10 to the same for S O; s11–s15 to deposition concentrations in the background year 1895 for Ca^{2+}, s16–s20 to the same for S O; s21–s25 to the solubility constant for $Al(OH)_3$, s26–s30 to sulfate adsorption capacity, s31–s35 to cation exchange capacity; and s36-s40 to weathering rates of Ca^{2+} and Mg^{2+}, see Table 4.

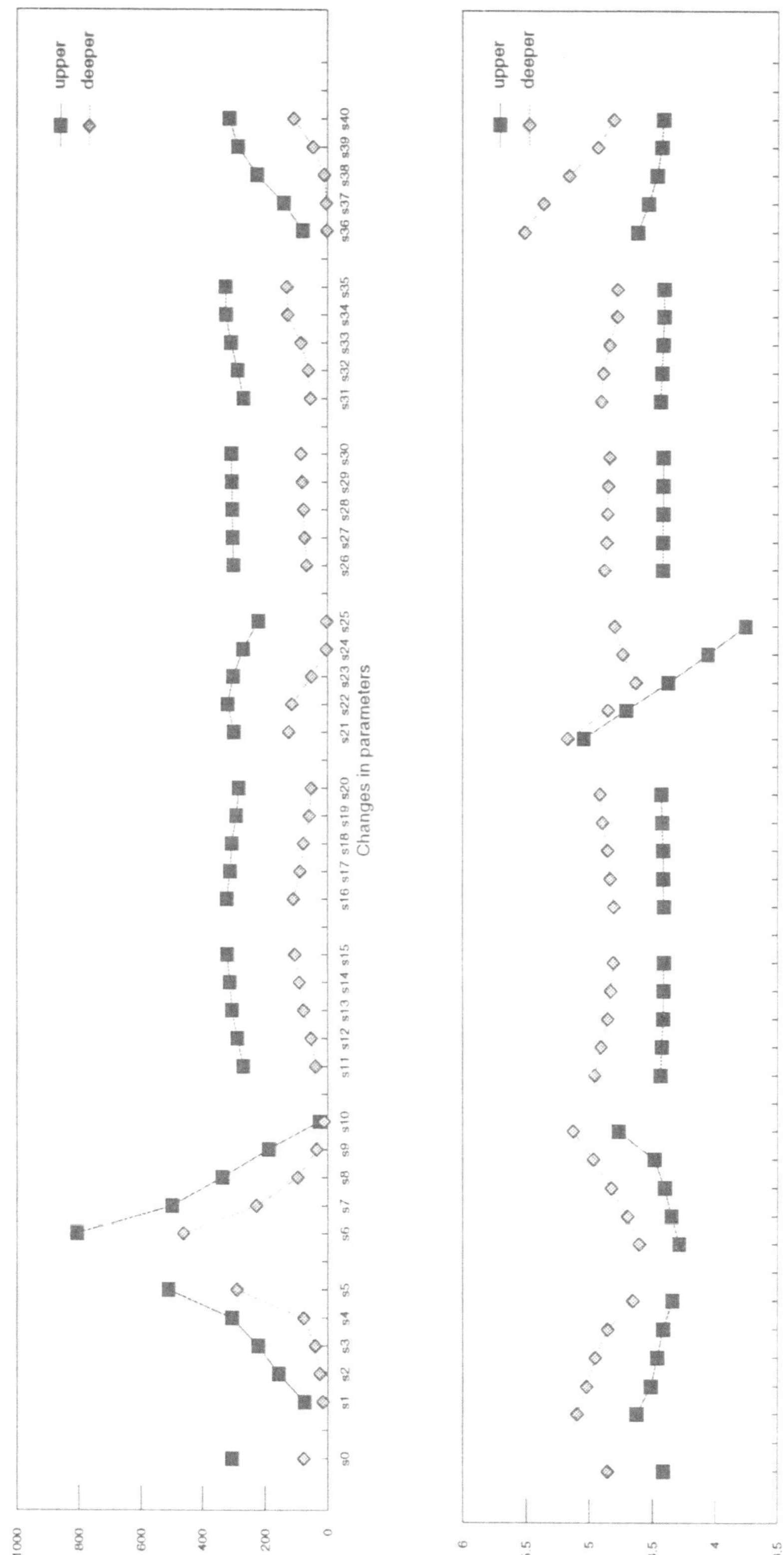

Figure 4. *(Continued)*

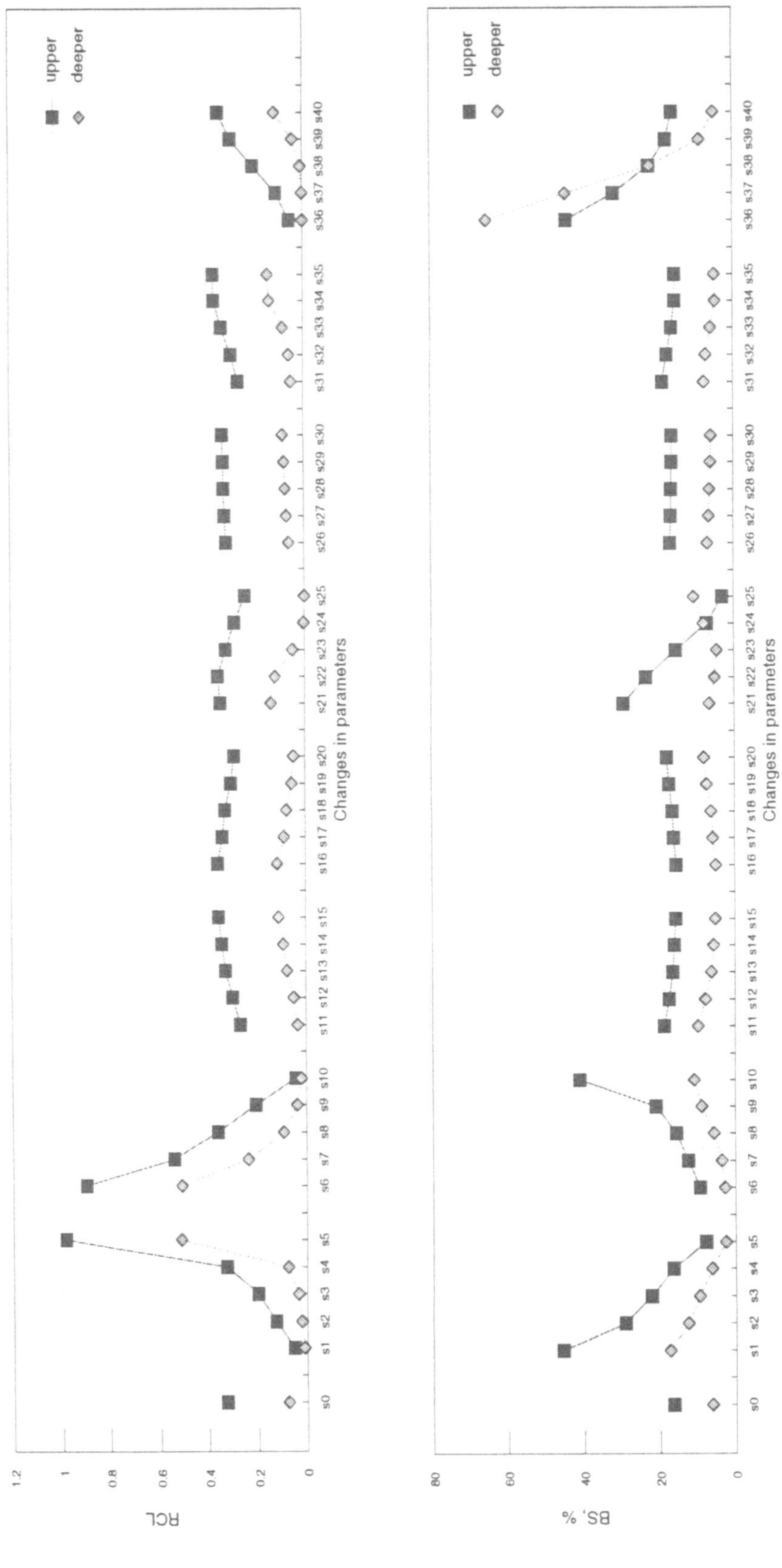

Figure 4. *(Continued)*

ranges chosen are large, i.e. for weathering rates the ratio between the highest and the lowest value is 10 and for the dry deposition factor 4. The solubility constant for $Al(OH)_3$ (s21–s25) is particularly important for the resulting BS and pH. When this constant is decreased, also the selectivity constants, e.g. S_{AlCa}, will change. This results in increases in the sum of base cation concentrations and the H^+ concentration, and, in most cases, a decrease in the concentration of Al^{3+}. Since more base cations are leached out, BS and pH decrease in the upper horizon. In the deeper horizon these properties may increase due to higher input of base cations from the upper horizon. (The increase in base cation concentrations when the solubility constant for $Al(OH)_3$ decreases is not clear for the upper horizon in Fig. 4, partly because the sum of only Ca^{2+} and Mg^{2+} is shown and partly because the difference has declined by the year 1995.) Within the chosen limits changes in the concentrations of Ca^{2+} and SO_4^{2-} in background deposition (s11–s20), in the maximum capacity for SO_4^{2-} adsorption (s26–s30) or in CEC (s31–s35) caused only minor changes.

It should be noted that if aqueous concentrations for the background year are calculated by assuming steady state (input equals output), one may get strange results. If the base saturation for the background year is set to the same value in all calculations for a given plot, an *increase* in the weathering rate may result in a *lower* base saturation for 1995 because the model gives a higher flux of base cations out from the system. The results in Fig. 4 were obtained by using the same initial concentrations in all calculations.

Because of the great effect of varying the dry deposition factors for Ca^{2+} and Mg^{2+}, we present two additional hindcasts with higher and lower values for these factors, see Table 5. Weathering rates and initial conditions were also altered to obtain observed concentrations in soil waters. For case 2 (high dry deposition factors) the modelled base saturation in the upper soil increased from 9% in 1895 to 16 % in 1995 in spite of very low weathering rates. For case 3 (low dry deposition factors) it was necessary to assume higher weathering rates. Even then BS decreased from 47% to 16%, which seems an unrealistically large change. Although the uncertainties are large, these calculations seem to support our choice of dry deposition factors.

Table 5. Comparison of parameters used in three cases of the hindcast for plot E*

	Case 1	Case 2	Case 3
Dry deposition factors in 1995			
Ca^{2+}	2.0	2.5	1.5
Mg^{2+}	2.0	2.5	1.5
Weathering rates (meq/m²/yr) in the upper soil in 1995			
Ca^{2+}	12.0	0.0	20.0
Mg^{2+}	8.0	0.0	13.0
Na^+	0.0	0.0	0.0
K^+	2.0	3.6	1.3
Initial conditions (%) in the upper soil in 1895			
ECa	20.0	6.6	37.2
EMg	4.0	1.0	7.7
ENa	0.2	0.2	0.3
EK	1.5	1.5	1.5
BS	25.7	9.3	46.6

*Case 1 is the base. The percentages of vegetation uptake for cases 2 and 3 are the same as for case 1 (see Table 3), but the absolute values are changed slightly.

Responses to Assumed Future Deposition

We assumed six scenarios of possible future deposition for the model predictions during the period 1996–2045. In plan 1, there are no changes in deposition either for base cations or for anions. Plan 2 and plan 3 assume a 30% increase and a 30% decrease respectively in the total (wet plus dry) deposition of both Ca^{2+}, Mg^{2+} and SO_4^{2-}. In plan 4, it is assumed a 30% reduction only in deposition of Ca^{2+} and Mg^{2+} (no changes in SO_4^{2-} deposition). Plan 5 and plan 6 assume a 30% decrease and a 30% increase respectively in SO_4^{2-} deposition, but keep deposition of Ca^{2+} and Mg^{2+} at the present level. Figure 5 depicts the results for plot E.

If the deposition does not change over the next 50 years (plan 1) soil pH and base saturation in the catchment will continue to decrease, (0.001–0.104 pH units and 0.2–6.6% respectively) in both soil horizons, and R_{CL} will continue to increase according to these model calculations. However, the changes are very slow especially for the upper soil.

The most serious soil acidification is predicted for plan 6 (increase in SO_4^{2-} deposition). However, considerable further acidification is predicted also if only deposition of base cations decreases (plan 4). According to the model, R_{CL} will exceed 1.0 in the deeper soil at plot C and thus reach the assumed harmful level for plans 4 and 6. Some further acidification is also predicted if there are increases in deposition of both base cations and SO_4^{2-} (plan 2).

The model calculations predict only minor changes in soil acidification for plan 3 (30% reduction in deposition in both base cations and SO_4^{2-}). Plan 5 (30% decrease only in SO_4^{2-} deposition) is predicted to result in reduced soil acidification over the next 50 years. Significant increases in pH and BS, and decreases in R_{CL} and concentrations of base cations, SO_4^{2-}, and Al^{3+} are predicted for 2045, compared to the values in 1995, in all cases for the upper soils and in most cases for the deeper soils.

5. DISCUSSION

As shown above, removal of alkaline dust may lead to harmful effects on vegetation due to soil acidification. It is therefore very important to consider the trends in base-cation emissions and not only the sulfur emissions when discussing the future potential scenarios of acid deposition and the possible catchment responses. However, in spite of the possible negative effect of removal of alkaline dust on vegetation, a complete risk assessment may show that there are other, perhaps more important, positive effects, e.g., on human health[27].

The differences in present soil properties and in predicted values are striking within this small catchment. It is difficult to describe the whole catchment using only one plot or a median of several plots. We tried to construct a 'median catchment', but found it difficult to determine meaningful parameter values. This illustrates the problems in obtaining critical loads for larger regions.

The very low concentrations in soil waters at plot C led to difficulties in choosing input parameters. In Fig. 3(a) BS for this plot decreases only slightly from 1895 to 1995. It was possible to get a decline similar to those for the other plots, but only by using very high weathering rates and unrealistically high uptake rates. This illustrates a general problem in using a model like MAGIC if some important input values have not been measured. It is usually possible to get reasonable results by adjusting some values particularly for weathering and uptake.

A further problem is that the model uses only two soil layers. The assignment of actual soil horizons to the two layers used in MAGIC is not obvious. In addition to the divi-

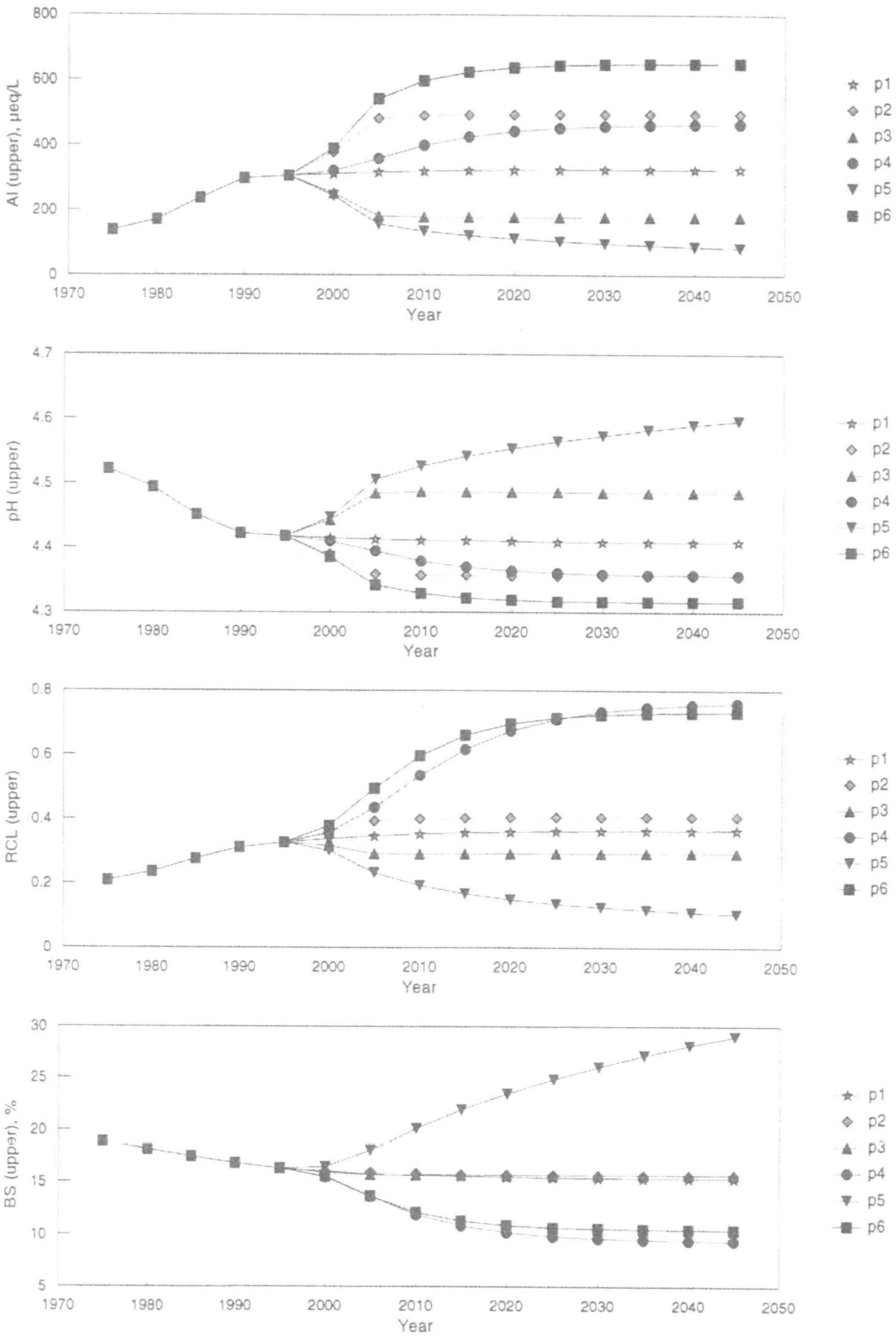

Figure 5. Responses of soils and soil waters for plot E to six assumed deposition scenarios for the period of 1996–2045.

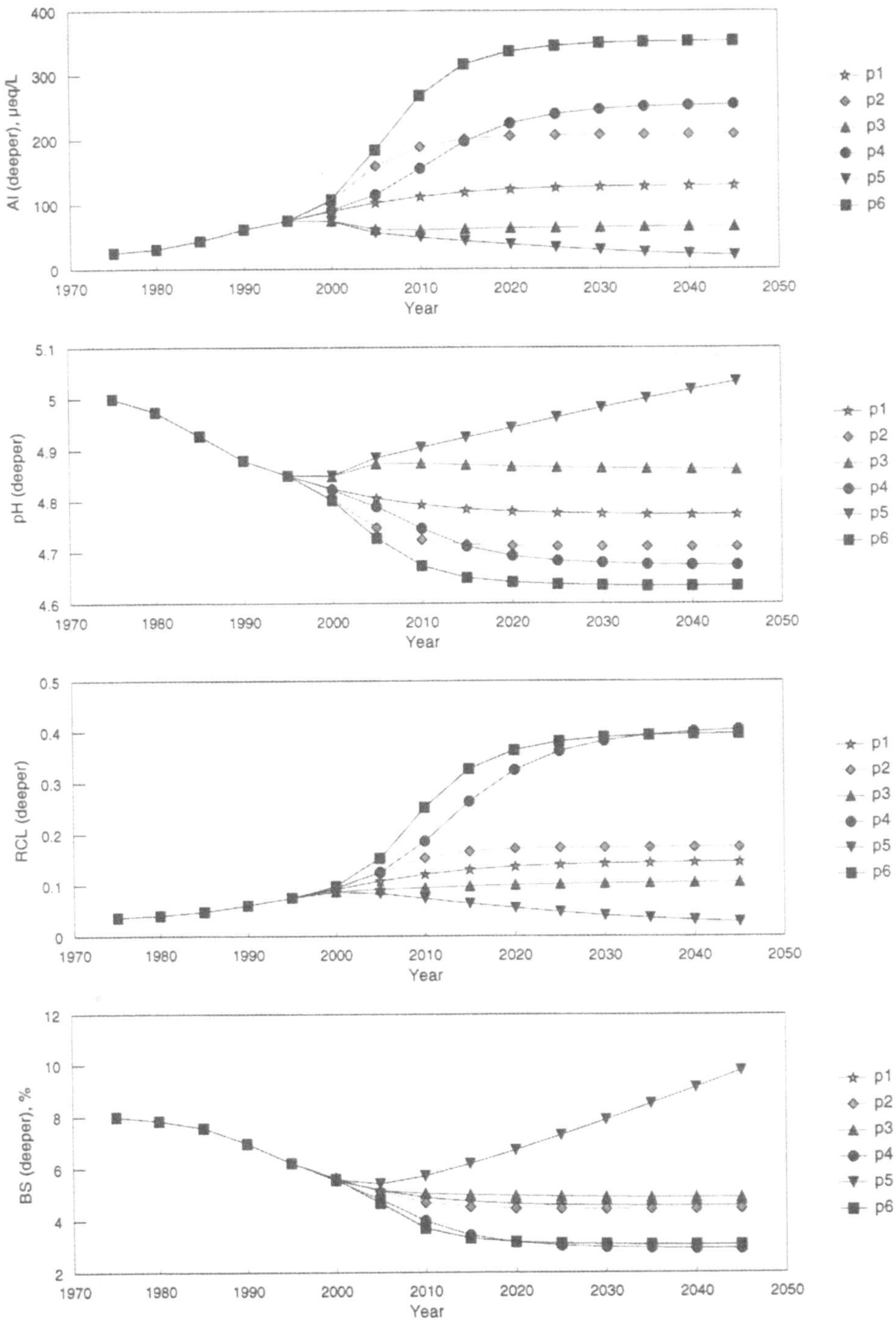

Figure 5. (*Continued*)

sion described in this paper, we tried to do it somewhat differently. This obviously changed the detailed results, but not the main conclusions.

Although the MAGIC model has been developed primarily for studying streamwater acidification, the emphasis in this work is on soil waters. Usually, the model outputs for streamwater are very close to the outputs for the soil waters for the deeper horizons. Since streamwater is a mixture of water from the whole catchment, comparison with modelling results using median parameter values is most reasonable. As mentioned we had difficulties in finding such values and for the parameter sets tried, the agreement for streamwater was not very good, probably mainly because the median parameters do not represent a suitable weighted average corresponding to the mixture of water types in the stream. The conclusion we can draw is that streamwater in the catchment is not likely to be acidified severely in the near future if deposition keeps at the present level, because of large acid neutralizing capacity in the deeper soils.

From the results shown in Table 2, we notice that most of the calculated concentrations are very close to the field observations. The largest deviations are for Al^{3+} and these cannot be remedied by simply changing one parameter. If the other soil variables have been selected, the outputs of Al^{3+} concentrations are determined by the solubility constant for $Al(OH)_3$ and the pH value achieved in the corresponding soil water, which is demonstrated clearly in the sensitivity analysis (Fig. 4). Put in another way, the deviation from an exact charge balance in the observations, which makes a perfect fit impossible, is mainly seen as a difference between observed and calculated Al^{3+} concentrations. An extensive trial and error might have distributed the necessary differences more evenly between the ions.

The aluminium submodel in MAGIC implies the relationship $K_{Al(OH)_3} = [Al^{3+}]/[H^+]^n$, with n equal to 3. Several studies show that the exponent n in field situations is often considerably lower than this value[10,28,29]; this is also the case for the Guiyang catchment[21]. It is therefore likely that MAGIC overestimates the shift in the aluminium concentrations corresponding to a given pH change. A modified version of the MAGIC model, in which the exponent (n) may be adjusted, has been described by Sullivan and Cosby[30]. However, it is, unfortunately, only possible to vary the exponent for streamwater. An approach assuming aluminium equilibrium with soil organic matter as the controlling mechanism for Al activity in soil water has recently been proposed[29,31,32]. It is also possible that equilibrium between Al^{3+} in solution and a solid phase is not always reached under field conditions, complicating the Al-modelling further.

Based on laboratory studies, Liao et al.[8] estimated weathering rates to be 46–103 $meqm^{-2}yr^{-1}$ for total base cations at plot A. This range is actually in better agreement with the values used for the other plots than for plot A. As mentioned earlier, the sulfate concentration in the deeper soil water at plot A is very high and so is the Ca-concentration. It was necessary to introduce S-weathering to obtain a reasonable fit with observations. The soil from the same plot used in the laboratory experiment did not show strong S-weathering and may not be representative for the soil controlling the water chemistry in the field.

The uptake rates used were generally near the upper end of the range mentioned in Section 3, except for plot C where higher values were necessary. This may seem somewhat high, since the highest value given by Cole and Rapp[23], 122 $meqm^{-2}yr^{-1}$, is for deciduous forests. However, considering the higher temperature in the subtropical Guiyang region compared to Europe and the US, we find the values acceptable.

In Fig. 6 we show modelled values for the year 2045 for a series of variables using different dry deposition factors, weathering and $Al(OH)_3$ solubility (cf. Table 4). Only results for plan 1 (no changes in deposition), plan 5 (30% decrease in S-deposition only) and

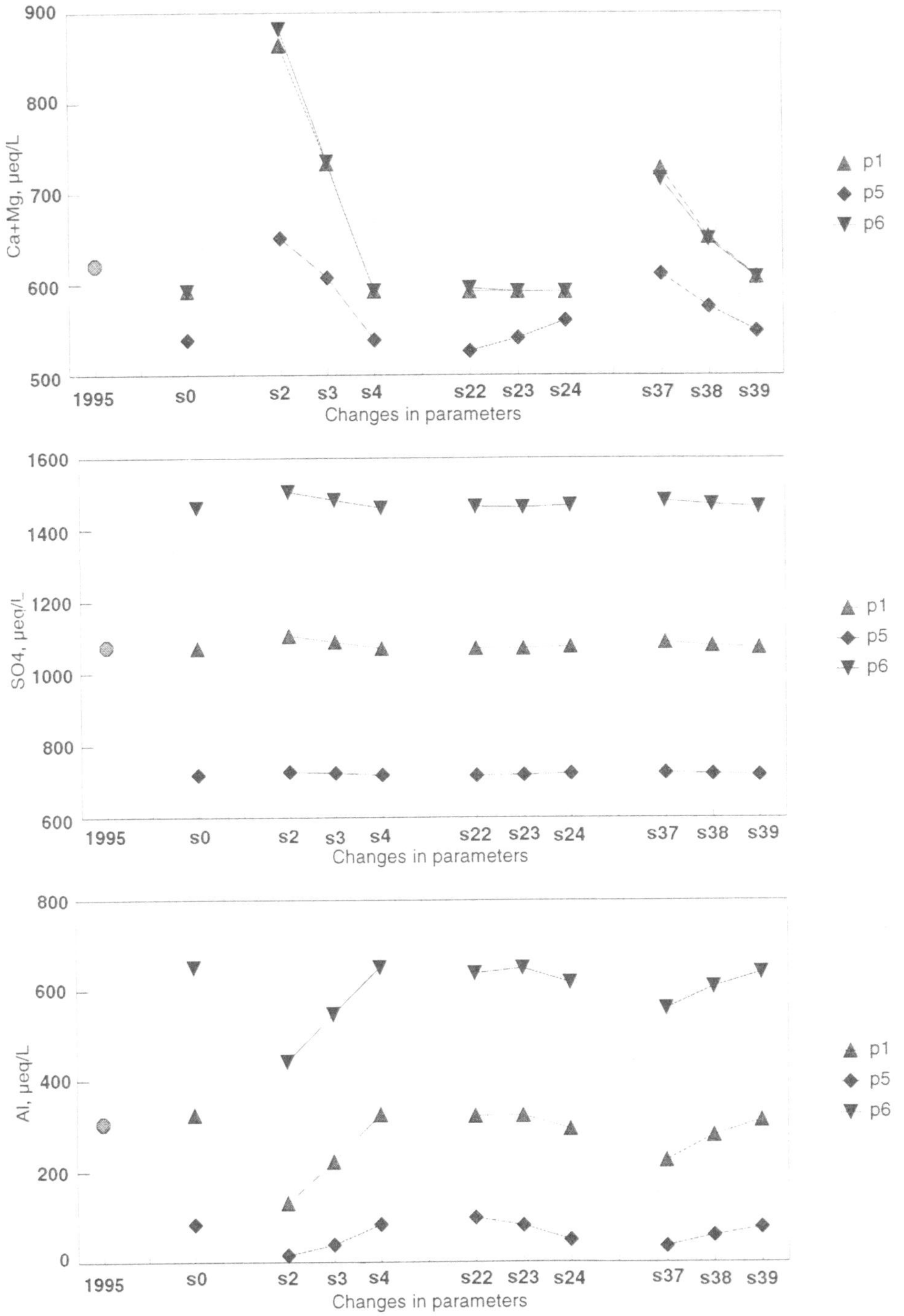

Figure 6. Output values for plot E for 2045 corresponding to the plans 1, 5 and 6 for different choices of dry deposion factors for Ca^{2+} and Mg^{2+} (s2–s4), logK[$Al(OH)_3$] (s22–s24) and base cation weathering (s37–s39). Present values (1995) are given to the left and s0 is the base case.

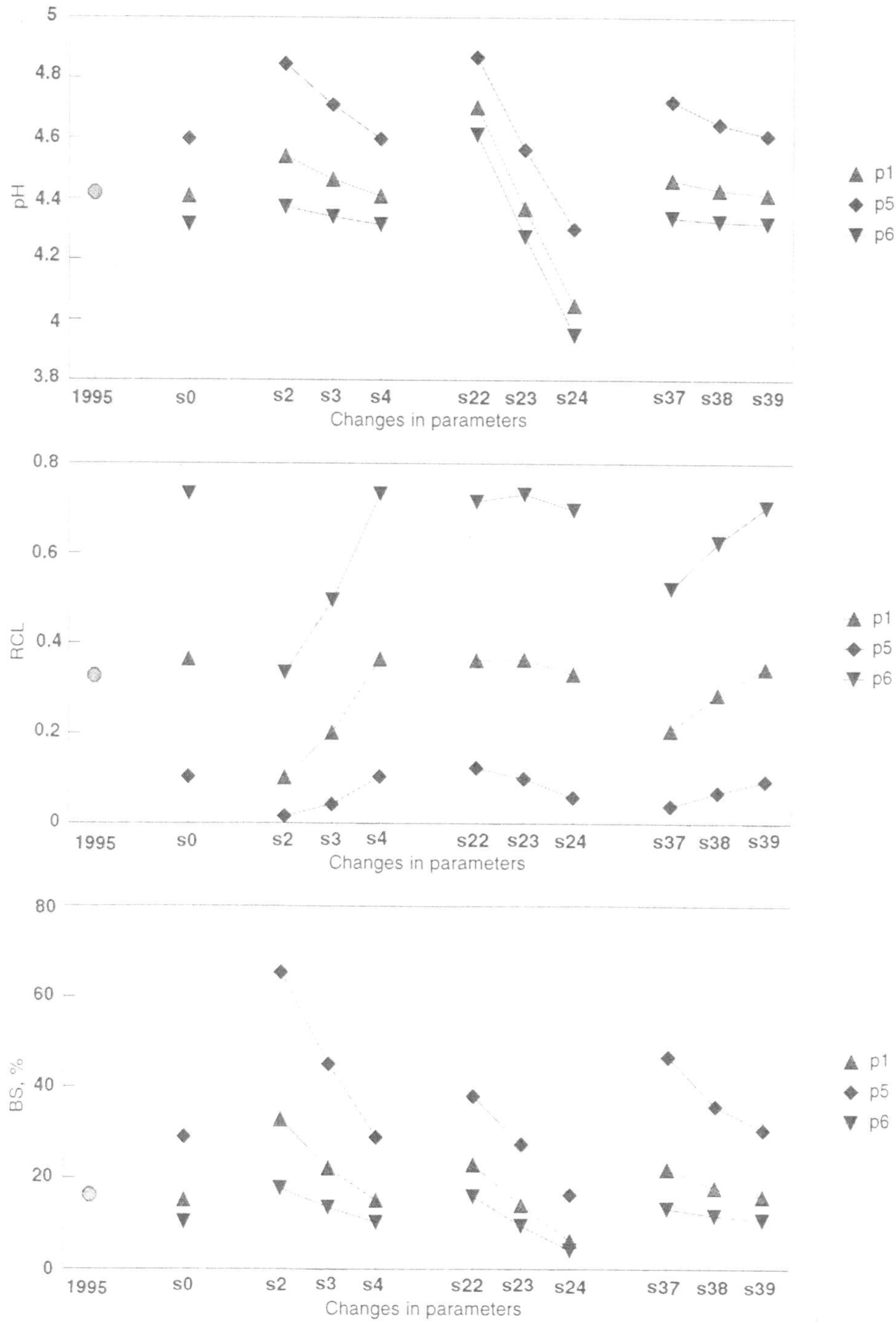

Figure 6. *(Continued)*

plan 6 (30% increase in S-deposition only) are given. The conclusion that plan 6 results in worse conditions seems quite robust looking at parameters as Al^{3+} or R_{CL}. The calculations also support the conclusion that plan 5 leads to improvements. However, the direction of the trends if the present deposition continues, is more uncertain. Our results thus indicate that dramatic changes in soil acidification status are unlikely for this scenario.

Although this modelling work give useful indications of future trends, improvements are needed. When several parameters are adjusted to obtain a reasonable fit for concentrations in soil waters, there is a danger that the results become biased in the direction of subjective opinions. To obtain more reliable predictions, more information on dry deposition factors, weathering and uptake is essential. Furthermore, the Al-submodel should be modified to make it in better agreement with recent studies.

6. CONCLUSIONS

1. Although the MAGIC model was developed originally to estimate the future trends in surface water chemistry, it can also be used to estimate the long-term changes of soils and soil waters in a catchment due to acid deposition, because the most important soil processes are taken into account in the model. However, some improvements, in particular in the Al-submodel, should be considered. Also, the limitation of only two soil layers represents a problem.
2. The model simulations for the Guiyang catchment indicate that soil acidification has occurred and probably is going on especially in the upper soils in this catchment due to long-term acid deposition.
3. The future trend in deposition of base cations is as important as the trend in sulfur deposition. According to the most likely model calculations, soil acidification will continue slowly if deposition continues at the present level. A 30% increase in S-deposition or a 30% decrease in deposition of base cations, seems to lead to deterioration of soils in this catchment. A 30% reduction in S-deposition is likely to result in improved soil conditions.
4. Partly due to variations in topography and vegetation, four different plots within this small catchment give quite different results, and show different sensitivity to acidification. To estimate parameters representative for the whole catchment is a very complicated, if not impossible, task.
5. The model results depend strongly on dry deposition factors, the solubility constant for $Al(OH)_3$, and weathering rates, which are difficult to determine either in the field or in the laboratory. These uncertain results are both because some important parameters chosen for the MAGIC model are uncertain and because of model deficiencies. It is recommended to use the results of field and laboratory studies as far as possible in determining input values and to use comparison between modelled and observed present soil water-concentrations for final adjustments.

ACKNOWLEDGMENT

We thank Professor Jerry L. Schnoor at University of Iowa (U.S.A.) for his helpful comments and suggestions on the manuscript.

REFERENCES

1. Zhao, D., Zhang, X., Yang, J., Mao, J. and Xiong, J., 1995. Critical load of sulfur deposition for ecosystem and its application in China. J. Environ. Sci., **7**: 325–337.
2. Dai, Z., Liu, Y., Wang, X. and Zhao, D., 1997. Changes in pH, CEC and exchangeable acidity of some forest soils in southern China during the last 32–35 years. Submitted to Water Air Soil Pollution.
3. Zhao, D. and Seip, H. M., 1991. Assessing effects of acid deposition in southwestern China using the MAGIC model. Water Air Soil Pollut., **60**:83–97.
4. Xue, H.B. and Schnoor, J.L., 1994. Acid deposition and lake chemistry in southwest China. Water Air Soil Pollut., **75**: 61–78.
5. Wang, R., Yang, J. and Tang, H., 1996. Impacts of acid deposition on ecosystems in China. Research Report, Research Ceter for Eco-Environmental Sciences, Chinese Academy of Sciences, P.O.Box 2871, Beijing 100085, China.
6. Liao, B., Larssen, T., Seip, H.M. and Vogt, R.D., 1994. Anion adsorption and aluminium release from Chinese soils treated with different concentrations of $(NH_4)_2SO_4$ and NH_4NO_3. J. Ecol. Chem., **3**:281–301.
7. Liao, B., Larssen, T. and Seip, H.M., 1997. Response of five Chinese forest soils to acidic inputs: batch experiment. Submitted to Geoderma.
8. Liao, B., Seip, H.M. and Larssen, T., 1997. Response of two Chinese forest soils to acidic inputs: leaching experiment. Geoderma, **75**:53–73.
9. Seip, H.M., Zhao, D., Xiong, J., Zhao, D., Larssen, T., Liao, B. and Vogt, R. D., 1995. Acidic deposition and its effects in southwestern China. Water Air Soil Pollut., **85**:2301–2306.
10. Stone, A. and Seip, H.M., 1990. Are mathematical models useful for understanding water acidification? Science of the Total Environment, **96**:159–174.
11. Sullivan, T.J., 1997. Ecosystem manipulation experimentation as a means of testing a biogeochemical model. Environmental Management, **21**:15–21.
12. Xie, S., Hao, J., Zhou, Z., Qi, L. and Yin, H., 1995. Assessment of critical loads in Liuzhou, China, using static and dynamic models. Water Air Soil Pollut., **85**: 2401–2406.
13. Cosby, B.J., Hornberger, G.M., Galloway, J.N. and Wright, R.F., 1985. Modeling the effects of acid deposition: assessment of a lumped-parameter model of soil water and streamwater chemistry. Water Resour. Res., **21**:51–63.
14. Cosby, B.J., Wright, R.F., Hornberger, G.M. and Galloway, J.N., 1985. Modeling the effects of acid deposition: estimation of long-term water quality responses in a small forested catchment. Water Resour. Res., **21**:1591–1601.
15. Cosby, B.J., Hornberger, G.M., Rastetter, E.B., Galloway, J.N. and Wright, R.F., 1986. Estimating catchment water quality response to acid deposition using mathematical models of soil ion exchange processes. Geoderma, **38**:77–95.
16. Neal, C., Whitehead, P., Neale, R. and Cosby, J., 1986. Modelling the effects of acidic deposition and conifer afforestation on stream acidity in the British uplands. J. Hydrol., **86**:15–26.
17. Hornberger, G.M., Cosby, B.J. and Wright, R.F., 1989. Historical reconstructions and future forecasts of reginal surface water acidification in southernmost Norway. Water Resour. Res., **25**:2009–2018.
18. Wright, R.F., Cosby, B.J. and Hornberger, G.M., 1991. A regional model of lake acidification in southernmost Norway. Ambio, **20**:222–225.
19. FAO, 1978. FAO/UNESCO Soil map of the world, 1:5 000 000. UNESCO, Paris.
20. FAO, 1994. FAO/UNESCO Soil map of the world, Revised legend, with corrections. ISRIC Technical Paper 20, Wageningen.
21. Larssen, T., Xiong, J., Vogt, R.D., Seip, H.M., Liao, B. and Zhao, D., 1997. Studies of soils, soil water and stream water at a small catchment near Guiyang, China. Water Air Soil Pollut., in press.
22. Jenkins, A. and Cosby, B.J., 1989. Modelling surface water acidification using one and two soil layers and simple flow routing. In: Regional Acidification Models, J. Kämäri, D. Brakke, A. Jenkins, S. Norton, and R. Wright (eds.). Springer-Verlag, Heidelberg. pp. 253–266.
23. Cole, D.W. and Rapp, M., 1981. Elemental cycling in forest ecosystems. In: Dynamic properties of forest ecosystems, D.E. Reichle (ed.). Cambridge University Press, Cambridge. pp.341–409.
24. De Vries, W., 1993. Average critical loads for nitrogen and sulfur and its use in acidification abatement policy in the Netherlands. Water Air Soil Pollut., **68**:399–434.
25. Prietzel, J. and Feger, K.H., 1992. Dynamics of aqueous aluminium species in podzol affected by experimental $MgSO_4$ and $(NH_4)_2SO_4$ treatments. Water Air Soil Pollut., **65**:153–173.
26. Sverdrup, H. and de Vries, W., 1994. Calculating critical loads for acidity with the simple mass balance method. Water Air Soil Pollut., **72**:143–162.

27. Bown, W., 1994. Dying from too much dust. New Scientist, **141**:12–13.
28. Reuss, J.O., Walthall, P.M., Roswall, E.C. and Hopper, R.W.E., 1990. Aluminium solubility, calcium-aluminium exchange, and pH in acid forest soils. Soil Sci. Soc. Am. J., **54**:374–380.
29. Berggren, D. and Mulder, J., 1995. The role of organic matter in controlling aluminium solubility in acidic mineral soil horizons. Geochim. Cosmochim. Acta., **59**:4167–4180.
30. Sullivan, T.J. and Cosby, B.J., 1995. MAGIC model applications for surface and soil waters as input to the Tracking and Analysis Framework (TAF). Model Documentation. Report for U.S. Department of Energy, DOE/ER/30196–7. E&S Environmental Chemistry, Inc., Corvallis, Oregon.
31. Tipping, E., Berggren, D., Mulder, J. and Woof, C., 1995. Modelling the solid-solution distributions of protons, aluminium, base cations and humic substances in acid soils. Europ. J. Soil Sci., **46**:77–94.
32. Wesselink, L.G., van Breemen, N., Mulder, J. and Janssen, P.H., 1996. A simple model of soil organic matter complexation to predict the solubility of aluminium in acid forest soils. Europ. J. Soil Sci., **47**:373–384.

28

THE RELATIVE IMPORTANCE OF ALUMINUM SOLID-PHASE COMPONENT IN AGRICULTURAL SOILS TREATED WITH OXALIC AND SULFURIC ACIDS

Xiao Ping Zhu,[1] Marek Kotowski,[2] and Lucjan Pawłowski[2]

[1]Institute of Soil Science
Academia Sinica
Nanjing, 210008, China
[2]Department of Environmental Protection Engineering
Technical University of Lublin
40 Nadbystrzycka Str.
20-618 Lublin, Poland

ABSTRACT

Our ability to predict the toxic amount of labile aluminum in acidic soils is limited by our understanding the interaction between different solid sources of aluminum in soils and soil solution pH. A soil acidification experiment, consisting of five consecutive equilibration of brown soils and peat soil with sulfuric and oxalic acid solutions was presented in this paper to determine the relative contribution of solid-phase aluminum to the labile aluminum in soil solutions.

The distribution of solid-phase aluminum pools in soils showed the predominance of organic-bound aluminum in all the studied soils, whereas the amounts of exchangeable and inorganic amorphous aluminum were relatively small. The reduction of the solid-phase aluminum pools by the simulated leaching with sulfuric or oxalic acids showed that organic-bound, amorphous, and free aluminum forms could act as aluminum sources. Organic-bound aluminum played the predominant role in soil acid neutralization. Generally, organic acid mobilized much more aluminum from mineral aluminum fraction, while inorganic acid prefer to act with organically bound aluminum fraction in mineral soil. Aluminum mobilization is a kinetic process.

Key words: Aluminum mobilization, soil.

Chemistry for the Protection of the Environment 3, edited by Pawłowski *et al.* Plenum Press, New York, 1998

1. INTRODUCTION

Recently, much more attention has been paid to the physical forms of Al in soil and their transformation concerning the kinetic release of Al to soil water. It has been suggested that the competing reactions between Al pool phases may be responsible for regulating aqueous Al activity[1,2]. Theoretically, Al in soils can be physically divided into:

1. crystalline Al (primary and secondary minerals);
2. non-crystalline Al such as amorphous Al oxides and hydroxides;
3. organic bound Al (Al-humus complexes);
4. interlay Al and polymer Al;
5. exchangeable Al (specifically and non-specifically adsorbed Al)[3,4].

There are significant differences in amounts and forms of solid-phase aluminum in soils, depending mostly on age of soil, parent material, climate (amount of rain, temperature) and topography. The complex interactions between the aqueous aluminum and the various solid-phase aluminum pools are a source of much confusion. It is therefore useful to establish the responses in distribution of solid-phase aluminum sources to increased soil acidity. Several studies have considered solid-phase aluminum forms in relation to the potential mobility of aluminum in soils[1,2,5–8]. However, the chemical nature of reactive aluminum in the soil (i.e. easily mobilized aluminum fraction) has not been related to the various pools of Al in soils. Most researchers have applied selective dissolution procedures to identify the solid sources of aluminum.

The aim of this paper was to determine which soil aluminum fraction was primarily responsible for the labile aluminum leached into solution, and to compare the solubility of different aluminum solid phases affected by inorganic or organic acids.

2. MATERIAL AND METHODS

2.1. Soil Characteristic

Brown soils (0–40 cm) were sampled from Elizówka (7 km north of Lublin) long-term agricultural experimental field[9]. This field station was established in 1971 by the Agricultural University of Lublin. Two soil treatments were investigated. One has received no fertilizers since the establishment of the station, and the other is annually fertilized with N, P and K (80–360 kgN/ha, 22–40 kgP/ha, and 100–249 kgK/ha). The soil is developed from loess, and classified as brown loam earth. Cultivated plants were changed from time to time. Winter wheat was grown at time of sampling. A peat soil (20–40 cm) was collected from Ludwin located 30 km northeast of Lublin. The site has wild grass (*Poa palustris*), sedges (*Carex fusca, Carex lasiocarpa*) and some birches (*Betula pubescens*), no agricultural activity has been applied to this soil.

Soil samples were air-dried at room temperature, gently crushed, and passed through a 2-mm sieve to remove plant roots and coarse fragments. Selected chemical-physical properties of the soil used in this study are described in Table 1. The X-ray diffraction analysis of soils revealed the mixed mineral composition with a predominant amount of feldspar and a small amount of illite in the brown soils. Organic matter in the peat soil was the predominant component, followed by feldspar and illite. Total contents of the major oxide components in the parent material are given in Table 2. The chemical composition

Table 1. Selected soil physical-chemical characteristics

Code	Site	Parent material	Depth (cm)	pH (H_2O)	Exch. Al mM/kg.	Organic matter (%)	Illite (%)	Feldspar (%)
F1	Elizówka	Loess	0–20	4.08	1.87	2.3	2.1	95.6
F2	Elizówka	Loess	20–40	4.47	1.97	1.6	2.3	94.2
UF1	Elizówka	Loess	0–20	5.90	0.24	1.8	6.4	89.0
UF2	Elizówka	Loess	20–40	6.00	0.24	1.8	5.3	92.9
P2	Ludwin	Peat	20–40	5.46	1.90	49.4	20.2	30.4

F—fertilized brown soil, UF—unfertilized brown soil, P—peat soil; 1—topsoil, 2—subsoil.

of all the studied soils were characterized by high concentrations of Si and low concentrations of Al and basic cations (Ca, Mg, K, Na).

2.2. Analytical Procedures

Analytical grade reagents and deionized water were used throughout all experiments. Dilute sulfuric acid and oxalic acid were selected because sulfure has been found to be the predominant anion in acid deposition, while oxalic acid is the predominant species of low-weight-molecular organic acids exuded by plant roots and micro-organisms in natural soil system. Both inorganic acid from acid deposition and organic acid from natural resource have very important effects on Al mobilization. Calcium chloride was used to keep ionic strength constant, and minimized the variation of solute concentrations in time.

The soil samples were acidified 40 mL aliquots of 0.2, 5.0 mM sulfuric acid or 5.0 mM oxalic acid at ratio of soil to solution of 1:100 for 24 hours. After that the solutions were separated by centrifuging at a speed of 4000 r/m for 15 min. Next 40 ml aliquot of acidic solution was added to soil samples and the process was repeated 5 times in order to remove all the labile Al from the soils. After these treatments, the soil samples were carefully recovered in order to measure aluminum solid phases together with the un-acidified soil samples.

The Al pools in the both the acidified and un-acidified soil samples were determined by using selective dissolution with 0.1 M sodium pyrophosphate ($Na_4P_2O_7$) for 16 h; acid ammonium oxalate for 4 h at pH 3.0 in the dark; and citrate-dithionite buffer for 16 h. The soils were sequentially extracted because of sample limitation. After each extraction, soils were very carefully washed two times with deionized water in order to remove residual extracting solution. The general relationships among the extractants used and the solid phases of Al believed to be extracted in the soils are summarized in Table 3[10–12]. Labile Al

Table 2. Soil properties and the total contents of major oxide components in the parent material

Code	SiO_2 (%)	Al_2O_3 (%)	Fe_2O_3 (%)	CaO (%)	MgO (%)	Na_2O (%)	K_2O (%)	CEC meq/kg	BS (%)	AIS (%)	HS (%)
F1	82.26	8.40	2.13	1.49	2.86	0.62	0.43	43	69	13	18
F2	82.77	8.48	2.24	1.38	2.97	0.65	0.47	36	71	16	13
UF1	79.96	8.82	2.66	1.29	3.19	0.67	0.69	27	88	3	9
UF2	81.07	8.77	2.80	1.24	3.15	0.73	0.66	24	90	3	7
P2	35.78	3.60	1.48	0.76	1.46	0.23	2.67	138	94	4	2

F—fertilized brown soil, UF—unfertilized brown soil, P—peat soil; 1—topsoil, 2—subsoil.

Table 3. Determination of solid-phase Al pools by selective dissolution procedures[10–12]

Methods used for pool extraction or determination	Description of solid phase Al
1M KCl extraction	Salt-extractable or exchangeable Al
0.1 M Na-pyrophosphate extraction	Al associated with organic carbon
0.1 M acid oxalate extraction	Poorly ordered to amorphous, pedogenic Al
Dithionite-citrate-buffer extraction (DCB)	Crystalline Al

in the extracted solutions was measured by complexation with 8-hydroxyquinoline at pH 8.3, followed by 20 seconds extraction period in methyl isobutyl ketone (MIBK) to estimate labile Al. A spectrophotometer was used to quantify adsorption at a 395 nm wavelength. The adsorption at a 600 nm wavelength was measured to counteract the possible interference of iron[13]. Aluminum in the MIBK extracts was measured by inductively coupled plasma spectrometer.

The computer program AlCHEMI 4 was used to speciate aluminum and to calculate ion activities, and ion activity products. The main species of inorganic Al considered here were complexes with hydroxide, sulfate, and silicate. To evaluate the potential for equilibrium with Al mineral phases, saturation indices (SI) were also calculated by comparing the measured solution activity with solubility product data. The measured solution activity expressed as ion activity products (IAP) was divided by the temperature-corrected Al solubility products (K): SI = log(IAP/K) where the positive, negative and zero SI values indicate that a solution is oversaturated, undersaturated, or in equilibrium, respectively, with reference to solubility of the mineral phases concerned. The Al mineral considered were amorphous gibbsite, natural gibbsite, microcrystalline gibbsite, kaolinite, halloysite, imogolite, smectite, jurbanite ($AlOHSO_4 \cdot 5H_2O$), alunite ($KAl_3(SO_4)_2(OH)_6$) and basaluminite ($Al_4(SO_4)(OH)_{10} \cdot 5H_2O$), as expressed in Table 4.

3. RESULTS AND DISCUSSION

3.1. Distribution of Solid-Phase Aluminum Fractions

Fig. 1 shows the values of Al pools in un-acidified brown soil and peat soil. Both soils exhibited organic-bound as the highest amount of Al pool, and the lowest amount as

Table 4. Thermodynamic data used in Al solubility calculation

Mineral	Reaction	$LogK_{sp}$ (298K)
Gibbsite	$Al(OH)_3 \Leftrightarrow Al^{3+} + 3OH^-$	−33.00
Jurbanite	$Al(OH)SO_4 \Leftrightarrow Al^{3+} + SO_4^{2-} + OH^-$	−17.80
Alunite	$KAl_3(SO_4)_2(OH)_6 \Leftrightarrow 3Al^{3+} + 2SO_4^{2-} + 6OH^- + K^+$	−85.32
Basaluminite	$Al_4(SO_4)(OH)_{10} \Leftrightarrow 4Al^{3+} + SO_4^{2-} + 10OH^-$	−117.50
Imogolite	$Al_2SiO_3(OH)_4 + 6H^+ \Leftrightarrow 2Al^{3+} + 3H_2O + H_4SiO_4$	12.00
Halloysite	$Al_2Si_2O_5(OH)_4 + 7H_2O \Leftrightarrow 2Al(OH)_4^- + 2H_4SiO_4 + 2H^+$	−38.82
Kaolinite	$Al_2Si_2O_5(OH)_4 + 7H_2O \Leftrightarrow 2Al(OH)_4^- + 2H_4SiO_4 + 2H^+$	−36.91
Smectite	$(H,Na,K)_{0.28}Mg_{0.29}Fe(III)_{0.23}Al_{1.58}Si_{3.93}O_{10}(OH)_2 + 10.04H_2O \Leftrightarrow$ $0.28(H,Na,K)^+ + 0.29Mg^{2+} + 0.23Fe^{3+} + 1.58Al(OH)_4^- + 3.93H_4SiO_4$	−34.91

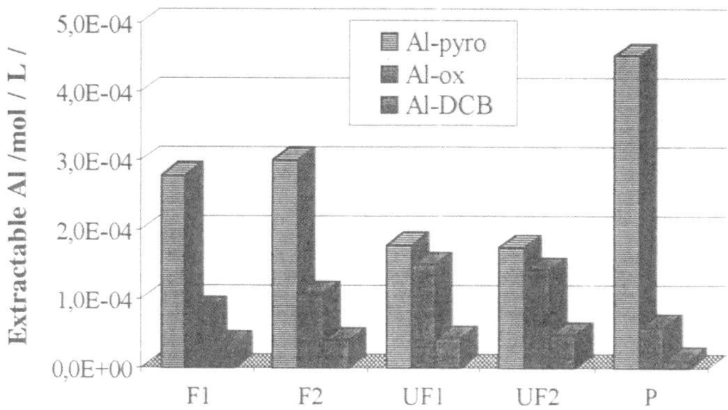

Figure 1. Concentrations of aluminum extracted by sodium pyrophosphate, ammonium oxalate and citrate-dithionite buffer in the brown soil and peat soil. F—fertilized brown soil; UF—unfertilized brown soil; P—peat soil. 1—topsoil, 2—subsoil.

DCB-extractable aluminum. That indicates that most of the extractable aluminum was strongly bound with organic matter in all the soil horizons.

The most noteworthy feature in Fig. 1 is the relative aluminum distribution in brown soil affected by long-term fertilization relative to the reference plot. The fertilized plot was furthermore strongly acidified as indicated by low pH_{H_2O} values (Table 1). The extractable aluminum pool reaches 0.44 mM, which is 22% higher than in the unfertilized plot. Examination of relative distribution of solid aluminum pools (Table 5) showed that organically bound aluminum in (0–20 cm) fertilized plot accounted for 70% of total extractable aluminum. While in the reference plot, organic bound aluminum represented

Table 5. The relative contribution of each solid-phase Al pools to the total released Al (%)

Code	Al-pyro	Al-oxalic	Al-DCB
F1	70	20	10
F2	68	23	9
F1-H-S	81	16	3
F2-L-S	88	08	4
F2-H-S	76	20	4
F2-OX	60	30	10
UF1	50	39	11
UF2	48	38	14
UF1-H-S	54	43	3
UF2-L-S	36	50	14
UF2-H-S	52	41	7
UF2-OX	43	44	13
P2	86	11	3
P2-H-S	90	10	0
P2-OX	89	11	0

F—fertilized brown soil, UF—unfertilized brown soil, P—peat soil; 1—topsoil, 2—subsoil; L-S—0.2 mM H_2SO_4, H-S—5.0 mM H_2SO_4, OX—5.0 mM oxalic acid.

50% of total extractable aluminum (0–20 cm). The amorphous aluminum pool showed the reverse case. The relative distribution of amorphous aluminum from the unfertilized plot was 39%, which was nearly two times as much as that in the fertilized plot. This result indicates that the changes in the size of various aluminum pools after long-term fertilization mainly occur in organically-bound aluminum pool.

Many investigations have suggested that organically bound aluminum in the solid phase is very active and can be the major neutralizer for acid deposition[5,14,15]. Compared to the reference brown soil, the little lower amorphous aluminum pool in the fertilized plot may reflect the transformation of solid-phase aluminum from amorphous aluminum fraction to organically bound aluminum fraction in soils. Although sub-horizon soil has somewhat more labile aluminum, there was no significant difference in relative distribution of solid phase aluminum pools between the horizons within the soil profiles. Due to its high content of organic matter (50%) the peat soil, had about 90% of its extractable aluminum complexed with organic matter while DCB extractable aluminum was eligible.

3.2. Source of the Released Aluminum in Solution

In order to understand the relative importance of the various solid-phase aluminum pools in acid neutralization, the dissolution behavior of the solid-phase aluminum fractions in acidic conditions were examined. The relative amounts of the Al pools in different the acid treated soils were almost the same as in the non-acid treated soil; i.e. higher amounts of organic bound Al than amorphous and crystalline Al were found in all the horizons except the fertilized subsoil horizon (Fig. 2). On an average, the relative distributions of organic-bound Al, amorphous Al and free Al were 44%, 37%, 19% respectively in the acid-pretreated soils.

The degree to which each of the Al pools decreased due to the acid pretreatment varied with the soils as shown in the Table 6. For example, in fertilized brown soil pretreated by 5.0 mM H_2SO_4, the amount of organic-bound Al in the topsoil and subsoil accounted

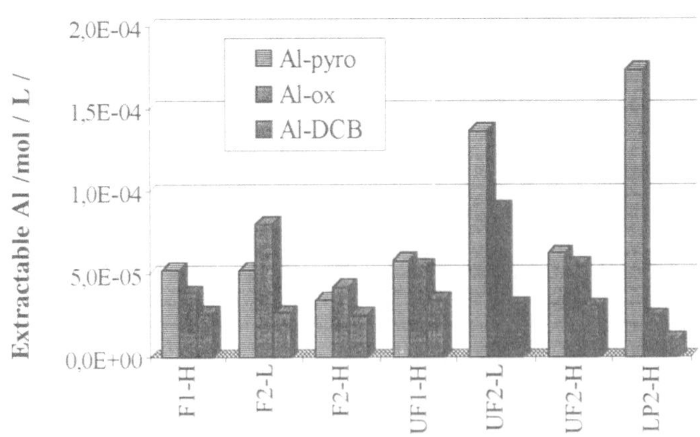

Figure 2. Aluminum concentrations extracted by sodium pyrophosphate, ammonium oxalate and citrate-dithionite buffer in the brown soil and peat soil pretreated by the sulfuric acid. F—fertilized brown soil; UF—unfertilized brown soil; LP—peat soil. 1—topsoil, 2—subsoil; H—5.0 mM H_2SO_4; L—0.2 mM H_2SO_4.

Table 6. Comparison between the released Al amount in solution and the changes of solid-phase Al fractions when treated by inorganic or organic acids

Soil	Al released to solution (Al_{rel})	Changes in solid-phase Al fraction (mol/dm^3)				
		$Al_{Org.}$	$Al_{Amor.}$	DBC-Al	SUM	SUM/$Al_{rel.}$
F1-H-S	5.43E–04	2.24E–04	4.41E–05	5.11E–06	2.74E–04	0.50
F2-H-S	6.44E–04	2.47E–04	6.33E–05	1.33E–05	3.23E–04	0.50
F2-OX	9.60E–04	1.54E–04	7.74E–05	2.32E–05	2.55E–04	0.27
UF1-H-S	5.24E–04	1.19E–04	9.44E–05	6.15E–06	2.19E–04	0.42
UF2-H-S	5.60E–04	1.11E–04	8.63E–05	1.48E–05	2.12E–04	0.38
UF2-OX	9.48E–04	1.09E–04	1.12E–04	3.13E–05	2.53E–04	0.27
P2-H-S	4.19E–04	2.78E–04	3.22E–05	—	3.10E–04	0.74
P2-OX	4.44E–04	3.16E–04	3.76E–05	—	3.53E–04	0.80

F—fertilized brown soil, UF—unfertilized brown soil, P—peat soil;
1—topsoil, 2—subsoil; H-S—5.0 mM H_2SO_4, OX—5.0 mM oxalic acid.

for 19% and 18% of that in the non-acid pretreated topsoil and subsoil, respectively. The loss of organic-bound Al accounted for 80% of total Al release. Whereas the amorphous Al represented 47% and 40% of that in the topsoil and subsoil untreated by acids, respectively. The loss of the amorphous Al made a 16–20% contribution to the total amount of released Al. The "free" Al forms estimated by DCB also showed a decreased trend after acid treatment. It decreased by 16% and 34% in the topsoil and subsoil respectively compared to the counterpart in the acid-untreated soil. The contribution of this free Al release to the total Al was very small, less than 5%. These results indicate that all the forms of Al pools in the brown soils are involved in the Al release process, and contribute to acid neutralization. But the major proportions of liable Al forms are associated with organic matter. In peat soil, the release of organic-bound Al dominated the neutralization of acid deposition. Also we should remember that not all of the Al extracted by pyrophosphate is necessarily associated with organic matter. Some amounts of the Al extracted may have been derived from inorganic sources.

Fig. 2 also shows that although the acid pretreatment of soil with 5.0 mM sulfuric acid removed large amount of aluminum from soil, there were still some amounts of solid-phase Al left. This may indicate that the same Al pools may have different characteristic of Al release. Parts of them are easily released than other parts. For example, Al could be weakly or strongly complexed by organic matter. The weakly organic-bound Al would be easily released at the beginning of acidification. This suggests that acid neutralization by Al mobilization in soil systems is a gradual process, and the duration needed for this process (capacity) is dependent on the soil characteristic.

The amounts of solid phases aluminum removed by the acid treatment depend also on the acid concentrations applied in the pretreatment (Fig. 2). Higher acid deposition removed more aluminum from soil and caused a larger decrease of reactive aluminum pools in the soil. In the subsoil of fertilized and unfertilized brown soil, the concentrations in extracts of organic bound aluminum were 34 μM and 62 μM when pretreated by 5.0 mM H_2SO_4. This amounts to only 65% and 45%, respectively, of what was found after pretreatment with 0.2 mM H_2SO_4. The same trend was found for the changes of amorphous aluminum and free aluminum. Furthermore, after pretreatment with sulfuric acid, the unfertilized brown soil showed relatively higher amounts of organic-bound aluminum, amorphous aluminum and free aluminum than fertilized brown soil. This may indicate that application of large amounts of fertilizer have made soil aluminum more active.

3.3. Role of Inorganic and Organic Acids in Releasing Solid-Phase Aluminum

The comparison of changes in solid-phase Al pools in soils pretreated with sulfuric or oxalic acids shows the different characteristic of solid-phase aluminum solubility. Complexing organic acid, such as oxalic acid, mobilized more Al than inorganic acid. More organically bound Al was therefore removed from the soil samples by the oxalic acid than by the sulfuric acid during the pretreatment soil acidification process (Table 6). As shown in Fig. 3, the amount of organically bound aluminum in the fertilized brown subsoil, pretreated by 5 mM oxalic acid, was nearly 2.8 times greater than that treated by the same concentration of sulfuric acids. Whereas the amount of amorphous and free Al showed the opposite trend., The values of amorphous Al and free Al in the sulfuric treatment were 43 and 25 μM, which was 1.5 and 1.7 times greater than for the oxalic acid treatment. The organically bound aluminum in the subsoil of the fertilized plot accounted for 76% of the acid neutralization when pretreated by 5.0 mM sulfuric acid, while it only reached 60% in the pretreatment of 5.0 mM oxalic acid (Table 5). The unfertilized plot show the same trend. More organic-bound Al was released by sulfuric acid than by oxalic acid. The high amount of amorphous and free Al observed in the sulfuric treatment could be the result of the Al transformation from a organic-bound phase to amorphous Al; i.e. some of the Al released during the pretreatment may precipitate by forming Al amorphous minerals as Al hydroxides or Al hydroxysulfate. To determine if some mineral phase may have been formed during the pretreatment, mineral saturation indices were calculated from ionic activities. All solutions were found to be undersaturated in respect to the Al minerals. Furthermore, Figure 4 shows that the IAP with respect to gibbsite decreased in the sub-horizons of the fertilized plot, unfertilized plot, and peat soils when treated by 5.0 mM H_2SO_4. This implies that the formation of Al hydroxides was impossible. Therefore, the changes in solid-phase aluminum pools in the acid-pretreated soils imply instead differences in the affinity of organic acid and inorganic acid to the solid-phase aluminum pools. Natural-derived (weak organic) acid prefers to react with the Al from inorganic fractions, while anthropogenic (strong mineral) acids mobilize more organic-bound Al. This result has important environmental implication.

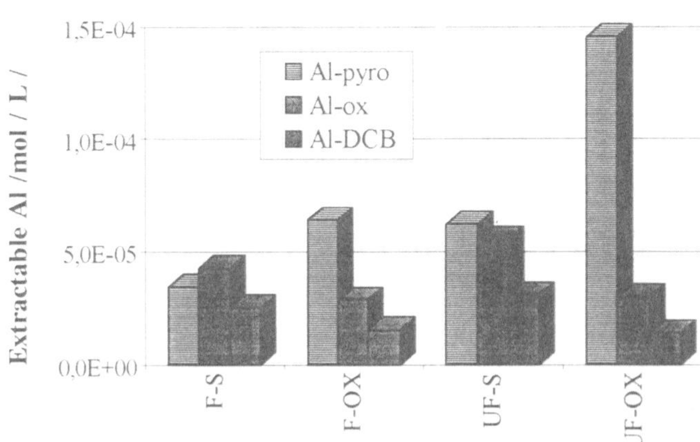

Figure 3. The comparison of solid bound aluminum fractions in brown soils pretreated by the sulfuric acid and oxalic acid. F—fertilized plot; UF—unfertilized plot; S—H_2SO_4; OX—oxalic acid.

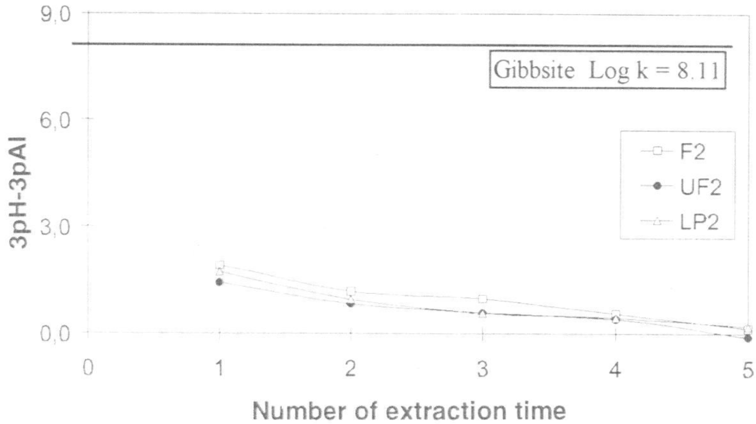

Figure 4. The changes of Al solubility with respect to gibbsite in the subsoil horizon of fertilized plot (F), unfertilized plot (UF) and peat soils (LP).

A budget of the amount of Al released against the amount of Al found in the soil pools show that the depletion of the three Al pools in the soils were significantly less than the amount Al released into the solutions. When the brown soil samples were treated by 5.0 mM sulfuric acid, the total changes of solid-phase aluminum fractions were only 40–50% of amounts of the released aluminum in solution. This indicates that some aluminum is relocated in the soil, from strongly bound mineral phases to more easily reacting phases. In the peat soil, organic matter is the major component. The changes of solid-phase aluminum fraction accounted for 70–80% of the released aluminum in solutions. But when the mineral soil was treated by complexing organic acids, the depletion of aluminum pools was significantly less than the amounts of aluminum released to solution (Table 6). This further confirms that organic acid prefers to act with mineral aluminum fraction.

The large difference between the amount of aluminum released to solution and the decreases of solid-phase aluminum fraction has not been understood very well. It may reflect the kinetic process of aluminum mobilization. Furthermore, according to Merino et al.[16], some of the released Al during the experiment may be adsorbed onto the surface of clay minerals or to organic matter[17]. This would inhibit the sodium pyrophosphate, acid oxalate solution, and dithionite citrate buffer solution to release further Al from the soils. Merino et al.[16] suggested that other mineral forms may act as the source of Al. All these factors indicate that aluminum release is a gradual process in soil. The release of aluminum to the solution under these conditions seems to be restricted by kinetic constraints.

4. CONCLUSION

The distribution of solid-phase aluminum pools in soils shows the predominance of organic-bound aluminum in all the studied soils. Long-term fertilization greatly increased organically bound aluminum fraction. The reduction of the solid-phase aluminum pools observed in all soil horizons when acidified suggested that organic-bound, amorphous, and DCB extractable aluminum forms could act as aluminum sources. Organic-bound aluminum plays the predominant role in soil acid neutralization. Oxalic acid had higher affinity to react with mineral aluminum, while sulfuric acid removed more aluminum from organically bound aluminum fraction.

ACKNOWLEDGMENT

Authors would like to express their gratitude to Dr R.D. Vogt for his providing valuable comments and corrections.

REFERENCES

1. Dahlgren, R.A., Driscoll, C.T., McAvoy, D.C.: Al precipitation and dissolution rates in spodosol Bs horizons in the northeastern USA. *Soil Sci. Soc. Am. J.*, 1989, *55*, 1382–1390.
2. Walker, W.J., Cronan, C.S., Bloom, P.R.: Al solubility in organic soil horizons from northern and southern forested watersheds. *Soil Sci. Soc. Am. J.*, 1990, *54*, 369–374.
3. Tan, K.H.: Degradation of soil minerals by organic acids. In: P.M.Huang and M.Schnitzer (eds) *Interactions of soil minerals with natural organics and microbes*, Spec. pub. 17. SSSA, Madison, WI., 1–28, 1986.
4. Dahlgren, R.A., Ugolini, F.C.: Distribution and characterization of short-range-order minerals in spodosols from the Washington Cascades. *Geoderma*, 1991, *48*, 391–413.
5. Dahlgren, R.A., Walker, W.J.: Al release rates from selected spodosol Bs horizons: Effect of pH and solid-phase Al pools. *Geochim. Cosmochim. Acta.*, 1993, *57*, 57–66.
6. David, M.B., Driscoll, C.T.: Al speciation and equilibrium in soil solutions of a Haplorthod in the Adirondack Mountains. *Geoderma*, 1984, *33*, 297–318.
7. Hoges, S.C.: Al speciation: a comparison of five methods. *Soil Sci. Soc. Am. J.*, 1987, *51*, 57–64.
8. Wright, R.J., Baligar, V.C., Weight, S.F.: Estimation of phytotoxic Al in soil solution using three spectrophotometric methods. *Soil Sci.*, 1987, *144*, 224–232.
9. Zhu, X.P.: Environmental behavior of aluminum chemistry due to acidification. Doctor thesis, Technical University of Lublin.
10. Mckeague, J.A., Day, J.H.: Dithionite and oxalate-extractable Fe and Al as aids in differentiating various classes of soils. *Can. J. Soil Sci.*, 1966, *46*, 13–22.
11. Mckeague, J.A.: An evaluation of 0.1 M pyrophosphate and pyrophosphate-dithionite in comparison with oxalate as extractants of the accumulation products of podzols and some other soils. *Can. J. Soil Sci.*, 1967, *47*, 95–99.
12. Mehra, O.P., Jackson, M.L.: Iron oxide removal from soils and clays by dithionite-citrate system buffered with sodium bicarbonate. Proc. 7th Conf. *Clays and Clay minerals*. Pergamon Press, New York. NY., 1960, 317–327.
13. Sullivan, T.J., Seip, H.M., Muniz, I.P.: A comparison of frequently used methods for the determination of aqueous aluminum. *Inter. J. Environ. Chem.*, 1986, *26*, 61–75.
14. Mulder, J., Breemen, N.V., Eijck, H.C.: Depletion of soil Al by acid deposition and implications for acid neutralization. *Nature*, 1989, *337*, 247–249.
15. Mulder, J., Stein, A.: The solubility of Al in acid forest soils: Long-term changes in acid water: the need for a reappraisal. *Geochim. Cosmochim. Acta*, 1994, *58*, 85–94.
16. Merino, A., Alvarez, E., garciarodeja, E.: Response of some soils of Galicia (NW Spain) to H_2SO_4 acidification. *Water Air Soil Pollut.*, 1994, *74*, 1–2.
17. Furrer, G., Zysset, M., Charlet, L., Schindler, P.W.: Mobilization and fixation of Al in soils. In: E.Merian et al., (eds) *Metal Compounds in Environment and Life (Interaction between chemistry and biology)*, Science and Technology Letters. P O B. 81, Northwood, Middlesex, UK, 1991.

THE ROLE OF ORGANIC MATTER AND ALUMINUM IN ZINC AND COPPER TRANSPORT THROUGH FOREST PODSOL SOIL PROFILES

Marek Kotowski

Department of Environmental Protection Engineering
Technical University of Lublin
40 Nadbystrzycka Str., 20-618 Lublin, Poland

ABSTRACT

Studies of mobilization of zinc and copper from three sites in Poland with moderate, high and very high levels of acid anions in precipitation (Janów Forest, Izerskie Mountains and forest of Pulawy region) are presented in this paper.

It was found that concentrations of heavy metals in podzol soil solutions from upper layers (O, A, E, and B) were highly depended on dissolved organic carbon whereas in B/C and C levels on aluminum concentration. In podzolized ranker loam soils concentrations of Zn and Cu depends mostly on DOC content.

Key words: zinc mobilization, copper mobilization, soil, DOC.

1. INTRODUCTION

Acidification of soil causes leaching of a number of toxic metals such as Al, Zn, Cu or Pb to soil solutions. Particularly sandy soils (e.g. podzol soils), containing low concentration of base cations, where mobilization of Al has become the major acid neutralizing process, seem to be liable to acidification.

Investigations on metal sorption by soil organic matter[1,2] indicate that in acidic soils (pH < 5.5) affinity of heavy metals to humic acids decreases as follows: Hg >> Pb > Cu >> Zn > Co. In mineral layers without any organic substances the mobilization power of metal ions changes as follows Co ~ Zn > Cu > Pb ~ Hg.

Chemistry for the Protection of the Environment 3, edited by Pawłowski *et al.* Plenum Press, New York, 1998

Reddy et al.[3] tested acidification of soils containing pyrite resulting from its oxidation. They found that content of dissolved organic carbon (DOC) decreased and concentration of soluble compounds of Zn and Cu increased together with reduction of the soil solution pH. They calculated activity of Zn^{2+}, Cu^{2+}, and Pb^{2+} ions by means of the GEOCHEM program[4] based on the reaction equilibrium constants. Activity of the mentioned ions was distinctly greater for DOC = 0.

In acid fresh waters heavy metals are mobilized by acid deposition and their transport is controlled by high-molecule weight organics[5].

Results of the mentioned research indicate that pH is the main parameter responsible for heavy metals concentration. The quantity of organic substances dissolved in soil solutions also makes an essential factor.

This paper presents the results from monitoring studies at the three sites in Poland with moderate, high and very high levels of acid anions in precipitation. Data were performed in the years 1994–96. Author attempt to describe mechanisms controlling heavy metals concentrations in acid soil solutions.

2. SITES AND METHODS

A field studies were carried out in the Janów Forest, in the Ciekonek stream catchment (the Izerskie Mountains) and in the forest of Pulawy region.

Mostly litogenic soils of the Izerskie Mountains of gneiss and granite origin, poor of basic materials are highly sensitive to acidification to the great degree. The sampling site, in the Ciekonek stream catchment, is located 7 km from Szklarska Poreba and 5 km from the Czech Republic border. The region is situated at about 770 m a.s.l., the average annual temperature is 6.1°C (about 3°C in January and 14°C in July), and precipitation is about 1200 mm per year.

Pulawy is situated on the right bank of the Vistula River, at the elevation of 160 m a.s.l. The average annual temperature is 7.4°C with about 500 mm of precipitation per year. Surroundings of Pulawy because of their industrial character (factory which manufacture nitrogen fertilizers) are systematically exposed to emission of acid precipitation, ammonia and fertilizers dust. The sampling site was located about 2 km north-east of industrial zone.

The Janów Forest area is an example of moderate polluted region[6–8] (Vogt et al., 1994; Larsen at al. 1996; Kotowski, 1997). Imission of sulfur and nitrogen compounds is the lowest in comparison to other experimental areas. The site is located about 80 km south of Lublin at the elevation of about 200 m a.s.l. The climate is continental with average annual temperature of 7.6°C (January—3.8°C, July—18.4°C). Average precipitation is about 600 mm per year and evapotranspiration about 450 mm.

At the Janów site, soil samples were taken from five profiles however at Ciekonek catchment and Pulawy area three profiles were sampled. The average standard deviations calculated for all of samples, on basis of three replicates were less than 5% in most of cases.

In this paper only heavy metals, iron oxides and organic matter content in the soils is discussed. More detailed informations about monitoring sites, soils and soil waters are given by Kotowski[8]. Soil water samples have been taken from the monitoring sites systematically in the years 1994–1996.

Dissolved organic carbon (DOC), concentration of anions, base cations, heavy metals, total aluminum (Al_{tot}), and total monomeric aluminum (Al_a), as well as pH were determined in the samples of soil waters.

Values of pH were measured with a pH-meter, produced by Orion company, model 231. Dissolved organic carbon was determined by a Shimadzu analyzer, model TOC-5000. Concentrations of metals in the tested solutions were determined by ICP method using model 850B (Hilger Analytical, UK). Anions were analyzed by standard procedures with the application of an ion-exchange chromatograph, Waters Division of Millipore, Action Analyzer 625. Total monomeric aluminum concentration was determined by the Barnes-Driscoll extraction-spectrophotometric method with 8-oxyquinoline[9,10].

3. RESULTS AND DISCUSSION

Soil in all studied horizons have pH_{H2O} < 5.0[8]. Effective cation exchange capacity (CEC_E) of organic levels (for 3 studied sites) is considerably high and ranges from 180.4 to 189.2 meq/kg of dry soil. CEC_E of A, E and B horizons are two times lower for Ciekonek soils (75.1–88.3 meq/kg) and five–ten times lower for Janów and Pulawy soils (14.2–39.1). The other horizons have a very low CEC_E (3.0–6.1 meq/kg).

The mineral cation exchanger is dominated by Al although in horizons A and E aluminum and hydrogen ions play similar role. In organic horizon acidic groups are dominated by base cations, but total content of Al in organic matter is the highest in all the studied soils.

Significant amounts of heavy metals were found only in upper soil horizons. Contents of heavy metals, iron oxides and organic matter in every genetic horizons of the examined soils are presented in Table 1.

In soils of the Janów Forest site, mainly zinc and lead were found. In O horizon their contents are 72.4 and 68.7 mg/kg of soil, respectively, while in A horizon they are 2–3 times lower. Zn and Pb contents decreased with the depth and in C horizon never exceeded level of 2.5 mg/kg. Copper, nickel and chromium in quantities of few mg per kg of soil occur only in organic and alluvial horizons.

Table 1. Content of organic matter, iron and heavy metals in the soils

Sampling site	Organic matter [%]	Fe_2O_3 [%]	Cu mg/kg	Zn mg/kg	Pb mg/kg	Ni mg/kg	Cr mg/kg	Co mg/kg	Cd mg/kg
Janów O	85.8	—	9.1	72.4	68.7	4.7	8.9	0.9	0.62
Janów A	5.1	—	2.8	20.8	32.1	1.3	3.1	0.05	0.15
Janów E	0.43	—	0.05	1.8	2.2	0.05	0.7	0.05	0.03
Janów Bhs	—	2.28	1.4	7.1	6.9	1.1	3.9	0.07	0.05
Janów Bs	—	0.93	0.7	4.8	4.9	1.7	1.9	0.12	<0.01
Janów B/C	—	—	0.05	3.8	3.4	0.9	0.9	<0.05	<0.01
Janów C	—	—	0.05	1.9	1.1	0.5	0.8	<0.05	<0.01
Pulawy O	88.7	—	6.2	83.6	96.4	1.6	4.9	3.6	1.12
Pulawy E	1.16	0.15	4.6	29.4	36.8	0.6	2.6	1.4	0.13
Pulawy B	—	1.27	2.1	14.3	8.6	0.5	2.8	2.4	0.08
Pulawy B/C	—	0.09	0.4	8.9	7.3	0.1	1.3	0.3	0.06
Pulawy C	—	—	0.6	4.3	1.2	0.06	2.4	0.12	0.06
Ciekonek O	91.0	—	18.4	42.0	121	7.6	16.3	2.7	0.55
Ciekon. A/E	6.4	—	16.6	23.2	51.6	4.2	14.6	0.16	0.32
Ciekon. Bhs	2.25	3.14	4.6	11.9	37.2	2.8	7.8	1.2	0.11
Ciekon. B	1.29	0.23	3.1	14.6	39.4	3.3	9.8	1.4	0.09

Higher quantities of Zn and Pb were found in soils from Pulawy region. In O horizon contents of these metals are 83.6 and 96.4 mg/kg, respectively, and they are decrease with the depth to the level of 4.3 and 1.2 mg/kg in C horizon. Copper in significant amounts (2.1–6.2 mg/kg) was found only in upper levels (O, E and B).

In the soil of Ciekonek catchment high Pb content (average 121 mg/kg) was determined in the organic horizon. The horizon contains also Zn, Cu and Cr at concentrations of 42.0, 18.4, and 16.3 mg/kg, respectively. Pb content sharply decreases with depth, but in B horizon is relatively high and equals 39.4 mg/kg. Concentrations of Zn diminish to the level of about 15 mg/kg in B horizon. The other heavy metals concentrations in deeper horizons were on the level of few mg/kg.

High concentrations of heavy metals in upper soil horizons (O, A or E) are found to be related to the high content of organic matter. The metals distribution within the soil profile seems to be depended mostly on processes of anthropogenic deposition. Although transport of heavy metals from lower soil levels by tree roots, and then their re-deposition with plants decay onto the soil surface can also play significant role.

Content of heavy metals in Polish forest soils were determined by Andersen at al.[11]. They sampled the organic and mineral soil horizons at 14 locations in Poland including Janów Forest, Pulawy area and Izerskie Mountains. Data from the Janów Forest are in good agreement with these found by author. Contents of heavy metals in the soil from Pulawy area found by Andersen at al. are much lower, probably because their samples were taken from the area situated much further from the factory. Results marked by Anderson at al. as Izerskie, differ mostly at organic horizon, they found lower Zn and much higher Pb and Ni contents.

In general founded levels represent normal heavy metal contents in the soils, and correspond well with those from Kabata-Pendias and Dudka[12], but some authors[13,14] have reported higher heavy metal concentrations for Polish soils.

In natural conditions pH value of soil solutions increases along with increasing depth in all studied soils[8]. The acidification degree of upper soil layers (O and A) in Janów Forest reaches average pH values of 3.78 and 3.82; in the E level the mean pH value is 4.01, and in the B, B/C and C horizons and groundwaters ranges 4.35–4.50. In Pulawy region the pH values in soil waters are similar to that in Janów Forest. In Ciekonek catchment soil waters at all horizons are more acidic and in organic horizon pH goes to 3.67 while in E horizon to 3.84.

Correlations between pH values, concentrations of cations, anions, organic carbon and various aluminum forms and concentrations of heavy metals as Zn and Cu were searched. Linear correlation between concentrations of the examined heavy metals and dissolved organic carbon in soil solutions of O, A, E, Bhs levels and aluminum concentration in the C horizon were found. STATISTICA 4.0 program was performed for linear regression analysis.

In the upper soil horizons (O, A, and E) at the Janów Forest and Pulawy site containing organic matter even in small quantities (E), concentrations of heavy metals linearly decrease with the DOC increase. In the organic horizon at Janów Forest site Zn concentration is reduced from 16.7 to 3.1 μM, with DOC increase from 13.8 to 38.2 mg/L and in the alluvial level, zinc concentration changes within the range of 18.4–2.5 μM with changes of DOC from 4.7 to 33.6 mg/L (Fig. 1). The widest range of copper concentration changes from 0.79 to 0.08 μM with DOC increase from 7.42 to 36.4 mg/L was found in the elluvial horizon from Pulawy site (Fig. 2).

In the B horizons of Janów and Pulawy soils containing the significant quantities of iron compounds the correlation between Zn and Cu concentrations, and the DOC content

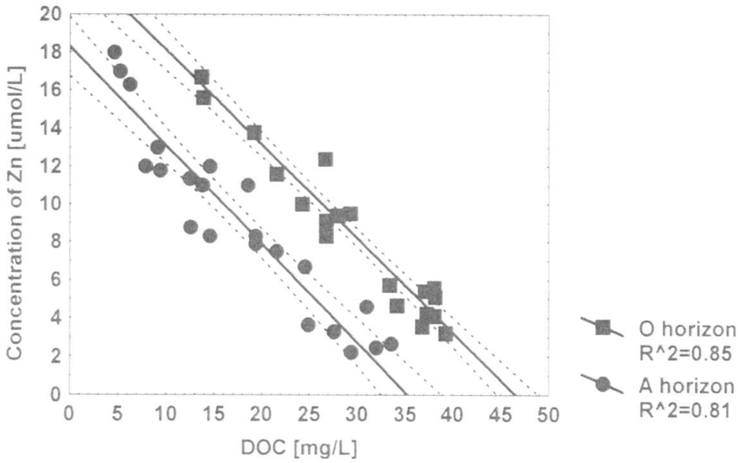

Figure 1. Concentration of Zn versus DOC concentration in soil solutions of organic (O) and alluvial (A) horizons at Janów site.

in soil solution is also linear, but unexpectedly increasing one (Figs 3 and 4). DOC values observed in these genetic levels do not exceed 14 mg/L and concentrations of zinc and copper fluctuate within the ranges of 0.63–11.6 μM and 0–4.86 μM, respectively.

At the B/C and C levels DOC can be found only in insignificant concentrations (up to 4.3 mg/L) and no correlation between Zn and Cu concentrations and DOC was found. All the parameters mentioned above were examined and a distinct correlation between concentrations of total monomeric aluminum (Al_a) and concentrations of heavy metals was found. Increase of Zn and Cu concentrations is clearly correlated with increase of Al_a (Fig. 5). In the soil solutions of the C genetic horizon, aluminum concentration fluctuates within the range of 78.9–346.9 μM, Zn concentration: 0.46–3.28 μM., and Cu concentration: 0.06–0.41 μM.

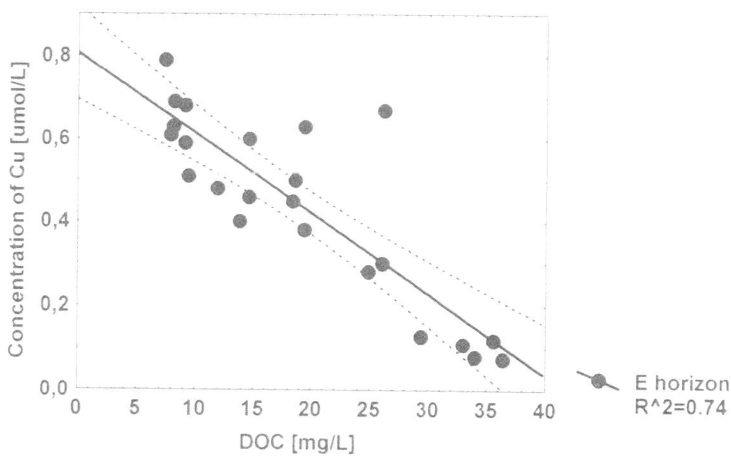

Figure 2. Concentration of Cu versus DOC concentration in soil solutions of elluvial (E) horizon at Pulawy area.

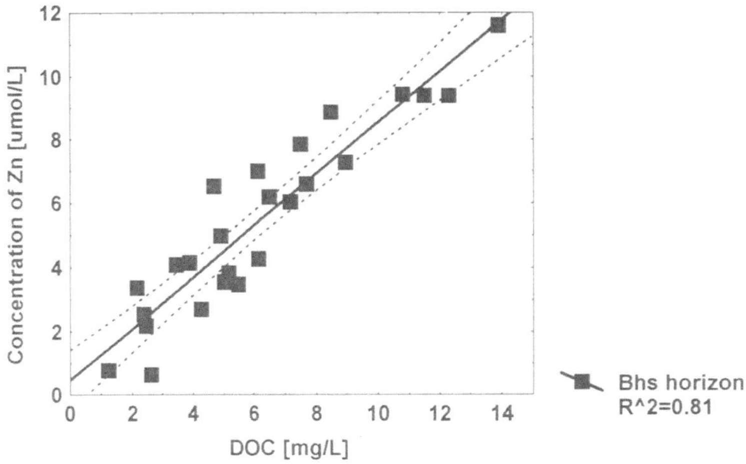

Figure 3. Concentration of Zn versus DOC concentration in soil solutions of illuvial (Bhs) horizon at Janów site.

Groundwater of the Janów Forest soils contain minimal quantities of DOC (up to 3.3 mg/L), significant concentrations of aluminum, and some quantities of Zn and Cu. However, no positive correlation between those quantities has been found.

At the Ciekonek catchment all of the studied soil horizons (O, E, B and B/C) contain organic matter and for all of them concentrations of heavy metals decrease with the DOC increase. Concentrations of zinc and copper are similar to that found in Janów and Pulawy sites. In O horizon zinc concentration decreases from 19.8 to 6.2 μM, with DOC increase from 18.4 to 42.6 mg/L and in the B horizon, zinc concentration changes within a range of 28.4–5.86 μM with changes of DOC from 6.8 to 17.3 mg/L (Fig. 6). Copper concentration ranges from 1.37 to 0.11 μM with DOC increase in the O horizon.

In acidified soils, the affinity of zinc ions to humic acids is very low[1] and their mobilization power is higher in comparison with that of Cu and Pb, what might explain rela-

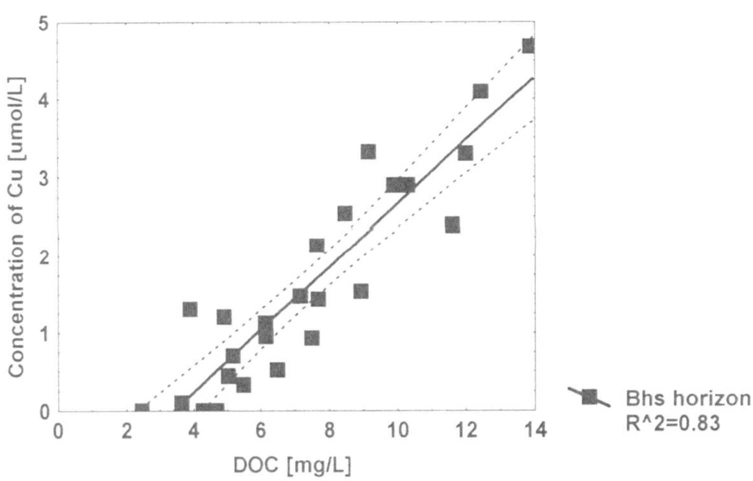

Figure 4. Concentration of Cu versus DOC concentration in soil solutions of illuvial (Bhs) horizon at Janów site.

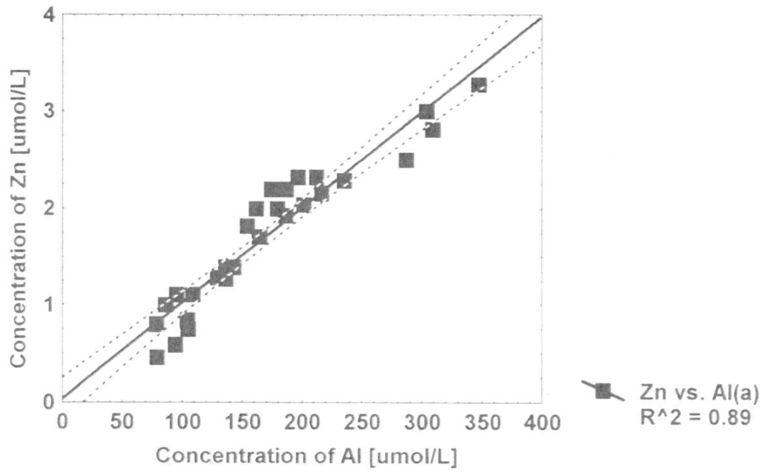

Figure 5. Concentration of Zn versus Al concentration in soil solutions of C horizon at Pulawy area.

tively high zinc concentrations in soil solutions of organic and alluvial layers. Affinity of copper to humic acids in acidic soils is much higher than in the case of zinc[1] and that is why copper concentration in soil solutions increases minimally. Decrease of heavy metals concentrations occurring along with the increase of dissolved organic carbon is probably related with the quality of the substances forming DOC. In the upper soil layers organic acids form complexes which might be sorbed on the soil organic matter.

It is possible that the B horizons of sandy soils are reached mainly by those organic acids that did not get sorbed by humic acids of the O, A, and E layers and form with Zn and Cu well soluble complex compounds. The content of heavy metals in the solid phase of B/C and C horizons was minimal and appears that zinc and copper found in soil solutions of this layers came from the higher levels as a result of ion exchange reaction with Al ions. Such a mechanism of the release of heavy metals and aluminum can be confirmed

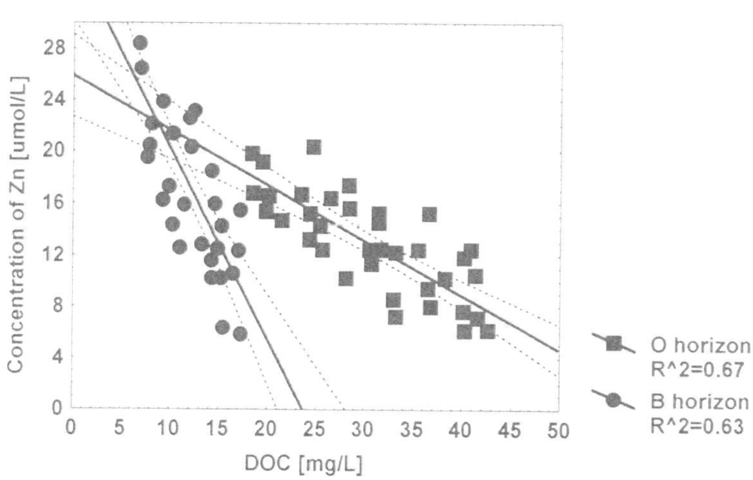

Figure 6. Concentration of Zn versus DOC concentration in soil solutions of O and B horizons at Ciekonek catchment.

by the results of Kotowski et al.[15] It has been established that 70–80% of aluminum present in podzol groundwater come from upper soil layers, mainly from B level.

4. CONCLUSIONS

Achieved results allowed to conclude:

- Concentrations of Zn and Cu in O, A, E and B podzol soil horizons are controlled by content of dissolved organic carbon. In the levels containing organic substances (O, A, E) concentrations of zinc and copper in soil solutions linearly decrease along with the increase of DOC and depend directly on the affinity of organic metal complexes to humic matter. In the B horizons concentrations of zinc and copper grow linearly with increase of DOC leached from higher soil levels.
- In the B/C and C podzol soils horizons, concentrations of heavy metals are proportional to the concentration of aluminum migrations from upper soil layers, mainly from the B horizons.
- In podzolized ranker loam soils containing organic matter in significant quantities in all horizons concentrations of Zn and Cu depends mostly on DOC content and decrease linearly with the DOC increase.

More advanced conclusion require further research.

ACKNOWLEDGMENT

Author would like to express their gratitude to Prof. H.M. Seip for helpful comments, and to Dr M.R. Dudzinska for reviewing manuscript.

REFERENCES

1. Kabata-Pendias, A.: Behavioral properties of trace metals in soils. *Applied Geochemistry*, 1993, *2*, 3–9.
2. Kabata-Pendias, A., Adriano, D.C.: Trace metals. In: Rechcigl, J.E., (ed.) *Soil amendments and environmental quality.* Lewis Publishers, 1995, 139–167.
3. Reddy, K.J., Wang, L., Gloss, S.P.: Solubility and mobility of copper, zinc and lead in acidic environments. *Plant and Soil*, 1995, *171*, 53–58.
4. Sposito G., Mattigod S.V.: GEOCHEM: A computer program for the calculation of chemical equilibria in soil solutions and other natural water systems. The Kearney Foundation of Soil Science, University of California, Riverside, California: USA, 1980.
5. Vesely, J.: Effects of acidification on trace metal transport in fresh waters. In: Steinberg, C.E.W. and Wright, R.F. (eds.) *Acidification of freshwater ecosystems: implications for future.* John Wiley & Sons Ltd., 1994.
6. Vogt, R.D., Godzik, S., Kotowski, M., Niklinska, M., Pawlowski, L., Seip, H.M., Sienkiewicz, J., Skotte, G., Staszewski, T., Szarek, G., Tyszka, J., Aagard, P.: Soil, soil water and stream water chemistry at some Polish sites with varying acid deposition. *Journal of Ecological Chemistry*, 1994, *3*, 325–356.
7. Larssen, T., Vogt, R.D., Seip, H.M.: A comparison of soil- and water chemistry in a cachment in China with sites in Poland and Norway. In: Pawlowski, L. at al. (eds.) *Chemistry for the protection of the environment, 2.* Plenum Press, 1996, 421–434.
8. Kotowski, M.: Soil and soil water chemistry at some Polish sites with acid podzol soils, 1998, This volume.

9. Barnes, R.B.: The determination of specific forms of aluminum in natural water. *Chem. Geol.*, 1975, *15*, 177–191.
10. Driscoll, C.T.: A procedure for the fractionation of aqueous aluminum in dilute acidic water. *Inter. J. Environ. Anal. Chemistry*, 1984, *16*, 267–283.
11. Andersen, S., Odegard, S., Vogt, R.D., Seip, H.M.: Background levels of heavy metals in Polish forest soils. *Ecol. Eng.*, 1994, *3*, 245–253.
12. Kabata-Pendias, A., Dudka, S.: Baseline data for cadmium and lead in soils and some cereals in Poland. *Water Air Soil Pollut.*, 1991, *57–58*, 723–731.
13. Pawlowski, L.: Chemical threat to the environment in Poland. *Sci. Total Environ.*, 1990, *96*, 1–21.
14. Kucharski, R., Marchwinska, E., Gzyl, J.: Agricultural policy in polluted areas. *Ecol. Eng.*, 1994, *3*, 299–312.
15. Kotowski, M., Pawlowski, L., Seip, H.M., Vogt, R.D.: Mobilization of aluminium in soil columns exposed to acids or salt solutions. *Ecol. Eng.*, 1994, *3*, 279–290.

30

ALUMINUM MOBILIZATION BY SULFURIC AND NITRIC ACIDS FROM SOME POLISH SOILS

Xiao Ping Zhu,[1] Marek Kotowski,[2] and Lucjan Pawłowski[2]

[1]Institute of Soil Science
Academia Sinica, Nanjing, 210008, China
[2]Department of Environmental Protection Engineering
Technical University of Lublin
40 Nadbystrzycka Str., 20-618 Lublin, Poland

ABSTRACT

Study was conducted to determine the influence of sulfuric and nitric acids on the Al solubility for brown, black, peat and calcareous soils. The results show that the amounts of labile aluminum mobilized depend significantly on the soil types and horizons. Much more labile aluminum is mobilized in the brown soil than in the black, peat and calcareous soils. The greatest difference in labile aluminum mobilization is between agriculturally managed soils and unfertilized soils. It is apparent that long-term ammonium fertilization significantly increase soil aluminum solubility. The mobilization of labile aluminum in the soil profile increases also with soil depth.

The trivalent aqueous aluminum activity relative to the hydrogen ion activity does not comply with a gibbsite or gibbsite-like mineral solubility control.

In most cases, aluminum mobilization by nitric acid is higher than by sulfuric acid. Nitric acid mobilized less aluminum than sulfuric acid only at high acid concentrations (> 5.0 mM) and in soils with high organic matter content.

Key words: aluminum mobilization, soil, aluminum speciation.

1. INTRODUCTION

The increased concentration of inorganic aluminum in soil solution is probably the most important effect of acid deposition in recent forest decline in Europe[1,2]. Therefore, consider-

Chemistry for the Protection of the Environment 3, edited by Pawłowski *et al.* Plenum Press, New York, 1998

able efforts have been made in understanding the effect of sulfates on aluminum mobilization mechanisms in forest soils. Few comparisons have been made between chemical reactions of HNO_3 and H_2SO_4 in soils. Since the acidification models are widely used to predict effects of acid deposition, it is of prime importance to improve our understanding of interactions the different components of acid deposition have with soils of various landscape covers. Changes in land use or vegetation type may lead to soil acidification with subsequent aluminum mobilization[3]. Furthermore, for a proper evaluation of the effects of elevated aluminum concentrations, it must be considered that aluminum in soil solutions exists as various species of different phytotoxicity. Reactions of aluminum with naturally-occurring organic complexing agents such as oxalic and citric acids are very important factor in detoxifying aluminum. It is necessary to quantitatively evaluate the distribution of aluminum species in soil solution influenced by inorganic and organic acids.

Although the understanding of aluminum release in acid soils has been improved over the last decades, most of our knowledge is based on forest soil. These soils, however, differ considerably from agriculture soils. Certain agricultural activity has also resulted in an intense aluminum mobilization in forest soils[4]. In a laboratory study, Kotowski et al.[5] reported high aluminum concentration by extraction with fertilizer salts (NH_4NO_3 and $(NH_4)_2SO_4$) from forest sandy soil. Few studies have quantified the variability of aluminum mobilization characteristics impacted by acid deposition in both forest and agricultural soils.

The aim of this report is to examine the responses of aluminum chemistry to the two prevailing strong inorganic acids (H_2SO_4, HNO_3) from agricultural and forest soils, and to determine the mineral phase controlling soil Al solubility using a solubility product-equilibrium approach.

2. METHODS AND MATERIALS

Six soil profiles were sampled from four different soils in Lublin, Poland. The characteristics concerning parent material, soil type, vegetation, and annual precipitation at the sampling sites are gathered in Table 1. Soil samples were air-dried at room temperature, and passed through a 2-mm sieve. Subsoil and deep soil of calcareous soil contained 14.5% and 26.6% of calcite chips respectively, which were removed before sieving. Selected physico-chemical properties of the soil used in this study are presented in Table 2. The pH_{H_2O} values in the topsoil vary from 4.08 in fertilized plot to 7.45 in the calcareous soils. Below the root zone, in the deep soil, pH is higher and differs from 5.60 to 8.08. The X-ray diffraction analysis of soils reveals a mixed mineral composition with a predominant amounts of the primary mineral feldspar and a small amounts of illite in all mineral soils. The calcareous soil contain also less than 3.3% of calcium carbonate. Organic matter in the peat soil is the predominant component, followed by illite and feldspar. Total contents of the major oxide components in the mineral soil material are given in Table 3. The chemical composition of all the studied soils is characterized by a high amounts of Si and low amounts of Al and very little basic cations (Ca, Mg, K, Na).

Batch extraction studies were conducted on different soil to determine the solubility of Al as a function of pH. The acidity of solutions was adjusted by sulfuric or nitric acids. Air dried and sieved soil samples of 0.400 g were added to 50-ml polypropylene centrifuge tubes containing 40 ml aliquot of acidic solutions, and shaken at room temperature for 24 hours. The suspension was then centrifuged at a speed of 4000 r/m for 15 min. Labile Al in the centrifuge was measured by complexation with 8-hydroxyquinoline at pH = 8.3, followed by rapid extraction in methyl-isobutyl ketone.[6]

Table 1. Characteristic of selected soil profiles

Code	Site	Depth (cm)	Precipitation (mm)	Parent material	Vegetation	Soil classification	Land use type
F	Elizówka	0–20	566	Loess	Wheat	Brown earth	Arable
	Elizówka	20–40	566	Loess	Wheat	Brown earth	Arable
	Elizówka	40–60	566	Loess	Wheat	Brown earth	Arable
UF	Elizówka	0–20	566	Loess	Wheat	Brown earth	Arable
	Elizówka	20–40	566	Loess	Wheat	Brown earth	Arable
	Elizówka	40–60	566	Loess	Wheat	Brown earth	Arable
LF	Dolhobczów	0–20	580	Loess	Birch	Black soil	Forest
	Dolhobczów	20–40	580	Loess	Birch	Black soil	Forest
	Dolhobczów	40–60	580	Loess	Birch	Black soil	Forest
LA	Dolhobczów	0–20	580	Loess	Wheat	Black soil	Arable
	Dolhobczów	20–40	580	Loess	Wheat	Black soil	Arable
	Dolhobczów	40–60	580	Loess	Wheat	Black soil	Arable
P	Ludvin	0–20	580	Peat	Grass	Peat	Pasture
	Ludvin	20–40	580	Peat	Grass	Peat	Pasture
	Ludvin	40–60	580	Peat	Grass	Peat	Pasture
LC	Sielec	0–20	580	Calcite	Wheat	Calcareous	Arable
	Sielec	20–40	580	Calcite	Wheat	Calcareous	Arable
	Sielec	40–60	580	Calcite	Wheat	Calcareous	Arable

F—fertilized brown soil; UF—unfertilized brown soil; LF—black forest soil; LA—black arable soil; P—peat soil; LC—calcareous soil.

Table 2. Selected soil physical-chemical characteristics

Code	Depth (cm)	pH (H_2O)	Exch. Al mM/kg	Subst. org. (%)	$CaCO_3$ (%)	Kaolinite (%)	Illite (%)	FeOOH (%)	Feldspar (%)
F									
	0–20	4.08	1.87	2.3	/	/	2.1	Trace	95.6
	20–40	4.47	1.97	1.6	/	/	2.3	0.6	94.2
	40–60	6.41	0.19	0.5	/	/	3.5	0.8	92.4
UF									
	0–20	5.90	0.24	1.8	/	/	6.4	/	89.0
	20–40	6.00	0.24	1.8	/	/	5.3	/	92.9
	40–60	6.16	0.18	0.8	/	/	8.0	/	91.2
LF									
	0–20	7.01	Trace	2.5	/	/	1.8	/	95.7
	20–40	7.04	0.42	1.1	/	/	3.4	/	95.5
	40–60	7.26	0.12	0.5	/	/	4.6	/	94.9
LA									
	0–20	6.44	0.17	5.5	/	/	2.4	/	92.1
	20–40	6.56	0.12	1.2	/	/	0.6	/	98.2
	40–60	6.64	0.09	0.6	/	/	2.5	/	96.9
P									
	0–20	5.41	Trace	68.4	/	/	22.0	/	9.6
	20–40	5.46	1.90	49.4	/	/	20.2	/	30.4
	40–60	5.60	0.18	55.6	/	/	20.6	/	23.8
LC									
	0–20	7.45	Trace	3.7	3.25	/	21.4	/	71.7
	20–40	7.64	Trace	2.0	1.80	/	15.6	/	80.6
	40–60	8.08	Trace	0.3	1.05	/	3.4	/	95.3

F—fertilized brown soil; UF—unfertilized brown soil; LF—black forest soil; LA—black arable soil; P—peat soil; LC—calcareous soil.

Table 3. The total contents of major oxide components in the parent material

Code	Site	SiO_2 (%)	Al_2O_3 (%)	Fe_2O_3 (%)	CaO (%)	MgO (%)	Na_2O (%)	K_2O (%)
F-1	Elizówka	82.26	8.40	2.13	1.49	2.86	0.62	0.43
F-2	Elizówka	82.77	8.48	2.24	1.38	2.97	0.65	0.47
F-3	Elizówka	78.00	11.89	4.83	1.08	3.43	1.18	0.89
UF-1	Elizówka	79.96	8.82	2.66	1.29	3.19	0.67	0.69
UF-2	Elizówka	81.07	8.77	2.80	1.24	3.15	0.73	0.66
UF-3	Elizówka	77.77	12.09	4.02	1.30	3.26	1.17	0.75
LF-1	Dolhobczów	81.88	7.73	1.81	1.08	2.54	0.64	0.77
LF-2	Dolhobczów	77.70	11.12	3.32	0.69	2.88	1.00	0.91
LF-3	Dolhobczów	78.53	11.67	2.95	1.18	2.81	0.95	0.90
LA-1	Dolhobczów	78.11	5.96	1.58	0.94	2.57	0.44	1.86
LA-2	Dolhobczów	82.84	6.47	1.08	1.44	2.52	0.42	0.53
LA-3	Dolhobczów	82.54	7.62	1.71	1.22	2.80	0.57	0.65
P-1	Ludwin	19.32	2.48	2.37	0.00	0.91	0.19	3.12
P-2	Ludwin	35.78	3.60	1.48	0.76	1.46	0.23	2.67
P-3	Ludwin	29.10	3.37	2.27	0.00	1.35	0.18	4.80
LC-1	Sielec	74.51	9.02	2.38	2.73	3.19	0.86	2.55
LC-2	Sielec	76.98	9.33	2.40	2.29	3.19	0.84	2.67
LC-3	Sielec	83.46	7.15	1.22	1.79	3.17	0.52	0.75

F—fertilized brown soil; UF—unfertilized brown soil; LF—black forest soil; LA—black arable soil; P—peat soil; LC—calcareous soil, 1—topsoil; 2—subsoil; 3—deep soil.

The computer program AlCHEMI4[7] was used to calculate ion specie activities. Saturation indices (SI) were also calculated to evaluate dissociation of Al minerals as a mechanism for controlling the mobilization of Al from the soils by nitric and sulfuric acids,. The measured solution activity expressed as ion activity products (IAP) was divided by the temperature-corrected Al solubility products (K_{sp}): $SI=\log(IAP/K_{sp})$ where the positive, negative and zero SI values indicate that a solution is oversaturated, undersaturated, or in equilibrium, respectively, with reference to solubility of the mineral phase concerned; I.e. natural gibbsite, microcrystalline gibbsite, kaolinite, halloysite, imogolite, smectite, jurbanite, alunite and basaluminite.

3. RESULT AND DISCUSSION

3.1. Effect of Sulfuric Acid Concentration

In general, the concentration of labile Al in sulfuric acid treatments increased with increasing concentration of H_2SO_4 (Fig. 1). It may be seen that there is a short range at low sulfuric acid concentrations that yield large Al mobilization, followed by an extended range of H_2SO_4 concentration of less Al release. The Al release in solution with no addition of acid is negligible in all the studied soils. Comparison of the curves in figure 1 shows that an increase of sulfuric acid concentration substantially increases mobilization of Al from brown soils, especially from soil that where ammonium fertilized, while Al mobilization was low in the peat and calcareous soils. Plots of labile Al vs. pH in extractant solution show the responses of labile Al to the H_2SO_4 addition (Fig. 2). The concentrations of labile Al decrease with increasing extractant pH. Fig. 2 also shows a

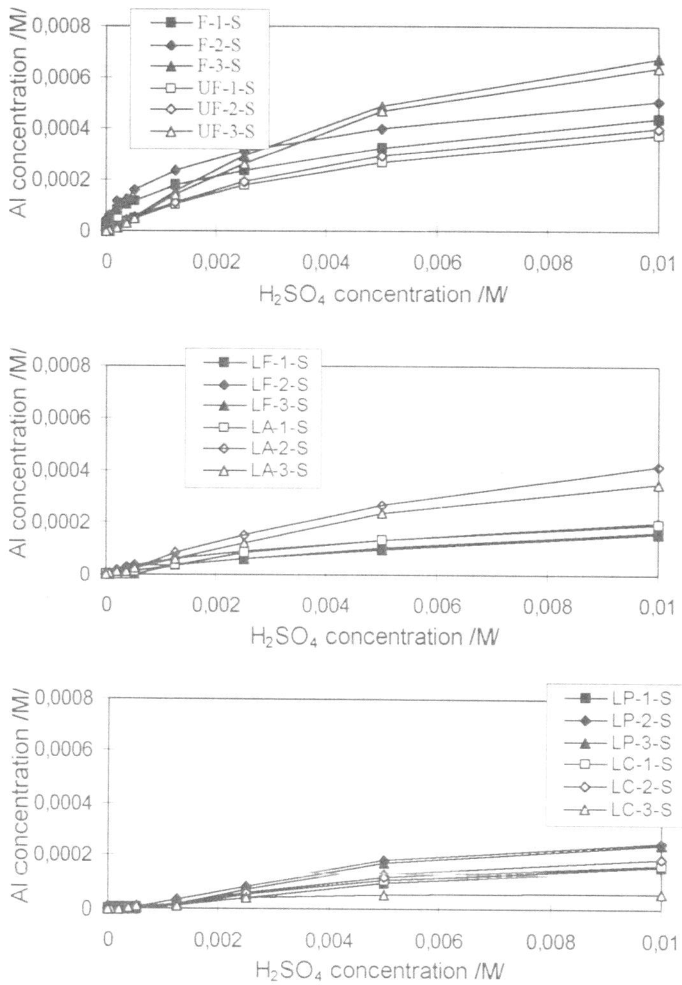

Figure 1. The dependence of the labile Al in soils on H_2SO_4 concentrations. (F—fertilized brown soil, UF—unfertilized brown soil, LF—black forest soil, LA—black arable soil, LP—peat soil, LC—calcareous soil; 1—topsoil, 2—subsoil, 3—deep soil).

differentiation of soil sensitivity to H_2SO_4 input. For example, a small addition of acid (0.05 mM) could mobilize 62.6 μM labile Al in fertilized brown subsoil. While in the black forest soil, 1.25 mM sulfuric acid was needed to reach the same level of labile Al. It is easy to see that the influence of acid deposition on Al mobilization was more significant in brown soil than in peat, black and calcareous soils. The main mechanism of acid neutralization in brown soil was by Al mobilization. In black, peat and calcareous soils, which are less sensitive to acid deposition, large amounts of added acid may be buffered by non-Al mobilization until concentration of acid inputs came to 5.0 mM or higher.

3.2. Responses of Soils

Fig. 3 shows the differences in the amount of Al mobilized by 5.0 mM H_2SO_4 in the studied soils. This acidity is nearly 80 times greater than the average acidity of precipita-

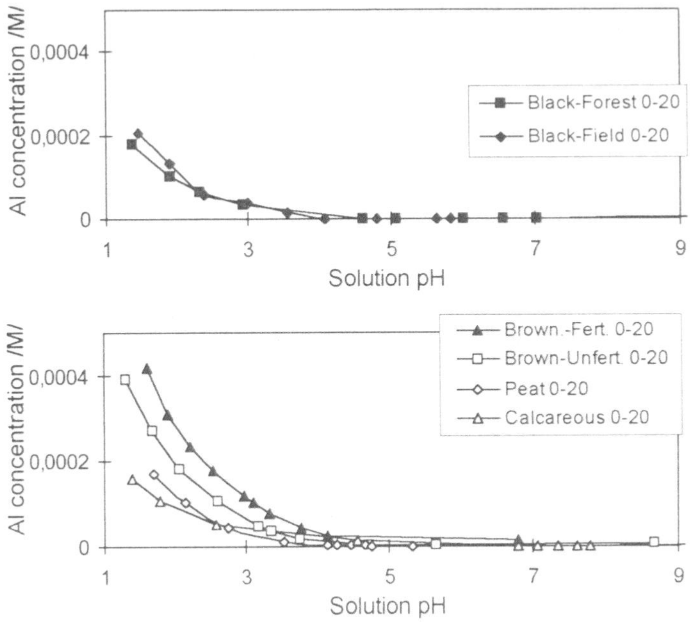

Figure 2. Concentration of labile Al as a function of pH (H_2SO_4) in the topsoil horizons.

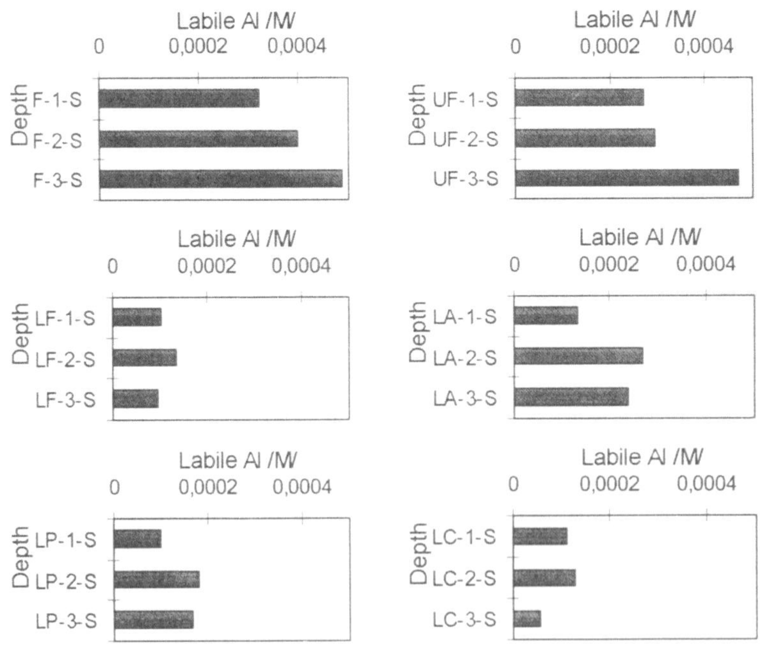

Figure 3. The changes of labile aluminum with soil treated by 5.0 mM H_2SO_4. (F—fertilized brown soil, UF—unfertilized brown soil, LF—black forest soil, LA—black arable soil, LP—peat soil, LC—calcareous soil; 1—topsoil, 2—subsoil, 3—deep soil; S—H_2SO_4).

tion at the sampling site. Such extreme experimental conditions are necessary in order to compare soil interactions. As expected, the studied soil responded differently. Comparison of the curves in figure 1 shows that an increase of sulfuric acid concentration substantially increases mobilization of Al from brown soils, especially from soil that where ammonium fertilized, while Al mobilization was low in the peat and calcareous soils.

Aluminum mobilization greatly increases with soil depth, but the changes of labile Al concentrations vary for different soils (Fig. 3). For the convenience of comparison, all the studied soils are divided into three groups: group I—brown soil from fertilized and unfertilized plot; group II—black arable and forest soil; group III—peat and calcareous soils.

In group I, both soils profiles show an increased mobilization of Al with depth (Fig. 3). Furthermore, the figure 3 show that the ammonium fertilized soil samples, have a greater ability of releasing Al. The labile Al concentration in the extracted solutions reaches 310 μM for the topsoil, 387 μM for the subsoil, and 487 μM for the deep soil, which is greater by 15.4%, 31.2%, 4.0% than that in counterpart horizon of unfertilized plot soil, respectively. These results indicate that under the same conditions, long-term ammonium fertilization highly changes chemical state of Al in soils. The influence of long-term fertilization on Al mobilization was also confirmed by Porebska and Mulder[9].

In the soil group II, black soils, the differences of labile Al concentrations reflect the strong influence of land use on Al mobilization. The labile Al concentrations came to 101, 136, and 97 μM, respectively, in the topsoil, subsoil and deep horizons of forest soil. Whereas the cultivated soils with single agricultural cropping gave 133, 267, and 239 μM labile aluminum respectively. I.e. the cultivated soil had 31.6%, 96.7% and 146% higher Al mobilization than that in the corresponding horizons of forest soil. The labile Al concentration in topsoil of the arable soil accounted only for 50%, and 55.6% of that in subsoil and deep soil. As to the forest soil, there were only minor changes of labile Al concentrations in comparison with the arable soil. The difference of labile Al between the soil profiles support the hypothesis that although natural soil acidification has brought about Al mobilization, agricultural management such as ammonium fertilization and harvest strongly intensify the processes of soil acidification.

The Al concentrations in the extract from the black soils were much lower compared to from the brown soils sampled from the same region (Fig. 3). The fertilized brown soil sample mobilized 1.5–2.4 times more labile Al than did black soil. Even the unfertilized brown soil mobilized more labile Al than the black soil. On the average, the concentrations of mobilized labile Al in the topsoil, subsoil and deep black soil horizons were 117, 202, 168 μM, respectively, only accounting for 40.2%, 59.2%, 35.1% of that in the corresponding brown soil horizons. As pointed out above, the amount of Al mobilized from brown soil increased with soil depth; the highest Al concentration being in the deep soil. The black soil showed instead the highest concentration of labile Al in the subsoil horizon, followed by deep soil and topsoil horizons.

In soil group III, peat and calcareous soils, the patterns of labile Al concentration with soil depth are quite similar to the black forest soil. The highest labile Al concentration occurs in the subsoil horizon, followed by the deep soil and topsoil horizons.

Peat soil shows a somewhat higher concentration of mobilized Al in the soil profile. The concentrations of labile Al in peat soil were 105, 180, and 166 μM in the topsoil, subsoil and deep soil horizons respectively; i.e. there was only minor difference in the Al mobilization between the subsoil and deep soil horizons. The amounts of labile Al extracted from subsoil and deep peat soil were by 32.4% and 71.1% higher than that in the black forest soil. Both soils have been used as wood land with different forest species and density. No agricultural activities have been applied. Compared to the brown soils, the

Table 4. The relationship between soil pH and labile Al

Soil horizon	Equation	R
Topsoil	Labile Al = $5.08\times10^{-4} - 6.20\times10^{-5}$ * soil pH	−0.736
Subsoil	Labile Al = $6.05\times10^{-4} - 6.99\times10^{-5}$ * soil pH	−0.923
Deep soil	Labile Al = $1.01\times10^{-3} - 1.28\times10^{-4}$ * soil pH	−0.808
All horizons	Labile Al = $6.19\times10^{-4} - 7.13\times10^{-5}$ * soil pH	−0.647

amounts of Al mobilized in peat soil were relatively low, accounting for only 35–55% in the counterpart horizons. Calcareous soil profile exhibited the lowest labile Al concentration among the studied soils. The concentrations of labile aluminum in the extract solutions were: 110, 126, and 55 μM in the topsoil, subsoil and deep soil horizons, respectively. Subsoil showed only 15% higher concentration of labile Al than the topsoil and the labile Al mobilized in the deep soil horizon accounted for only 50% of that in the surface soil.

3.2.1. Soil pH. Figure 4 gives the relationship between labile Al mobilized by 5.0 mM sulfuric acid and the soil ph_{H_2O}. The soil pH dependence on labile Al gave significant linear regressions with slopes of $-6.2\cdot10^{-5}$ for the topsoil, $-7.0\cdot10^{-5}$ for the subsoil, and $-1.3\cdot10^{-4}$ for the deep soil horizons (Table 4). The relationship (R) between soil pH and the labile Al in the top soil was lower than in the subsoil and deep soil horizon. The low R value in the top soil horizon may be due to a higher content of the soil organic matter (cf. Table 2) which is found to caused a substantial reduction of exchangeable Al, especially at low pH values[10].

Although the amount of labile Al in the extract generally increased with soil acidity the concentration of labile Al also varied greatly at any given soil pH_{H_2O}. For example, in the brown and black soils the extract from the deep horizons had higher labile Al concentrations than the topsoil despite that they also have higher pH values (from 4.63 to 6.88). This is probably due to difference in solubility of different crystalline Al minerals and soil organic matter.

3.2.2. Agricultural Management. According to type of land use, these acid soils may be divided into two groups: group A of agricultural soils including brown soil, black arable soil and calcareous soil; group B of nonagricultural soils containing peat soil and black forest soil. Labile Al extracted from group A was higher than from group B (Fig. 5),

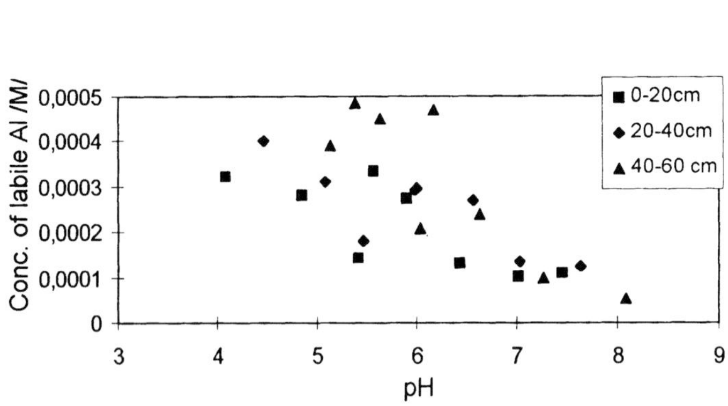

Figure 4. The relationships between labile Al in 5.0 mM H_2SO_4 and soil pH values.

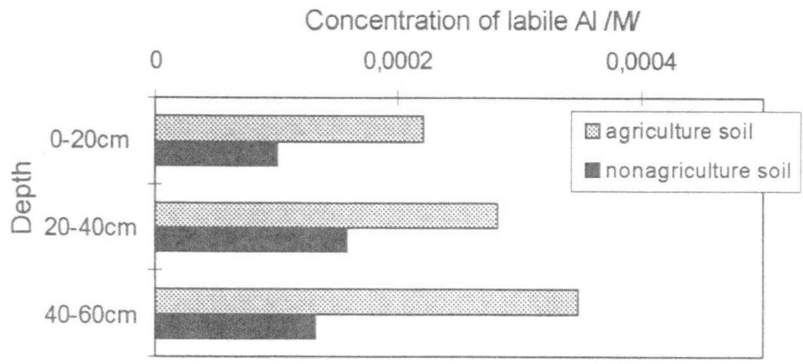

Figure 5. The comparison of average labile aluminum in agriculture soils and non-agriculture soils treated by 5.0 mM H_2SO_4.

except for the calcareous soil. The average of labile Al in the topsoil, subsoil and deep horizons of agricultural soil group were higher by 115.5%, 79.1% and 165.2% than the average values in the counterpart horizons of non-agricultural soil group. Within the agricultural soil group, the soils with long-term farming practice exhibited stronger tendency of Al mobilization to neutralize acid input. These result suggests that the activities of agricultural production affect soil chemical and physical properties, thus influence the processes of Al mobilization, especially in the areas where intensive agriculture management has been conducted.

3.3. Difference of HNO_3 and H_2SO_4 in Aluminum Mobilization

Treatment with nitric acid gave the same trend of labile Al change with acid concentration as was found for sulfuric acid in all the studied soils (Fig. 6; cf. Fig. 1). The concentrations of labile Al mobilized by HNO_3 were therefore all significantly correlated with the corresponding values obtained by sulfuric acid extraction.

Although both inorganic acids gave the same general trend of labile Al change labile Al concentrations in the extract from the nitric acid treatments was not identical to the sulfuric acid treatments (Fig. 7). Comparison of labile Al concentrations from all the soil horizons mobilized by nitric and sulfuric acids gave a rather complicated picture. Generally, nitric acid had lower ability in mobilizing Al than sulfuric acid when hydrogen ion concentrations was greater than 5.0 mM. While at lower H^+ concentration, the ability of aluminum mobilization between two acids was determined by soil type. In the peat soil, the Al mobilized by nitric acid was usually less than by sulfuric acid irrespective of acid concentrations, especially in the surface soil. On an average, the concentrations of labile Al in the nitric acid treatments make up 40%, 83%, and 80% compared to those treated by H_2SO_4 of these values in the topsoil, subsoil and deep horizon respectively. In the brown soils, no significant differences were observed between two acid treatments. In black arable soil nitric acid showed higher ability of mobilizing aluminum than sulfuric acid, and on an average, the labile Al concentrations in nitric acid treatments are 1.75, 1.41 and 1.44 times higher than in sulfuric acid treatments. Table 5 summarizes the differences in the studied soils treated by both inorganic acids.

The difference between sulfuric and nitric acid in mobilizing Al could be attributed to different mechanisms of Al release. This process is affected by soil pH, the characteristics of the adsorbent surface such as amounts and types of Al, iron oxides, soil clays, sulfate ion concentrations in solutions, the amounts of SO_4^{2-} already adsorbed, organic ions and other ion species in soil solutions and finally contact time between sulfate solution and adsorbents[13–16]. The differences in amounts of mobilized Al between sulfuric and nitric acids correspondingly relate to differences in all these factors in the tested soils.

In the peat soil, the organic matter content reaches 69%. The high concentrations of organic acids in solutions allow the existence of high amounts of organo-Al complexes. Sulfate ions furthermore form metastable sulfate complexes with the organo-Al chelates. In these complexes, Al could exist as polyvalent bridging cations[17,18]. The formation of such organo-Al-sulfate complexes could result in an increased release of Al. The dissolved organic carbon is reported to increase with the input very strong acidity. McColl and Pohlman[19] reported that the amounts of dissolved organic carbon remained fairly constant up to about 0.001 M H^+ input in both nitric and sulfuric acid treatments. With increasing H^+ ion up to 0.01 M, 2.5–3.0 fold more dissolved organic carbon was released in sulfuric acid treatment, while nitric acid treatment increased nearly 2-fold more dissolved

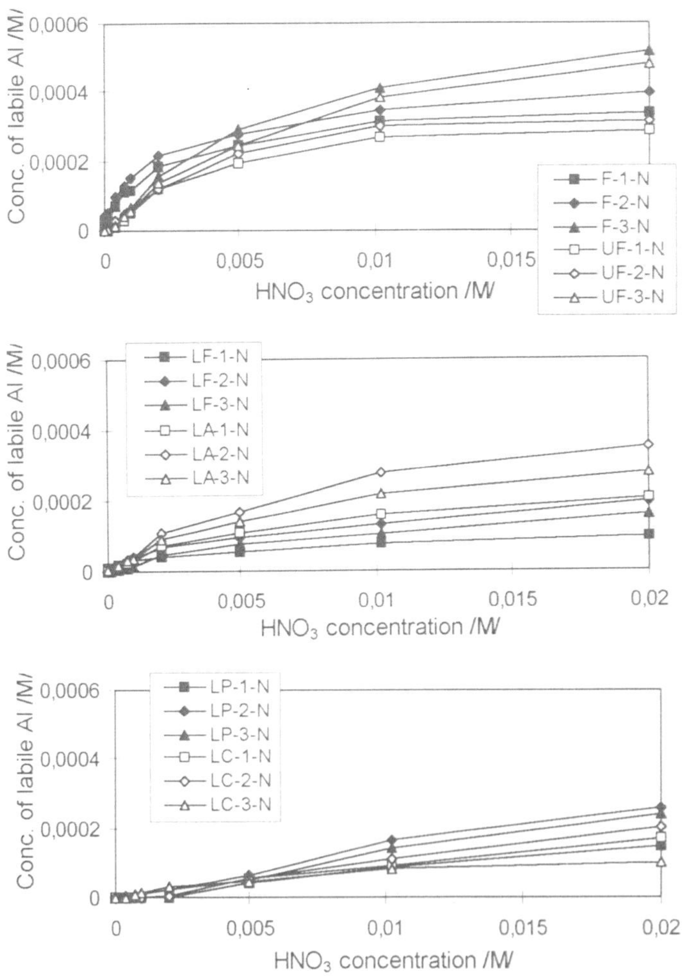

Figure 6. The changes of labile aluminum with the acid input concentrations (HNO_3). (F—fertilized brown soil, UF—unfertilized brown soil, LF—black forest soil, LA—black arable soil, LP—peat soil, LC—calcareous soil; 1—topsoil, 2—subsoil, 3—deep soil; N—HNO_3.)

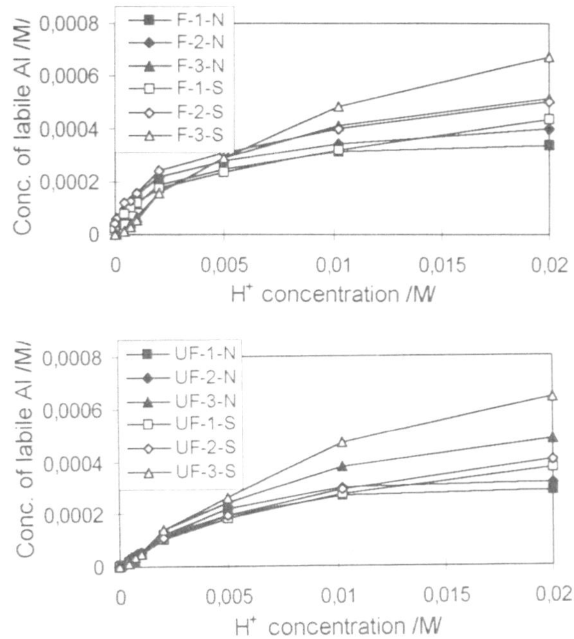

Figure 7. The amounts of mobilized labile Al versus H^+ concentrations in HNO_3 and H_2SO_4. (F—fertilized brown soil, UF—unfertilized brown soil; 1—topsoil, 2—subsoil, 3—deep soil; N—HNO_3, S—H_2SO_4.)

Table 5. The differences of labile Al between the nitric and sulfate at the same hydrogen concentrations*

Horizon	H^+ conc. [mol/L]	Brown fertilized	Brown unfertilized	Black forest	Black arable	Peat soil	Calcareous soil
0–20 cm							
	2.00E−02	0.81	0.85	0.54	0.99	1.03	0.92
	1.00E−02	0.94	0.99	0.76	1.19	0.84	0.82
	5.00E−03	1.04	1.09	0.88	2.38	0.40	1.01
	2.50E−03	1.04	1.14	1.19	1.90	0.12	0.32
	1.00E−03	0.95	1.02	/	2.27	0.26	/
	7.50E−04	1.05	0.77	/	/	0.23	/
	4.00E−04	0.88	0.49	/	/	0.20	/
	1.00E−04	0.91	1.18	/	/	0.12	/
20–40 cm							
	2.00E−02	0.82	0.68	0.48	0.84	1.04	1.80
	1.00E−02	0.89	1.01	0.57	1.03	0.93	0.88
	5.00E−03	1.06	1.14	0.82	1.38	0.72	0.97
	2.50E−03	1.03	1.08	0.89	1.58	0.74	1.19
	1.00E−03	0.94	1.08	0.91	1.54	0.93	/
	7.50E−04	1.01	1.10	1.05	2.17	0.87	/
	4.00E−04	0.81	1.36	0.94	1.34	0.93	/
	1.00E−04	0.85	2.11	/	/	0.47	/
40–60 cm							
	2.00E−02	0.77	0.77	0.90	0.80	1.06	1.59
	1.00E−02	0.84	0.81	1.38	0.90	0.93	1.50
	5.00E−03	0.99	0.92	2.33	1.13	0.56	1.37
	2.50E−03	1.02	0.98	1.84	1.56	0.76	1.13
	1.00E−03	1.21	1.07	1.05	1.72	0.91	0.78
	7.50E−04	1.32	1.16	0.99	2.74	0.98	0.59
	4.00E−04	1.19	1.04	0.73	1.83	0.68	/
	1.00E−04	0.86	0.95	/	0.83	0.49	/

*The values are the labile Al in the nitric treatments divided by labile Al in sulfate treatments.

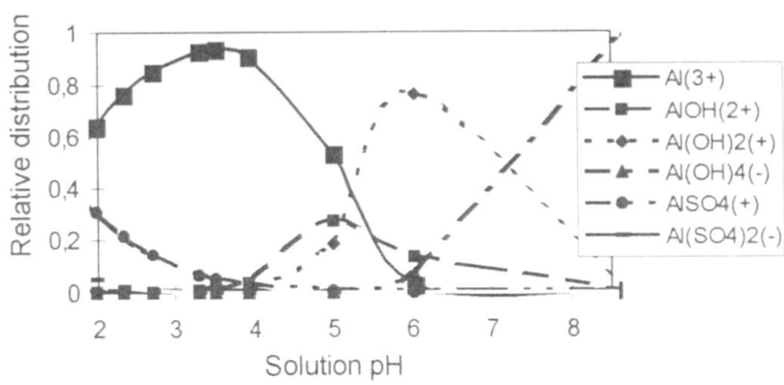

Figure 8. The relative distribution of aluminum species in unfertilized plot (deep soil) treated by H_2SO_4.

organic carbon. The quantities of labile Al released appeared to be related to the amount of dissolved organic carbon. Correspondingly, at strong acid input, sulfuric acid was enabled to holds more labile Al in soil solution than nitric acid.

3.4. Aqueous Chemistry of Mobilized Aluminum

Speciation of inorganic Al in sulfuric extracts exhibited marked changes associated with acidity (Fig. 8). The trivalent aluminum Al^{3+} was a predominant species at low pH values, followed by species bound to sulfate due to the greater concentrations of sulfate ions in suspensions. The other forms of Al species accounted for a minor contribution to the total Al. The species of $Al(SO_4)_2^-$, and $AlSO_4^+$ in the deep unfertilized plot soil made up 40% of the total labile Al when treated by 5.0 mM, and then declined to insignificant levels at pH higher than 5.0. The relative distribution of trivalent aluminum Al^{3+} reached the highest proportion (more than 90%) at pH 3.5, and became unimportant at pH greater than 6.0 (Fig. 8). The decrease of relative distribution of trivalent aluminum Al^{3+} at pH less than 3.5 was mainly due to the increase of $Al(SO_4)_2^-$, and $AlSO_4^+$ species. In the 10 mM sulfuric acid treatment, labile Al in all soil suspensions were composed of 50% $Al(SO_4)_2^-$, and $AlSO_4^+$ species. In the high pH range (more than 5.0), hydrolysis of trivalent aluminum became more prevalent.

Although all the soil extracts showed the predominance of Al^{3+} species at low acid input, the relative proportion of trivalent aluminum differ among the soil types and horizons depending on the degree of soil acidification (Fig. 9). The relative distribution of trivalent aluminum Al^{3+} in the fertilized brown soil is 97%, 93% in the topsoil and subsoil horizons, respectively. This was nearly 20% higher than in unfertilized brown soils. Trivalent aluminum Al^{3+} also held high proportion in the peat soil equilibrium solution (more than 60%), when treated by 0.05 mM sulfuric acid, although peat soil contains high amounts of organic matter. Fig. 9 also shows that there was a decreasing proportion of Al^{3+} with soil depth, while the relative amount of hydroxy Al species increased. A lower degree of soil acidification in deep soil cousing a higher soil pH changed the proportion of the individual species.

Alvarez et al.[20] did researches on the composition of soil solutions in a mining region. They pointed out that acidified soil solutions rich in Al and sulfates tended to exhibit a predominance of Al^{3+} and Al-sulfate complexes.

The distribution of inorganic monomeric Al species in nitric acid treatment is presented in figure 10. In contrary to the soils treated by sulfuric acid, Al sulfate complexes did not make an important contribution to the total Al fraction at lower pH value range (Fig. 10).

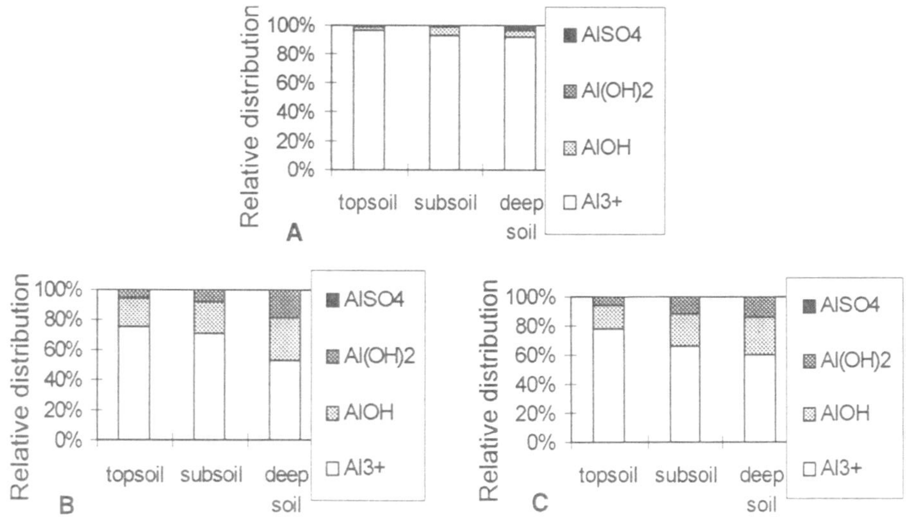

Figure 9. The relative distribution of aluminum species treated by 0.05 mM H_2SO_4. (A—fertilized brown soil; B—unfertilized brown soil; C—peat soil.)

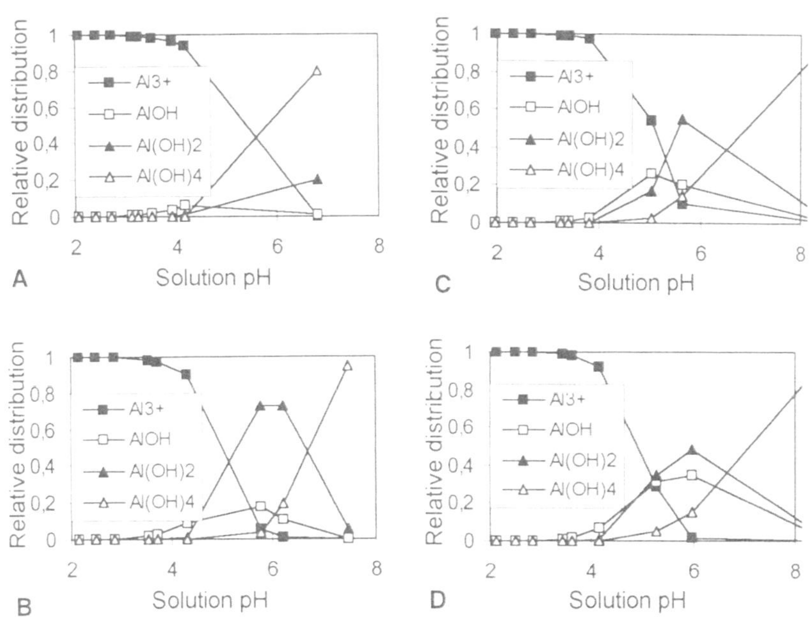

Figure 10. The relative distribution of Al species as function of solution pH in brown soil in nitric solutions. (A—fertilized brown soil, topsoil; B—fertilized brown soil, deep soil; C—unfertilized brown soil, topsoil; D—unfertilized brown soil, deep soil.)

Table 6. Saturation indices of equilibrium solution with respect to the selected Al solid mineral phase

Soil	H_2SO_4 conc. [mol/L]	SSG	SNG	SMG	SHAL	SIMOG	SJUR
Unfertilized brown soil							
20–40 cm	1.00E−02	−7.63	−8.29	−8.87	−12.16	−19.60	−1.41
	5.00E−03	−6.84	−7.50	−8.08	−11.37	−18.00	−1.42
	2.50E−03	−5.81	−6.47	−7.05	−10.34	−15.95	−1.38
	1.25E−03	−4.87	−5.53	−6.11	−9.40	−14.07	−1.44
	5.00E−04	−3.55	−4.21	−4.79	−8.09	−11.44	−1.57
	3.75E−04	−3.09	−3.75	−4.33	−7.63	−10.52	−1.62
	2.00E−04	−2.22	−2.88	−3.46	−6.76	−8.78	−1.78
	5.00E−05	0.03	−0.63	−1.21	−4.50	−4.27	−2.26
	0	1.74	1.08	0.50	−2.79	−0.85	−7.01
Fertilized brown soil							
20–40 cm	1.00E−02	−7.79	−8.45	−9.03	−12.32	−19.92	−1.46
	5.00E−03	−6.87	−7.53	−8.11	−11.40	−18.07	−1.36
	2.50E−03	−5.92	−6.58	−7.16	−10.45	−16.16	−1.34
	1.25E−03	−4.93	−5.59	−6.17	−9.46	−14.19	−1.29
	5.00E−04	−3.64	−4.30	−4.88	−8.18	−11.62	−1.33
	3.75E−04	−3.23	−3.89	−4.47	−7.76	−10.79	−1.34
	2.00E−04	−2.49	−3.15	−3.73	−7.03	−9.31	−1.37
	5.00E−05	−0.56	−1.22	−1.80	−5.12	−5.48	−1.52
	0	−0.24	−0.90	−1.48	−4.80	−4.84	−6.24

SSG—synthetic gibbsite; SNG—natural gibbsite; SMG—amorphous gibbsite; SHAL—halloysite; SIMOG—imogolite; SJUR—jurbanite.

3.5. Solubility Equilibrium with Mineral Phase

All the studied soils exhibited an increase in SI values with decreasing amount of acid in the extractant. Table 6 presents the saturation indices with respect to selected Al-containing minerals vs. concentration of sulfuric acid in soil extracts from subsoil horizons of the brown soil profiles. Fig. 11 shows the saturation indices with respect to gibbsite vs. pH for the fertilized brown soil. The deep soil showed a somewhat lower SI values compared to the topsoil and subsoil. Equilibrium with synthetic gibbsite could be reached at pH 4.3–5.0. But the increase in SI with pH shown in Fig. 11 showed that this mineral could not control the concentrations of dissolved Al. The calculation of pH vs. pAl relationships showed that for each soil horizon there existed a significant linear re-

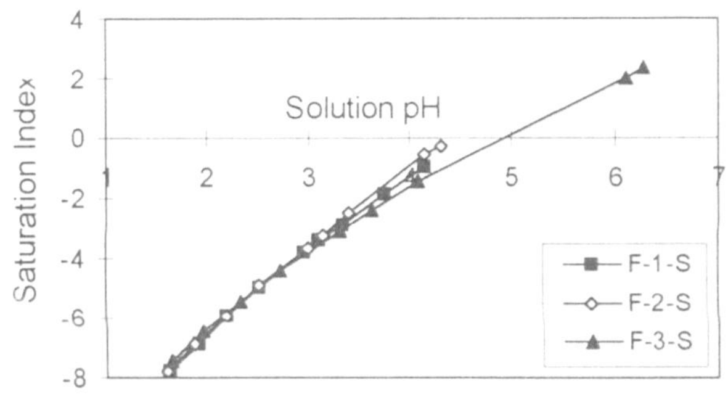

Figure 11. Saturation indices calculated for synthetic gibbsite in the fertilized brown soil.

Table 7. The relation between pAl-pH treated by H_2SO_4 and HNO_3

Soil types	Depth (cm)	SO_4-system Slope	SO_4-system R	NO_3-system Slope	NO_3-system R
Unfertilized brown soil					
	0–20	0.71	0.981	1.76	0.934
	20–40	0.72	0.982	1.80	0.936
	40–60	0.88	0.970	1.91	0.939
Fertilized brown soil					
	0–20	0.85	0.939	0.99	0.968
	20–40	1.05	0.892	1.27	0.939
	40–60	1.23	0.957	1.38	0.957
Peat soil					
	0–20	0.73	0.926	0.87	0.984
	20–40	0.76	0.945	0.86	0.964
	40–60	0.74	0.946	0.84	0.961
Black forest soil					
	0–20	0.71	0.981	0.995	0.851
	20–40	0.72	0.969	0.76	0.959
	40–60	0.88	0.959	1.56	0.955
Black arable soil					
	0–20	0.75	0.998	0.82	0.988
	20–40	0.79	0.988	0.65	0.990
	40–60	1.54	0.994	0.90	0.988

gression characterized by a distinctly different slope (Table 7). The slope was always significantly less than three.

The high amounts of sulfate ions together with the mobilized Al caused the solution to become supersaturated in regards to the aluminum-sulfate minerals alunite ($KAl_3(OH)_6SO_4$) and jurbanite ($Al(OH)SO_4$). It was therefore possible that sulfate ions may precipitate with Al in these soils. According to solubility equilibrium diagram almost all samples of the soils treated with sulfuric acid lied between the theoretical lines for alunite and jurbanite (Fig. 12). Thus, Al in the soil extracts appeared to be controlled by a mineral of this type. Probably the high input of sulfate ions (> 5.0 mM) resulted in a precipitation of Al with sulfate (cf. Table 5). The reaction of soils with the incoming acids could be expressed as follows:

Dissolution:

$$2Al(OH)_{3\,(s)} + 3H^+ + 3HSO_4^- \Rightarrow 2Al^{3+} + 3SO_4^{2-} + 6H_2O$$

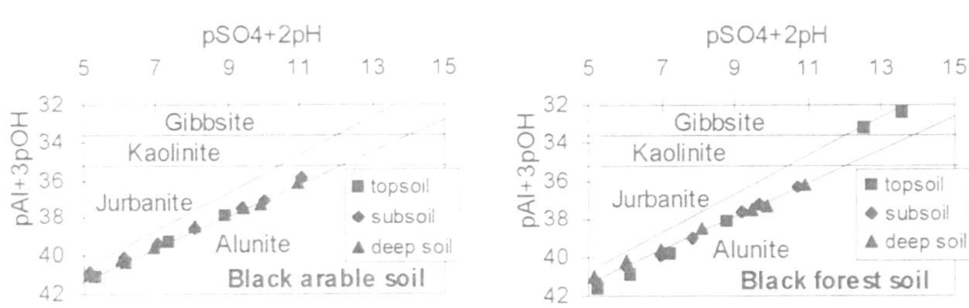

Figure 12. The relationships of pAl+3pOH and pSO_4+2pH with respect to the solubility lines for gibbsite, jurbanite, alunite and kaolinite (pH_4SiO_4 = 2.7).

Precipitation:

$$Al^{3+} + HSO_4^- + H_2O \Rightarrow Al(OH)SO_4 + 2H^+$$

$$3Al^{3+} + 2HSO_4^- + K^+ + 6H_2O \Rightarrow Al_3K(SO_4)_2(OH)_6 + 8H^+$$

The same pattern of solubility diagram in all the tested soil samples treated with sulfuric acid indicates that sulfate input was an important factor in determining Al solubility. Sjostrom[21] also found that there were positive correlation between the amount of soil exchangeable SO_4^{2-} and Al. Vogt et al.[22], studying soil and soil water interactions in the field at Polish sites with high deposition of acid rain, found that the Al^{3+} concentrations corresponded well to equilibrium with jurbanite. The less acidic soil solutions (low sulfuric acid concentration in the extractant) remained also between the theoretical lines for jurbanite and alunite, but were also close to the solubility line of gibbsite (Fig. 12). Thus precipitation of either phases may be possible in the high range of pH values.

In contrast all the extracts from soils treated by nitric acids were greatly undersaturated with respect to jurbanite and alunite. Saturation indices in regards to gibbsite clearly increased while SI for jurbanite remained constant with increasing pH values. Gibbsite was therefore apparently not operative in controlling aqueous Al in the studied soil solutions with pH less than 4.5 (Fig. 13). Where pH in the extractant was greater than 4.50, the SI values in regards to synthetic gibbsite of all treatments were near or larger than 0. At these pH values the solution was in equilibrium or oversaturated with respect to gibbsite. The constant but greatly negative SI for jurbanite may suggest an equilibrium with a more crystalline form of jurbanite (i.e. lager pK value) or a simplified and combined description of pH dependent Al dissolution and sulfate adsorption/desorption. Vogt et al.[8,22] did a comparison of potential Al mobilization between Janow, Poland and Birkenes, in Southernmost Norway. They found the same trend of saturation indices for gibbsite with pH values. The variation of saturation indices with soil solution pH values may indicate different mechanisms of controlling Al mobilization in different pH ranges. At higher pH values dissolution of gibbsite may account for most Al release in soil solutions. When pH decreases to less than 4.0, the mineral surface is saturated with adsorbed protons and previously mobilized Al, and this may inhibit further Al release[23].

In addition to the inhibition of clay mineral dissolution at lower pH ranges (less than 4.1), the organic matter may play an important role in controlling aqueous Al in acid soil solutions. Al activity could be controlled by Al mobilization from solid organic phase ma-

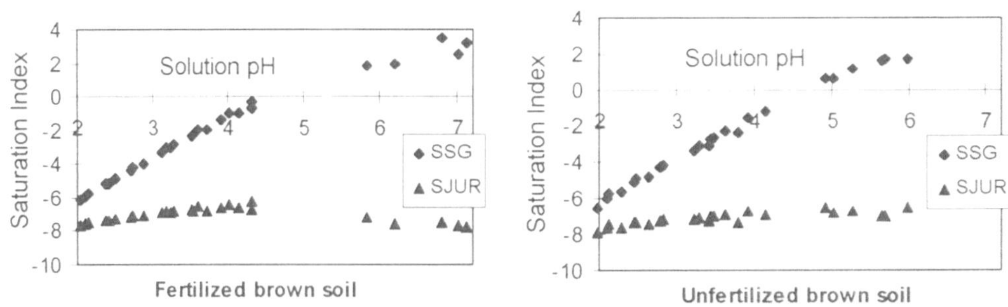

Figure 13. Saturation indices (SI) calculated for synthetic gibbsite (SSG) and jurbanite (SJUR) in the brown soil treated by nitric acid solution.

terial. If organic carbon was a major component of the soil matrix, Al^{3+} activity in the soil solution may be predominantly controlled by an equilibrium with organic complexes bound to solid phase. Such equilibrium systems may differ considerably depending upon the chemical structure of special humic substances. Thus soils containing high amounts of humic substances and low content of Al would be strongly undersaturated with Al containing minerals. Due to the complex nature of organic matter in soil system, organic–Al complexation in controlling Al activity has not been elucidated quite well.

4. CONCLUSION

The aluminum mobilization in soils was highly depended on acid input concentration, and significantly increased with solution acidity for pH < 4.1. Agricultural management, especially long-term ammonium fertilization, greatly increased soil sensitivity to acid deposition. In general, the sensitivity of the studied soils decreased as follows: brown soil > black soil > peat soil > calcareous soil. The difference of nitric and sulfuric acids ability to mobilize aluminum depended on acid concentration and soil type. At higher concentration (> 5.0 mM) or organic soil, sulfuric acid can mobilize more aluminum than nitric acid. Relationship between H^+ and Al^{3+} was far less than cubic in acid treatments (ranging from 0.65 to 1.91). Thus, Al in the sulfuric acid treatments appeared to be controlled by alunite and jurbanite, and in the nitric acid treatment can be in equilibrium with more crystalline form of jurbanite. At pH > 4.5 in both of acid treatments also gibbsite may count for Al release into soil solutions.

ACKNOWLEDGMENT

Authors would like to express their gratitude to Dr R.D. Vogt for his providing valuable comments and corrections.

REFERENCE

1. Ulrich, B.R., Mayer, C., Khanna, P.K.: Chemical changes to acid precipitation in loess-derived soil in central Europe. *Soil. Sci.*, 1980, *130*, 193–199.
2. Vogelman, H.W.: Catastrophe on Camels Hump. *Nat. Hist.*, 1982, *91*, 8–14.
3. Krug, E.C., Frink, C.R.: Acid rain on acid soil: a new perspective. *Sci.*, 1983, *221*, 520–525.
4. Prietzel, J., Feger, K.H.: Dynamics of aqueous aluminum species in a podzol affected by experimental $MgSO_4$ and $(NH_4)_2SO_4$ treatments. *Water Air Soil Pollut.*, 1992, *65*,153–173.
5. Kotowski, M., Pawlowski, L., Seip, H.M., Vogt, R.D.: Mobilization of Al in soil columns exposed to acids and salt solution. *Ecol. Engin.*, 1994, *3*, 279–290.
6. Driscoll, C.T.: A procedure for the fraction of aqueous aluminum in dilute acidic waters. *Int. J. Environ. Anal. Chem.*, 1984, *16*, 267–283.
7. Schecher, W.D., Driscoll, C.T.: An evaluation of uncertainty associated with Al equilibrium calculations. *Water Resour. Res.*, 1987, *23*, 525–535.
8. Vogt, R.D., Seip, H.M., Pawlowski, L., Kotowski, M., Odegard, S., Horvath, A., Andersen, S.: Potential acidification of soil and soil water: a monitoring study in the Janow Forest, southeastern Poland. *Ecol. Engin.*, 1994, *3*, 255–266.
9. Porebska, G., Mulder, J.: Effect of long term nitrogen fertilization on soil Al chemistry. *J. Ecol. Chem.*, 1994, *3*, 269–280, 1994.
10. Thomas, G.W., The relationship between organic matter content and exchangeable Al in acid soil. *Soil Sci. Soc. Am. J.*, 1975, *39*, 591.

11. David, M.B., Driscoll, C.T.: Aluminum speciation and equilibria in soil solution of a haplorthod in the Adirondack mountains (New York, USA). *Geoderma*, 1984, *33*, 297–318.
12. Johnson, N.M., Driscoll, C.T., Eaton, J.S., Likens, G.E., McDowell, W.H.: Acid rain, dissolved aluminum and chemical weathering at the Hubbard Brook Experimental Forest, New Hampshire. *Geochim. Cosmochim. Acta*, 1981, *45*, 1421–1437.
13. Courchesne, F.: Relationships between soil chemical properties and sulfate sorption kinetics in podzolic soils from Quebec. *Can. J. Soil Sci.*, 1992, *72*, 467–480.
14. Curtin, D., Syers, J.K.: Mechanism of sulfate adsorption by two tropical soil. *J. Soil Sci.*, 1990, *41*, 295–304.
15. Harrison, R.B., Johnson, D.W., Todd, D.E.: Sulfate adsorption and desorption reversibility in a variety of forest soils. *J. Environ. Qual.*, 1989, *18*, 419–426.
16. Prenzel, J.: Sulfate adsorption in soils under acid deposition: comparison of two modeling approaches. *J. Environ. Qual.*, 1994, *23*, 188–194.
17. Deconick, S.: Major mechanisms in formation of spodic horizons. *Geoderma*, 1980, *24*, 101–128.
18. Stevenson, F.J.: *Humus chemistry: genesis, composition, reactions.* John Wiley & Sons, New York, 1982.
19. McColl, J.G., Pohlman, A.A.: Soluble organic acids and their chelating influence on Al and other metal dissolution from forest soils. *Water, Air Soil Pollut.*, 1986, *31*, 917–927.
20. Alvarez, E., Petez, A., Calvo, R.: Al speciation in surface waters and soil solutions in areas of sulfide mineralisation in Galicia (N.W. Spain). *The Science of the Total Environment*, 1993, *133*, 17–37.
21. Sjostrom, J.: Al and sulfate in acid soils groundwaters on the Swedish west coast. *Water, Air and Soil Pollution*, 1994, *75*, 127–139.
22. Vogt, R.D., Godzik, S., Kotowski, M., Niklinska, M., Pawlowski, L., Seip, H.M., Sienkiewicz, J., Skotte, G., Staszewski, T., Szarek, G., Tyszka, J., Aagaard, P.: Soil, soil water and stream water chemistry at some Polish sites with varying acid deposition. In: Pawlowski, L., Seip, H.M., & Sullivan, T.J., (eds.), *Aluminum in the environment*, Special issue of *J. Ecol. Chem.*, 1994, *3*, 325–356.
23. Furrer, G., Zysset, M., Charlet, L., Schindler, P.W.: Mobilization and fixation of Al in soils. In: Merian, E., et al. (eds.) *Metal Compounds in Environment and Life (Interaction between chemistry and biology)*, Science and Technology Letters. P O B. 81, Northwood, Middlesex, UK, 1991.

31

SOIL AND SOIL WATER CHEMISTRY AT SOME POLISH SITES WITH ACID PODZOL SOILS

Marek Kotowski

Department of Environmental Engineering
Technical University of Lublin
40 Nadbystrzycka Str., 20-618 Lublin, Poland

ABSTRACT

Results of monitoring studies from three sites in Poland with moderate, high and very high levels of acid anions in precipitation (Janów Forest, Izerskie Mountains and forest of Pulawy region) are compared in this paper. Field observations of precipitation, throughfall, soil and soil water chemistry at the sites are presented.

The Pulawy area where is large factory which produces nitrogen fertilizers receives the highest deposition of nitrogen and sulfur compounds. Also the deposition of ammonium and sum of divalent cations (Ca^{2+} + Mg^{2+}) is the highest in this region. It was found that the molar critical load ratio, $Al_i/(Mg^{2+}+Ca^{2+})$, where Al_i is monomeric, inorganic aluminum, is very high in the all soils studied (up to 2.6). Al^{3+}, sulfate- and hydroxy-complexes are the dominating forms of aluminum in soil water. The observed aluminum concentrations can not be explained by equilibrium with a gibbsite ($Al(OH)_3$) mineral phase, as assumed in many acidification models.

Key words: soil chemistry, soil solution, monitoring, aluminum.

1. INTRODUCTION

Parts of Poland receive high deposition of acidifying compounds of sulfur and nitrogen[1,2]. In south-western Poland the total deposition of sulfur compounds is among the highest in Europe with background values above 7.0 g $S \cdot m^{-2} \cdot yr^{-1}$.[3] Significant parts of Polish soils in coniferous forest are podzolic, particularly vulnerable to acid deposition.

To estimate conditions for plant growth in forest ecosystems the most useful parameters are: pH, exchangeable base cations, and base saturation (BS) in the soils and pH,

Chemistry for the Protection of the Environment 3, edited by Pawłowski *et al.* Plenum Press, New York, 1998

concentrations of Al^{3+} and base cations in soil solutions. An acid soil with low base saturation is not easily acidified by acid deposition since the acid input must be very high to reduce the base saturation still further. However, the concentrations in soil water will change with the deposition. The molar critical load ratio in soil water ($R_{CL}=[Al_i]/[Ca^{2+}+Mg^{2+}]$), where Al_i is monomeric, inorganic aluminum, is often used to estimate biological sustainability[4,5]. However, the basis for the critical loads has been questioned[6].

The most detrimental result of soil acidification, in opinion of author, is mobilization of aluminum from organic matter and soil minerals, but mechanisms of Al mobilization are not well understood. In many published models (e.g. BIM[7], MAGIC[8], PROFILE[9]) Al^{3+} activity is regulated by equilibrium with a $Al(OH)_3$ (gibbsite) phase. However, many scientists have questioned this mechanism as a general model for mineral soil layers[10–13].

This paper presents the results from monitoring studies at the three sites in Poland with moderate, high and very high levels of acid anions in precipitation. Field observations of precipitation, throughfall, soil and soil water chemistry at the sites are presented. The results will be used for further discussions of critical loads, transport mechanisms and pathways of anthropogenic pollutants.

2. CHARACTERISTICS OF SAMPLING SITES

Field studies were carried out in the Janów Forest, in the Ciekonek stream catchment (the Izerskie Mountains) and in the forest of Pulawy region in Poland. Location of the sampling sites are shown in Fig. 1.

Izerskie Mountains

Most lithogenic soils of the Izerskie Mountains originate from gneiss and granite. They are poor in basic materials and effects of acid deposition are serious. The region is

Figure 1. Map showing location of the sampling sites in Poland discussed in this paper.

mainly forested by conifers, especially Norway spruce (*Picea abies*), which naturally cause decrease of soil pH to about 4.0. Additionally, the area has been seriously polluted by acidic precipitation for quite a long time. The site of sampling in the Ciekonek stream catchment is located 7 km from Szklarska Poreba and 5 km from the Czech Republic border. The area is partly covered by trees in varying conditions and is partly completely deforested. The region is situated at about 770 m a.s.l., the average annual temperature is 6.1°C (about 3°C in January and 14°C in July), and precipitation reaches about 1200 mm per year. At the sampling site, 12 soil water samplers were installed in 3 soil profiles (4 in each profile), one where the forest is in good condition, one where the forest is in poor condition and one profile in a deforested area. One precipitation collector and four throughfall collectors (2 in an area with healthy trees and 2 in an area with trees in poor conditions) were also installed.

Pulawy

A large factory in Pulawy, which produces nitrogen fertilizers, emits nitrogen oxides, ammonia and dust. Areas east and south of the industrial zone, with podzol soils and pine forests (*Pinus sylvestris*), are directly exposed to dry and wet deposition of acid gases and mineral acids. The sampling site was located about 2 km north-east of industrial zone. Pulawy is situated on the right bank of the Vistula River, at the elevation of 160 m a.s.l. The average annual temperature is 7.4°C with about 500 mm of precipitation per year. The site was equipped with 12 samplers in 2 soil horizons, but owing to lack of soil moisture, very often sampling of soil solutions was impossible. For that reason, soil water obtained by centrifugation of soil samples taken from each genetic horizons, was analyzed. A collector of precipitation and four throughfall samplers were also installed.

Janów Forest

The Janów Forest area is an example of moderately polluted region. Deposition of sulfur and nitrogen compounds is lower than in the two other studied areas. The vegetation is mainly pine (*Pinus sylvestris*) with blackberry (*Rubus hirtus*) and blueberry (*Vaccinium myrtillus*). Apparent changes of trees, bushes and undergrowth condition were not observed. The site is located about 80 km south of Lublin and 2 km west of Szklarnia village at the elevation of about 200 m a.s.l. The climate is continental with average annual temperature of 7.6°C. Average precipitation is about 600 mm per year and evapotranspiration about 450 mm. Two soil water collectors were installed in 1988, later 24 additional samplers were installed in 3 soil profiles. Throughfall was sampled in 20 collectors and bulk precipitation in one sampler about 500 m from the forest border. A lot of scientific work was done in this region in the years 1988–1994, results are presented by Vogt at al.[14,15] and Larsen at al.[16]

A summary of important characteristics of the sampling sites is given in Table 1.

3. CHARACTERISTICS OF SOILS

In Table 2, description of soil horizons at the sampling sites is presented, and in Tables 3 and 4 average values of soil properties and contents of all major chemical species in soils are given. At the Janów site samples of soils were taken from five profiles however at Ciekonek catchment and Pulawy area three profiles were sampled. All soil samples

Table 1. Characteristic of the sampling sites

Site	Elevation [m a.s.l.]	Parent material	Type of soil	Type of forest	Average temperature [°C]	Annual precipitation [mm]
Izerskie Mountain	740–800	gnejs	Podzolized rankers	spruce	6.1	1170
Janów Forest	200	sand	Podzol soil	pine	7.6	590
Pulawy	160	sand	Podzols	pine	7.4	500

were analyzed three times. The average standard deviations calculated for five or three number of samples, on basis of three replicates were mostly less than 5%, only in few cases were higher.

In the Janów Forest, typical podzol soils have developed with clear genetic horizons: O—organic, A—alluvial, E—elluvial, Bhs—illuvial, Bs—spodic, B/C—intermediate and C—bed rock. In the Pulawy region there are podzols with O, E, B, B/C and C horizons. The boundary between E and B horizons is diffusive and wavy. Soil parameters in the Pulawy region and the Janów Forest are very similar. Below the organic horizon the soils consist mostly (>95%) of sand. Organic and alluvial horizons are the most acidic ones at both sites.

Total Effective Cation Exchange Capacity (CEC_E) was analyzed according to the method developed by Hedershot and Duquette[17] in an unbuffered salt solution (0.1 M $BaCl_2$). In organic horizons CEC_E ranges from 180 to 190 meq/kg. The base saturation in organic horizons exceeds 65% and aluminum saturation ranges from 14.2 to 21.2% of total exchange capacity. Organic horizons of podzol soils are composed mainly of organic matter and content of mineral substances does not exceed 15% in any case (see Table 4). The alluvial horizon in the Janów Forest sites contains 5.1% of organic matter, and in the E horizons it decreases to 0.43% while in lower genetic horizons (B and C) organic matter content is negligible.

Table 2. Characteristic of soil horizons of the sampling sites

Site	Type of soil	Genetic levels	Colour	Thickness of level [cm]
Janów Forest	Podzol soil	O	Dark brown	5–10
		A	Dark gray	10–20
		E	Gray	3–17
		Bhs	Dark brown	18–20
		Bs	Yellowish brown	3–8
		B/C	Light yellowish brown	30–45
		C	Very pale brown	—
Pulawy	Podzols	O	Dark brown	10–20
		E	Gray	10–30
		B	Dark brown	8–17
		B/C	Yellowish brown	30–70
		C	Very pale brown	—
Ciekonek catchment	Podzolized rankers	O	Dark brown	5–13
		A/E	Gray-brown	4–7
		B1	Brown	5–9
		B	Brown	20–34
		C	Brown	—

Table 3. Average soil chemical and physico-chemical data

Sampling site	Genetic horizon	pH H_2O	pH $BaCl_2$	H^+ meq/kg	Na^+ meq/kg	K^+ meq/kg	Ca^{2+} meq/kg	Mg^{2+} meq/kg	Al^{3+} meq/kg	CEC meq/kg	BS %	Al %	HS %
Janów													
	O	3.82	2.72	25.2	3.1	8.9	112.4	12.7	26.9	189.2	72.5	14.2	13.3
	A	3.74	2.78	10.7	0.6	0.6	3.1	0.6	13.9	29.5	16.6	47.1	36.3
	E	4.22	3.44	2.3	0.4	0.1	0.8	0.2	10.4	14.2	10.6	73.2	16.2
	Bhs	4.41	4.14	0.4	0.4	0.2	0.9	0.1	19.6	21.6	7.4	90.7	1.9
	Bs	4.65	4.63	0.1	0.5	0.1	0.6	0.1	2.8	4.2	30.9	66.7	2.4
	B/C	4.58	4.59	0.1	0.4	0.1	0.7	0.1	3.3	4.7	27.7	70.2	2.1
	C	4.83	4.67	0.2	0.2	0.0	0.7	0.1	1.4	3.0	46.6	46.7	6.7
Puławy													
	O	3.61	2.71	23.7	4.1	6.7	126.0	16.1	38.2	180.4	65.7	21.2	13.1
	E	3.67	2.77	9.5	0.8	0.9	14.6	3.2	10.1	39.1	49.9	25.8	24.3
	B	4.28	3.81	0.9	0.3	0.2	0.9	0.3	16.4	19.0	9.0	86.3	4.7
	B/C	4.42	4.28	0.3	0.4	0.2	0.8	0.1	4.2	6.0	25.0	70.0	5.0
	C	4.63	4.42	0.2	0.3	0.2	0.8	0.1	1.6	3.2	43.7	50.0	6.3
Ciekonek													
	O	3.54	2.49	28.0	8.5	13.7	16.4	6.8	114.3	187.7	24.2	60.9	14.9
	A/E	3.69	2.84	19.1	1.0	0.8	6.4	1.4	48.1	76.8	12.5	62.6	24.9
	B1	3.38	2.56	21.3	0.4	1.1	1.5	1.2	49.6	75.1	5.6	66.0	28.4
	B	3.57	2.73	20.7	0.3	0.8	2.2	0.7	63.6	88.3	4.6	72.0	23.4

Table 4. Content of organic matter, iron and heavy metals in soil

Sampling site	Density [g/cm^3]	Sand [%]	Silt [%]	Clay [%]	Fe_2O_3 [%]	Organic matter [%]
Janów O	0.21	—	—	—	—	85.8
Janów A	1.34	95.5	4.2	0.3	—	5.1
Janów E	1.33	96.8	2.7	0.5	—	0.43
Janów Bhs	1.33	97.7	1.8	0.5	2.28	—
Janów Bs	1.32	97.7	1.9	0.4	0.93	—
Janów B/C	1.42	97.7	1.6	0.7	—	—
Janów C	1.65	99.1	0.6	0.3	—	—
Pulawy O	0.30	—	—	—	—	88.7
Pulawy E	1.29	95.2	4.1	0.7	0.15	1.16
Pulawy B	1.32	96.4	3.2	0.4	1.27	—
Pulawy B/C	1.32	96.2	2.8	1.0	0.09	—
Pulawy C	1.41	96.4	2.2	1.4	—	—
Ciekonek O	0.27	—	—	—	—	91.0
Ciekonek A/E	1.46	52.7	37.9	9.4	—	6.4
Ciekonek Bhs	1.78	56.3	36.6	7.1	3.14	2.25
Ciekonek B	1.68	49.1	44.7	6.2	0.23	1.29

Soils in the Ciekonek stream catchment are classified as podzolic rankers. The soil profile is rather clear and forms O, A/E, Bhs, B and C horizons. Rock chips in amounts of above 40% were found in B and C horizons. Small amounts of non-weathered bed rock exist in upper horizons. Silt and clay are the main components of mineral horizons of the Ciekonek catchment soils. Contents of organic substances in the mineral horizons A/E, Bhs and B decrease with depth (6.40; 2.25 and 1.29% respectively); these values are high in comparison with most podzol soils. Due to significant quantities of silt, clay and organic substance, the total exchange capacity of the soil mineral horizons is high and ranges from 75.1 to 88.3 meq/kg. Due to high fraction of Al (above 60%) in CEC_E and low content of exchangeable Ca and Mg, soil waters in these soils are very affected by acid deposition.

The above results confirm the fact that the studied soils are easily affected by acidification due to mobilization of great amounts of Al. Buffer capability of the studied soils is limited only to upper horizons, mainly the organic one.

4. PRECIPITATION

The Janów Forest and the Pulawy region receive similar amounts of precipitation, from 503 to 638 mm, while in the Izerskie Mountains the amount is nearly twice as much, from 1090 to 1170 mm annually. Monthly precipitation is highest in September and October. The amount of rain water falling directly on to the soil surface in forests is significantly less than the precipitation measured in open areas, only 65–80% of rain water reach the soil surface, 20–35% moisten trees, bushes and litter.

Annual quantities and chemical composition of precipitation and throughfall are presented in Table 5. Most concentrations are highest at Pulawy. The lowest concentrations of Me^{2+} are found in the Ciekonek catchment. The chemical composition of precipitation and throughfall is very different. Throughfall waters are enriched with substances deposited on trees by dry deposition.

Table 5. Average precipitation and throughfall chemistry

Sampling site	mm/yr	pH	H^+ µeq/l	NH_4^+ µeq/l	Me^+ µeq/l	Me^{2+} µeq/l	Cl^- µeq/l	NO_3^- µeq/l	SO_4^{2-} µeq/l	F^- µeq/l	Al_{tot} µeq/l
Janów											
1993											
precipitation	489	4,51	31	50	72	96	39	84	134	4	9.4
throughfall	346	4,17	67	128	136	248	94	142	372	9	34.3
1994											
precipitation	520	4.59	26	61	55	96	58	49	137	7	8.2
throughfall	323	4.20	63	202	124	204	116	123	379	12	27.6
1995											
precipitation	638	4.64	23	41	67	88	61	69	92	4	10.4
throughfall	479	4.16	69	82	153	166	138	103	239	9	21.3
Ciekonek											
1994											
precipitation	1090	4.28	53	40	84	62	52	36	150	—	2.1
throughfall	683	3.87	136	89	108	173	57	114	344	2	2.4
1995											
precipitation	1170	4.21	61	41	68	87	67	46	138	—	3.2
throughfall	694	3.81	154	97	143	138	67	126	361	3	3.4
Pulawy											
1994											
precipitation	503	3.41	387	159	218	408	27	427	778	9	—
throughfall	418	3.20	628	291	428	687	42	689	1542	24	—
1995											
precipitation	584	3.46	349	112	148	261	18	343	540	7	—
throughfall	474	3.24	573	294	232	423	36	471	1134	31	—

Table 6. Sulphur and nitrogen deposition for different catchment

	Deposition of sulphur [gS/m²/rok]	Deposition of nitrogen [gN/m²/rok]
Janów		
'93		
precipitation	1.05	0.92
throughfall	2.06	1.31
'94		
precipitation	1.14	0.80
throughfall	1.96	1.47
'95		
precipitation	0.94	0.98
throughfall	1.83	1.24
Ciekonek		
'94		
precipitation	2.62	1.16
throughfall	3.76	1.94
'95		
precipitation	2.57	1.43
throughfall	4.01	2.17
Pulawy		
'94		
precipitation	6.26	4.12
throughfall	10.31	5.73
'95		
precipitation	5.05	3.72
throughfall	8.60	5.08

Table 6 shows calculated values of S and N deposition. Deposition of sulfur in the Janów Forest region is the lowest, 1.05 $gS \cdot m^{-2} \cdot yr^{-1}$ in precipitation and 2.06 $gS \cdot m^{-2} \cdot yr^{-1}$ in throughfall. In the Izerskie Mountains amounts of sulfur compounds found in precipitation and in throughfall are 2–3 times higher. The Pulawy area receives not only the greatest quantity of sulfur—10.3 $gS \cdot m^{-2} \cdot yr^{-1}$ in 1994, but also N-compounds ($NH_4^+ + NO_3^-$—5.73 $gN \cdot m^{-2} \cdot yr^{-1}$) as well.

The high deposition of nitrogen and sulfur compounds in the Izerskie Mountains is the main cause of significant forest degradation of the region. Leaching of Ca^{2+} and Mg^{2+} ions out of soils diminishes their buffering capability. Deposition of sulfur oxides is much more hazardous for podzol soils than nitrogen oxides. The significant part of sulfate ions is immobilized or bound temporarily in organic horizons of soil (O and A), while nitrate ions exhibit considerably lower affinity to soil matter and they are taken up by vegetation.

The small forest areas in the industrial zone of the Pulawy region are not in good conditions, presumably because of the very high deposition of S and N compounds. A significant decrease in sulfur and nitrogen deposition from 1994 to 1995 is a reason for optimism, although a longer series is needed.

5. SOIL WATERS

Concentrations in soil solutions and groundwaters for the different sites are presented in Table 7. The concentrations given are average values.

Table 7. Chemistry of soil waters (average values)

	Horizon	N	DOC [mg/l]	H^+ [μeq/l]	K^+ [μeq/l]	Na^+ [μeq/l]	NH_4^+ [μeq/l]	Mg^{2+} [μeq/l]	Ca^{2+} [μeq/l]	Al_i [μeq/l]	F^- [μeq/l]	Cl^- [μeq/l]	NO_3^- [μeq/l]	SO_4^{2-} [μeq/l]
Janów														
	O	39	34.6	166	56	69	22	112	421	115	21	164	18	647
			18.3–94.6	8–275	21–197	24–94	7–46	55–234	164–1123	43–287	14–32	105–742	2–46	259–842
	A	47	23.9	151	38	54	34	67	236	162	16	121	21	446
			7.9–32.4	7–195	6–121	2–74	18–53	32–134	141–424	49–286	9–26	64–412	0–53	224–874
	E	23	13.6	98	43	61	21	45	126	236	15	136	18	392
			3.7–26.9	8–155	18–126	29–109	3–49	23–89	74–218	112–473	9–34	27–400	0–58	204–784
	Bhs	17	3.4	38	51	67	21	44	131	564	12	204	12	605
			2.4–10.9	18–98	24–95	31–104	3–51	31–258	28–200	243–1269	0–30	53–307	0–49	215–1491
	Bs	16	6.6	32	68	74	16	121	217	392	31	183	44	473
			0.9–25.6	16–71	49–145	21–93	2–41	63–189	93–416	183–873	20–47	104–544	0–73	343–1024
	B/C	26	3.8	42	74	51	18	146	306	506	28	232	66	662
			0.9–12.6	7–62	43–161	26–144	7–42	54–205	125–499	228–934	21–56	155–473	18–113	421–936
	C	49	4.0	42	52	52	16	54	121	486	18	134	9	643
			1.8–26.3	18–97	26–97	4–101	8–36	30–228	40–368	176–832	8–36	29–306	0–38	263–1151
	GW	72	3.3	44	47	64	13	74	257	627	21	193	5	826
			0.7–5.8	2–158	15–81	41–114	4–43	32–212	126–503	130–1380	8–54	116–563	0–45	412–1218
Pulawy														
	O	21	39.6	142	67	42	73	98	244	156	12	84	78	512
			23.7–68.6	28–287	21–145	21–84	41–183	54–183	84–393	48–368	4–34	46–148	34–196	293–712
	E	17	16.4	106	43	52	52	72	163	296	18	66	96	432
			10.2–32.4	41–224	18–163	22–112	23–143	46–128	71–224	98–438	5–29	26–121	61–218	263–642
	B	26	9.6	49	48	34	46	96	246	459	16	91	124	682
			1.9–28.7	12–87	14–152	18–87	18–63	43–161	112–399	286–678	0–39	39–146	63–236	382–834
	B/C	28	4.3	51	31	26	31	53	234	612	19	67	131	743
			1.2–8.4	14–112	21–89	9–61	17–62	37–143	137–402	218–1232	11–34	38–96	71–189	593–812
	C	37	4.1	43	33	21	23	56	148	587	22	59	82	696
			0.9–12.3	34–87	12–121	8–49	9–34	18–96	67–308	263–986	8–31	21–108	41–153	327–913
Ciekonek														
	O	41	41.3	212	31	46	18	59	112	234	12	81	63	418
			30.1–64.6	126–268	24–69	21–78	9–67	41–83	47–167	162–293	4–32	41–136	41–94	311–486
	E	19	21.0	144	42	59	46	48	120	506	17	74	122	567
			11.4–36.7	94–228	12–83	12–143	31–84	23–96	86–212	312–687	4–31	47–98	56–171	294–743
	B	46	12.3	51	38	48	23	31	136	643	11	62	107	624
			6.2–19.4	23–78	21–49	19–73	2–38	24–115	74–201	423–763	0–29	32–94	59–194	326–983
	B/C	46	10.2	48	32	41	8	42	168	696	9	53	84	606
			5.1–18.4	32–59	11–67	23–69	1–24	21–87	112–248	430–1246	0–31	24–101	42–161	312–897

Sulfate is the dominating anion in soil solutions in all genetic horizon. The highest concentrations were observed in the lowest mineral horizons (B and C) and in groundwaters. High concentrations of Cl^- ions (121–232 meq/L) and small quantities of nitrates occur in soil solutions in the Janów Forest region. In soil waters in the Pulawy region and the Izerskie Mountains, chloride concentrations are much lower than in Janów Forest but nitrates were found in high concentrations (up to 131 meq/L) in good agreement with the high deposition of nitrogen in these regions. Fluorides concentrations are rather low (max. 31 meq/L).

Ca and Al ions are the dominating cations in soil solutions of the studied soils. The organic horizons are the most acidic ones, with H^+ concentration 2–3 times higher than in the mineral horizons. In soil solution of deepest mineral horizons of all soils and in groundwater of the Janów Forest region concentration of inorganic monomeric Al is higher than of divalent ions.

6. ALUMINUM CONTROL

Very high concentrations of Al_i were found in soil water at all studied sites. The highest concentration of inorganic monomeric Al was found in B/C horizons at Pulawy and Ciekonek catchment, in Bhs horizon and ground water at Janów forest; the values are 1232, 1246, 1269 and 1380 µeq/L respectively. A molar ratio $Al_i/(Mg^{2+}+Ca^{2+})=1.0$ is the limit above which toxic effects on tree roots by inorganic monomeric Al forms increase dramatically. The value is found to be exceeded in some mineral horizons of the studied soils: Bhs, C and GW at Janów forest; B/C and C at Pulawy; E, B and B/C at Ciekonek catchment, as shown in Figure 2.

The concentrations of Al_i and SO_4^{2-} were highly correlated (Janów: $r^2 = 0.71$, n = 183; Pulawy: $r^2 = 0.62$, n = 46; Ciekonek: $r^2 = 0.74$, n = 87) implying that the sulfate concentration is one of the most important factors in determining Al-mobilization.

Due to lack of analytical methods to determine concentrations of inorganic Al forms in soil water, their concentrations were calculated using the chemical equilibrium model ALCHEMI 4[18]. ALCHEMI 4 includes temperature and activity correction (Debye-Hückel). Fig. 3 shows the results of the calculations. The dominating aluminum species are Al^{3+}, sulfate and hydroxy-complexes. Saturation indices (Fig. 4) for synthetic gibbsite show that this simple model does not work properly for any of the studied soils.

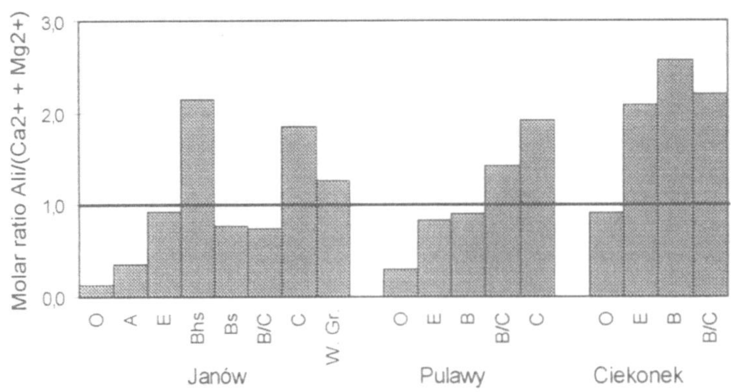

Figure 2. The molar ratio of $Al_i/(Ca^{2+}+Mg^{2+})$ for soil water from the Janów, Pulawy and Ciekonek catchment.

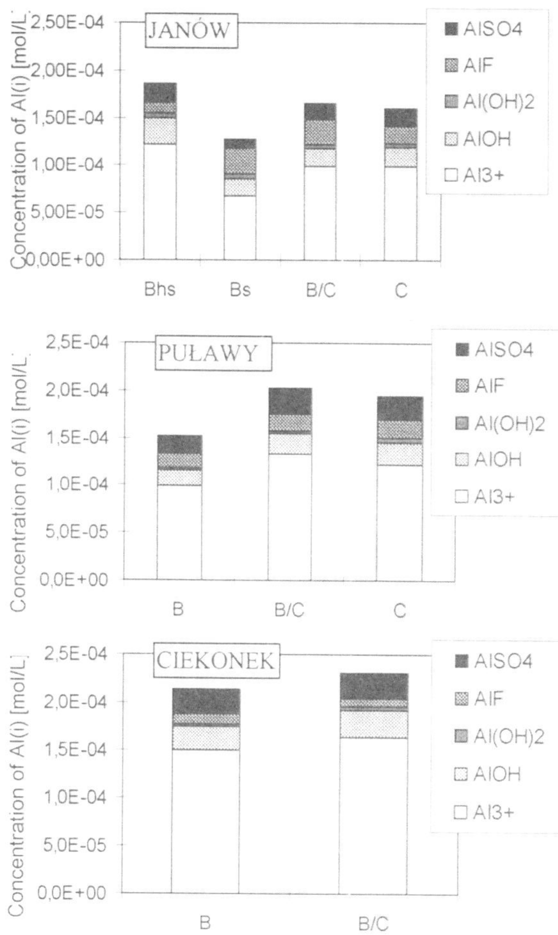

Figure 3. Aluminum inorganic species in soil water of B, B/C and C horizons of different soils.

7. SUMMARY

Among the studied sites, the Pulawy area receives the highest deposition of nitrogen and sulfur compounds. Also the deposition of ammonium and divalent cations (Ca^{2+} + Mg^{2+}) is the highest in this highly polluted region. The sandy soils (Janów Forest and Pulawy region) are acidified to pH_{H_2O} below 4.0 in two upper (O, A and O, E, respectively) horizons while in clay soils (Ciekonek) such low values are found in the whole soil profile.

Dominating cations and anions in soil water and ground water are Ca^{2+}, Al^{3+}, H^+ and SO_4^{2-}. As podzols and podsol rankers are poor in basic components, it may be assumed that deposition is a very important source of Ca^{2+} ions.

The molar ratio of $Al_i/(Mg^{2+}+Ca^{2+})$ is high in the all soils studied (up to 2.6) but it is found to exceed unity in all mineral horizons only at the Ciekonek catchment, where serious forest decline has occurred.

Future trends in soil properties and soil water concentrations will depend on deposition of sulfur and nitrogen oxides, but also on deposition of base cations.

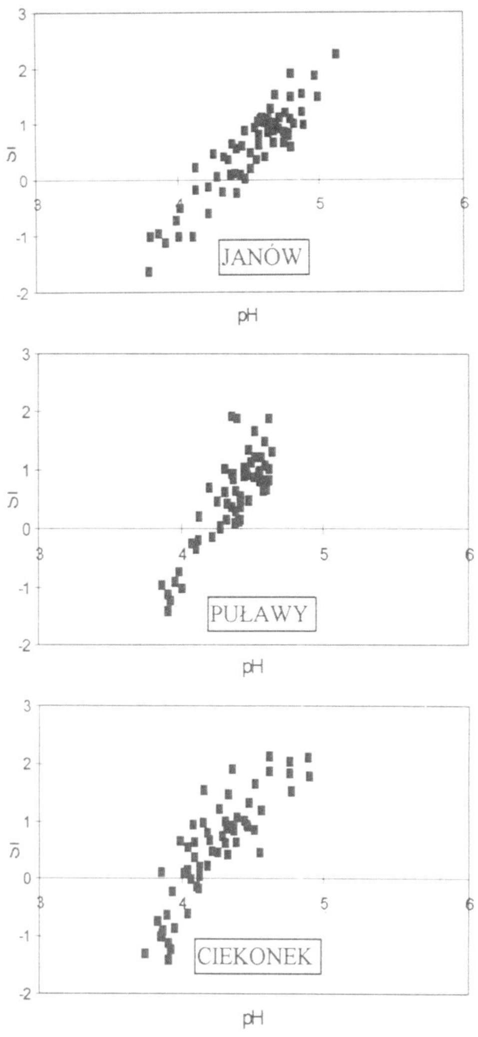

Figure 4. Saturation indices for syntetic gibbsite for soil water from mineral horizons from the Janów, Pulawy and Ciekonek catchment.

ACKNOWLEDGMENT

Author would like to express his gratitude to Prof. H.M. Seip and Dr M.R. Dudzinska for their providing valuable comments and corrections.

REFERENCES

1. Pawlowski, L., Dudzinska, M.R.: Environmental problems of Poland during economic end political transformation. *Ecol. Eng.*, 1994, *3*, 207–215.
2. Schaug, J.: Concentrations in air and deposition of acidic components in Poland. *Ecol. Eng.*, 1994, *3*, 217–224.
3. Tuovinen, J.P., Barret, K., Styve, H.: Transboundary acidifying pollution in Europe: Calculated fields and budgets 1985–93. *EMEP Data Report*, 1994.

4. Sverdrup, H., de Vries, W.: Calculating critical loads for the acidity with the simple mass balance method. *Water, Air and Soil Pollut.*, 1994, *72*, 143–162.
5. Rosseland, B.O., Eldhuset, T.D., Staurnes, M.: Environmental effects of aluminium. *Environ. Geochem. Health.*, 1990, *12*, 17–27.
6. Loekke, H., Bak, J., Falkengren-Grerup, U., Finlay, R.D., Ilvesniemi, H., Nygaard, P. H., Starr, M.: Critical loads of acidic deposition for forest soils: Is the current approach adequate? *Ambio*, 1996, *25*, 510–516.
7. Christophersen, N., Seip, H.M., Wright, R.F.: A model for stream water chemistry at Birkenes, Norway. *Water Resour. Res.*, 1982, *18*, 977–996.
8. Cosby, B.J., Hornberger, G.M., Galloway, J.N., Wright, R.F.: Times scales of catchment acidification. *Environ. Sci. Tech.*, 1985, *19*, 1144–1149.
9. Warfvinge, P., Sverdrup, H.: Calculating critical loads of acid deposition with PROFILE—a steady-state soil chemistry model. *Water, Air and Soil Pollut.*, 1992, *63*, 119–143.
10. Mulder, J., Breemen, N.V., Eijck, H.C.: Depletion of soil Al by acid deposition and implications for acid neutralization. *Nature*, 1989, *337*, 247–249.
11. Seip, H.M., Andersen, D.O., Christophersen, N., Sullivan, T.J., Vogt, R.D.: Variations in concentrations of aqueous aluminum and other chemical species during hydrological episodes at Birkenes, southernmost Norway. *J. Hydrol.*, 1989, *108*, 387–405.
12. Wesselink, L.G.: Time trends & mechanisms of soil acidification. Doctor Thesis, Wageningen—The Netherlands, 1994.
13. Vogt, R.D., Ranneklev, S.B., Mykkelbost, T.C.: The impact of acid treatment on soilwater chemistry at the HUMEX site. *Environ. Int.*, 1994, *20*, 277–286.
14. Vogt, R.D., Seip, H.M., Pawlowski, L., Kotowski, M., Odegard, S., Horvath, A., Andersen, S.: Potential acidification of soil and soil water: a monitoring study in the Janow Forest, southeastern Poland. *Ecol. Engin.*, 1994, *3*, 255–266.
15. Vogt, R.D., Godzik, S., Kotowski, M., Niklinska, M., Pawlowski, L., Seip, H.M., Sienkiewicz, J., Skotte, G., Staszewski, T., Szarek, G., Tyszka, J., and Aagaard, P.: Soil, soil water and stream water chemistry at some Polish sites with varying acid deposition. In: Pawlowski, L., Seip; H.M., & Sullivan, T.J., (eds.), *Aluminum in the environment*. Special issue of *J. Ecol. Chem.*, 1994, *3*, 325–356.
16. Larssen, T., Vogt, R.D., Seip, H.M.: A comparison of soil- and water chemistry in a catchment in China with sites in Poland and Norway. In: Pawlowski, L., Lacy, W.J., Uchrin, C.G., & Dudzinska, M.R., (eds.), *Chemistry for the protection of the environment*. Plenum Press, 1996, *51*, 421–434.
17. Hendershot, W.H., Duquette, M.: A simple barium chloride method for determining cation exchange capacity and exchangeable cations. *Soil Sci. Soc. Am. J.*, 1986, *50*, 605–608.
18. Schecher, W.D., Driscoll, C.T.: An evaluation of uncertainty associated with Al equilibrium calculations. *Water Resour. Res.*, 1987, *23*, 525–535.

THE ROLE OF CITRIC, LACTIC AND OXALIC ACIDS IN ALUMINUM MOBILIZATION FROM SOME POLISH AND CHINESE AGRICULTURAL SOILS

Xiao Ping Zhu

Institute of Soil Science
Academia Sinica
Nanjing, 210008, China
Department of Environmental Protection Engineering
Technical University of Lublin
40 Nadbystrzycka Str., 20-618 Lublin, Poland

ABSTRACT

The release of organic acids from plant roots and soil microorganisms can greatly increase aluminum solubility. Oxalic and citric acids have been identified as the major organic acids in soil solutions. Therefore, the objectives of this study were to examine the role of complexing and non-complexing organic acids in aluminum mobilization in different agricultural soil types from Poland and China. The concentration of labile aluminum in soils increases with the input of organic acid concentrations. According to batch experiments, the mobilization of Al in different soil by organic acids is as follows: ammonium fertilized brown soil > ammonium fertilized red soil > unfertilized brown soil > peat soil > calcareous soil. Aluminum mobilization is significantly influence by agricultural management. A lower amount of labile aluminum when using a non-complexing lactic acid illustrate the important role of complexation relative to only the H^+ ions released from the dissociation of carboxylic group in aluminum mobilization. Concentrations of Al mobilized by oxalic acid are 1.5~2.0 times and 8.7~22 times higher than mobilized by citric and lactic acids, respectively. Aluminum speciation in extracted soil solution using complexing organic acid shows the predominance of organically bound aluminum. When treating the sample with a non-complexing organic acid, the inorganic Al species account for more than 97% of labile aluminum.

Key words: aluminum mobilization, soil, organic acids.

Chemistry for the Protection of the Environment 3, edited by Pawłowski *et al.* Plenum Press, New York, 1998

1. INTRODUCTION

The geochemistry of aluminum in soils is strongly combined with natural organic matter. The release of organic acids from plant roots and soil microorganisms is of great importance for aluminum mobilization in acid soils. Many investigators discuss the undersaturation of Al concentration with respect to gibbsite and other minerals[1–6]. They found that complexation of aluminum by soil organic matter or aluminum adsorption on solid phase organic matters contributed to the control of Al solubility. The leaching results obtained by James and Riha[7,8] suggest that dissolved organic matter in forest soil significantly intensify aluminum mobilization by acid deposition. They observed a synergism between inorganic acid and dissolved organic matter in aluminum mobilization.

A number of authors have discussed an effect of organic acids on aluminum complexation in forest soils, however little attempt has been made to compare the mechanisms of Al solubility from agricultural soils by complexing and non-complexing organic acids. Therefore, the objectives of this study is to examine the role of complexing and non-complexing organic acids on the aluminum mobilization in some very different agricultural soils from Poland and China.

2. METHODS AND MATERIALS

Four soil profiles were sampled from Lublin, Poland and one profile from Yixing, China. The soils used in these batch experiments were sampled from 0–20 cm (topsoil) and 20–40 cm (subsoil) horizons.

The characteristics concerning parent material, soil type, vegetation, and annual precipitation at the sampling sites are presented in Table 1. Soil samples were air-dried at room temperature and passed through a 2-mm sieve. Physico-chemical properties of the soil used in this study are presented in Table 2. The thermogravimetric (using Paulik–Erdey system, model D-102, MOM–Hungary) analysis of soils revealed the mixed mineral composition with a predominant amounts of feldspar and a small amounts of illite in mineral soils. The calcareous soil contained less than 3.3% of calcium carbonate. Organic

Table 1. Characteristic of selected soil profiles

Code	Site	Precipitation (mm)	Parent material	Vegetation	Soil classification	Landuse type
CT-1	Yixing	1158	Q_4 red earth	Tea bush	Avrisols	Arable
CT-2	Yixing	1158	Q_4 red earth	Tea bush	Avrisols	Arable
F-1	Elizówka	566	Loess	Wheat	Brown earth	Arable
F-2	Elizówka	566	Loess	Wheat	Brown earth	Arable
UF-1	Elizówka	566	Loess	Wheat	Brown earth	Arable
UF-2	Elizówka	566	Loess	Wheat	Brown earth	Arable
LP-1	Ludvin	580	Peat	Grass	Peat	Pasture
LP-2	Ludvin	580	Peat	Grass	Peat	Pasture
LC-1	Sielec	580	Calcite	Wheat	Calcareous	Arable
LC-2	Sielec	580	Calcite	Wheat	Calcareous	Arable

CT—Red soil, F—Fertilized brown soil, UF—Unfertilized brown soil, LP—Peat soil, LC—Calcareous soil; 1—topsoil (0—20 cm depth) and 2—subsoil (20–40 cm depth).

Table 2. Selected soil physico-chemical characteristics

Code	Depth (cm)	pH (H_2O)	Exch. Al mM/kg	Subst. org. (%)	$CaCO_3$ (%)	Kaolinite (%)	Illite (%)	FeOOH (%)	Feldspar (%)
CT-1	0-20	4.85	1.88	2.10	/	10.20	5.30	1.20	79.00
CT-2	20-40	5.09	2.74	2.20	/	8.10	5.20	1.10	81.20
F-1	0-20	4.08	1.87	2.30	/	/	2.10	trace	95.60
F-2	20-40	4.47	1.97	1.60	/	/	2.30	0.60	94.20
UF-1	0-20	5.90	0.24	1.80	/	/	6.40	/	89.00
UF-2	20-40	6.00	0.24	1.80	/	/	5.30	/	92.90
LP-1	0-20	5.41	Trace	68.40	/	/	22.00	/	9.60
LP-2	20-40	5.46	1.90	49.40	/	/	20.20	/	30.40
LC-1	0-20	7.45	Trace	3.70	3.25	/	21.40	/	71.70
LC-2	20-40	7.64	Trace	2.00	1.80	/	15.60	/	80.60

CT—Red soil, F—Fertilized brown soil, UF—Unfertilized brown soil, LP—Peat soil, LC—Calcareous soil; 1—topsoil and 2—subsoil.

matter (measured as loss on ignition at 823K) in the peat soil was the predominant component, followed by illite and feldspar. Kaolinite was found only in the red soil. The soil pH_{H_2O} was measured in suspension of air-dried soils in water and exchangeable Al were extracted with 1M KCl[9]. Total contents of the major oxide components in the parent material analyzed using X-ray fluorescence spectrometer (made by Philips, model PW-2400) are given in Table 3. The chemical composition of all mineral soils is characterized by high content of Si (i.e. higher than 70%) and low contents of Al and basic cations (Ca, Mg, K, Na). The aluminum oxide content in Chinese soils is slightly higher than in the Polish soils.

Three kinds of short chained aliphatic organic acids (citric, oxalic and lactic) were used in batch experiments to examine their ability to mobilize Al. The concentrations of the organic acid was as followed: 0.2 mM, 1.0 mM and 5.0 mM. Subsamples of 400 mg of soil were shaken for 24 hours with 40 ml of organic acid solution and then centrifuged to

Table 3. The total contents of major oxide components in the parent material

Code	Site	SiO_2 (%)	Al_2O_3 (%)	Fe_2O_3 (%)	CaO (%)	MgO (%)	Na_2O (%)	K_2O (%)
CT-1	Yixing	74.38	14.11	5.38	0.00	1.67	0.65	0.18
CT-2	Yixing	73.85	14.01	5.55	0.00	1.75	0.63	0.16
F-1	Elizówka	82.26	8.40	2.13	1.49	2.86	0.62	0.43
F-2	Elizówka	82.77	8.48	2.24	1.38	2.97	0.65	0.47
UF-1	Elizówka	79.96	8.82	2.66	1.29	3.19	0.67	0.69
UF-2	Elizówka	81.07	8.77	2.80	1.24	3.15	0.73	0.66
LP-1	Ludwin	19.32	2.48	2.37	0.00	0.91	0.19	3.12
LP-2	Ludwin	35.78	3.60	1.48	0.76	1.46	0.23	2.67
LC-1	Sielec	74.51	9.02	2.38	2.73	3.19	0.86	2.55
LC-2	Sielec	76.98	9.33	2.40	2.29	3.19	0.84	2.67

CT—Red soil, F—Fertilized brown soil, UF—Unfertilized brown soil, LP—Peat soil, LC—Calcareous soil; 1—topsoil and 2—subsoil.

Table 4. Thermodynamic constants used in the species calculations

Reactions	log(K)
$Al^{3+} + H_2O \rightleftharpoons Al(OH)^{2+} + H^+$	−4.99
$Al^{3+} + 2H_2O \rightleftharpoons Al(OH)_2^+ + 2H^+$	−10.00
$Al^{3+} + 4H_2O \rightleftharpoons Al(OH)_4^- + 4H^+$	−23.00
$Al^{3+} + SO_4^{2-} \rightleftharpoons AlSO_4^+$	3.01
$Al^{3+} + 2SO_4^{2-} \rightleftharpoons Al(SO_4)_2^-$	4.90
$Al^{3+} + H_4SiO_4^0 \rightleftharpoons AlH_3SiO_4^{2+} + H^+$	−1.07
$Al(OH)_{3(solid)} + 3H^+ \rightleftharpoons Al^{3+} + 3H_2O$	9.35
Al^{3+} + Oxalate $\rightleftharpoons$ $AlOxalate^+$	7.8
Al^{3+} + 2Oxalate $\rightleftharpoons$ $Al(Oxalate)_2^-$	13.38
Al^{3+} + 3Oxalate $\rightleftharpoons$ $Al(Oxalate)_3^{3-}$	16.68
Al^{3+} + Citrate $\rightleftharpoons$ $AlCitrate^0$	9.72
Al^{3+} + Citrate $\rightleftharpoons$ $Al(Citrate)_2^{3-}$	14.98

Data are mainly taken from these listed in the ALCHEMI 4 program.

obtain clear solution. Labile Al was measured by complexation with 8-hydroxyquinoline at pH = 8.3, followed by rapid extraction in methyl-isobutyl ketone[10].

The computer program AlCHEMI4[11] was used to calculate ion activities. The main species of inorganic Al consider in this paper are complexes with hydroxide, sulfate, silicate and fluoride ions. The thermodynamic constants used in calculations of the Al species are gathered in Table 4[11,12]. Saturation indices (SI) were calculated to evaluate the mechanism controlling the concentration of Al mobilized from the soils by organic acids.

3. RESULTS AND DISCUSSION

3.1. Effect of Soil Type on Al Mobilization

Figures 1–3 show the influence of the short-chained, aliphatic, mono-, di- and tricarboxylic acids in mobilizing Al from soils. There are significantly positive and non-linear relationships between the concentrations of organic acids and labile Al. Oxalic acid with a concentration of 5.0 mM managed to mobilize 0.484 mM of labile Al from the red subsoil Already at 0.2 mM and 1.0 mM solution of oxalic acid the labile Al concentrations accounted for 12%, 56% of that in the 5.0 mM oxalic acid respectively. Significant mobilization of Al is observed for brown soil and red soil (Figs 1–3). Apparently ammonium fertilized brown soil has the highest content of labile aluminum. The lowest concentration of labile aluminum was as expected found in the calcareous soil. Subsoil horizon (20–40 cm) release more Al than the topsoil. These results indicate a downward movement of labile Al in all the soil profiles. According to the amounts of labile aluminum mobilized by organic acids, the relative sensitivity of soils rank as follow: ammonium fertilized brown soil > ammonium fertilized red soil > unfertilized brown soil > peat soil > calcareous soil.

The ammonium fertilized brown soil shows higher ability to neutralize the organic acids. The labile aluminum dissolved by oxalic acid (5.0 mM) reached 0.45 and 0.59 mM in the surface and subhorizon of fertilized brown soil, respectively, which is up to 15–25% higher than in the unfertilized brown soils. Apparently long-term ammonium fertilization significantly change chemical status of Al in soils. Much more Al is mobilized into soil

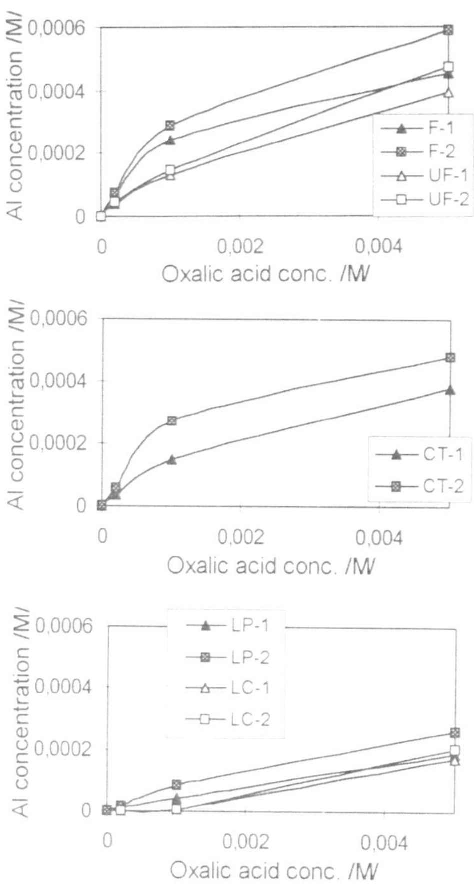

Figure 1. Labile Al concentrations versus oxalic acid concentration (M). (F—fertilized brown soil; UF—unfertilized brown soil; CT—red soil; LP—peat soil; LC—calcareous soil; 1—topsoil and 2—subsoil).

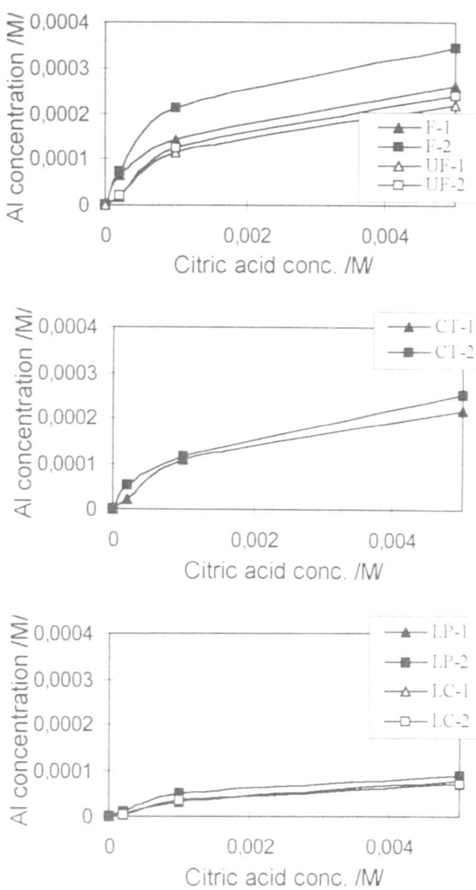

Figure 2. Labile Al concentrations versus citric acid concentration (M). (F—fertilized brown soil; UF—unfertilized brown soil; CT—red soil; LP—peat soil; LC—calcareous soil; 1—topsoil and 2—subsoil).

solution, i.e. existing as reactive form on the soil. Porebska and Mulder[13] also confirmed the influence of long-term fertilization on Al mobilization.

The mobilization of labile Al in peat soil and calcareous soil is lower. The amount of labile aluminum accounted for 43% and 35% of that in fertilized brown soil, respectively. The fertilized red soil also gave high labile aluminum concentrations in the oxalic acid treatment; i.e. 3.79 and 4.84 mM in the top- and subhorizon, respectively. Although red soil also was intensively fertilized, the amount of labile aluminum accounted for only 80% of that in fertilized brown soil. Apart from differences in physico-chemical properties of the soils this might be attributed to the type of fertilizers used. In Poland the most common nitrogen fertilizers are NH_4NO_3 and $(NH_4)_2SO_4$, while in China a large amount of NH_4HCO_3 is being applied. Application of mineral fertilizers resulted in an intense aluminum mobilization in forest soils[14]. In a laboratory study, Kotowski et al.[15] reported strongly elevated aluminum concentration by NH_4NO_3 and $(NH_4)_2SO_4$ from forest sandy soil. The Al mobilization by fertilizer salts was higher than by sulfuric acid.

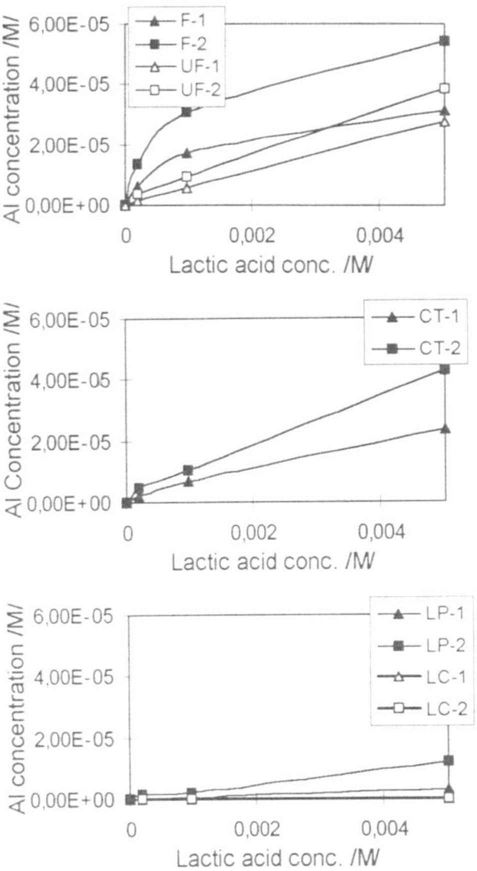

Figure 3. Labile Al concentrations versus lactic acid concentration (M). (F—fertilized brown soil; UF—unfertilized brown soil; CT—red soil; LP—peat soil; LC—calcareous soil; 1—topsoil and 2—subsoil).

3.2. Effect of Organic Acids on Al Mobilization

There were found significant relationships between the labile Al concentrations and the amounts of oxalic, citric and lactic acids in the leachates. But the ability of organic acids to mobilize aluminum vary. Figure 4 shows the labile Al concentrations with increasing concentration of organic acids in fertilized brown soil. The affinity of Al to oxalic acid

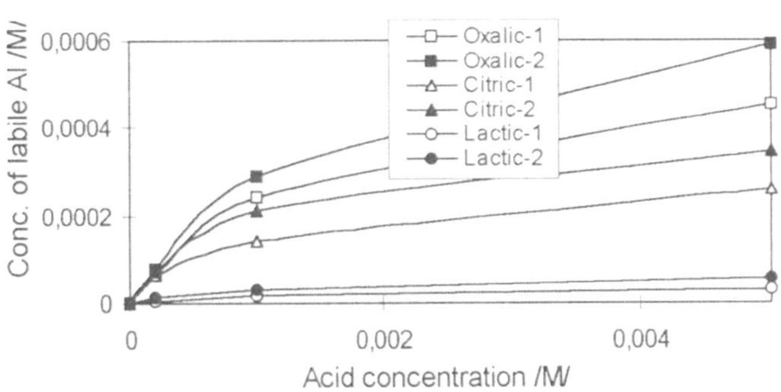

Figure 4. The comparison of three organic acids in mobilizing of aluminum from fertilized brown soil (1—top soil and 2—subsoil).

Table 5. The difference of labile Al between oxalic and citric acids

	Difference of labile Al for different organic acid concentrations		
Soil type	0.2 mM	1.0 mM	5.0 mM
CT-1	1.17E−05	4.30E−05	1.60E−04
CT-2	6.10E−06	1.54E−04	2.30E−04
F-1	5.56E−07	9.90E−05	1.91E−04
F-2	2.57E−06	7.70E−05	2.46E−04
UF-1	2.11E−05	1.80E−05	1.74E−04
UF-2	2.14E−05	1.80E−05	2.33E−04
LP-1	4.60E−06	9.29E−06	1.14E−04
LP-2	4.60E−06	3.51E−05	1.76E−04
LC-1	−3.70E−06	−2.67E−05	1.02E−04
LC-2	−4.44E−06	−2.96E−05	1.33E−04

CT—Red soil, F—Fertilized brown soil, UF—Unfertilized brown soil, LP—Peat soil, LC—Calcareous soil; 1—topsoil and 2—subsoil.

is the highest, followed by citric and lactic acids. These differences increase with organic acid concentration. Fertilized red subsoil treated by 0.2, 1.0 and 5.0 mM oxalic acid solutions gave labile Al concentrations of 58, 270, and 484 μM, which was 1.5, 1.4, 1.7 times higher than that at the same level of citric acid and 18.4, 22.7, 16.3 times higher than at the same level of lactic acid concentrations (see Tables 5 and 6). It is apparent that oxalic and citric acids have higher affinity for Al in soil solution. Generally, concentrations of Al mobilized by oxalic acid were 1.5~2.0 times and 8.7~22 times higher than by citric and lactic acids, respectively.

Organic-aluminum complexes in soil solution are usually formed by displacement of one or more protons in organic acid by Al ions. This reaction may be written as:

$$Al^{3+} + H_nL = [AlH_{(n-i)}L]^{(3-i)+} + iH^+ \ (n \geq i;\ n,i > 0)$$

Table 6. The difference of labile Al between citric and lactic acids

	Difference of labile Al for different organic acid concentrations		
Soil type	0.2 mM	1.0 mM	5.0 mM
CT-1	2.05E−05	1.01E−04	1.96E−04
CT-2	4.75E−05	1.06E−04	2.11E−04
F-1	5.87E−05	1.25E−04	2.30E−04
F-2	6.08E−05	1.81E−04	2.90E−04
UF-1	1.63E−05	1.08E−04	1.91E−04
UF-2	1.82E−05	1.16E−04	2.01E−04
LP-1	4.07E−06	3.26E−05	7.33E−05
LP-2	9.22E−06	4.82E−05	7.55E−05
LC-1	3.70E−06	3.00E−05	6.89E−05
LC-2	4.44E−06	3.41E−05	7.07E−05

CT—Red soil, F—Fertilized brown soil, UF—Unfertilized brown soil, LP—Peat soil, LC—Calcareous soil; 1—topsoil and 2—subsoil.

Organo-aluminum complexes can be formed through ligand exchange reactions on clay surfaces. For example, the formation of aluminum-oxalate complexes on the clay surface can be illustrated as:

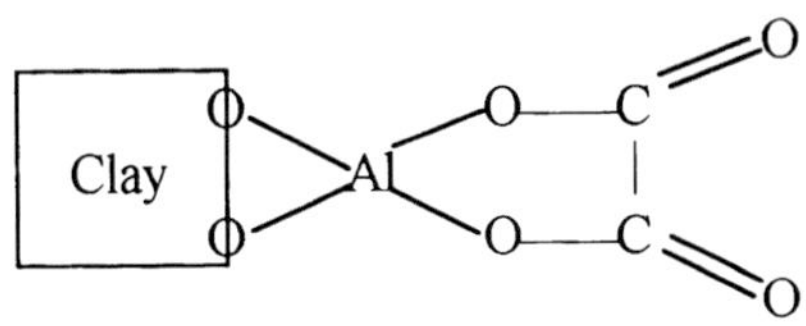

The weakening of O-Al bonds to the clay by surface-coordinated ligands allows the organic Al to dissolve. Organic complexation enhance dissolution of aluminum-containing minerals and allow high Al concentrations in soil solutions without precipitation.

The mobilization of Al is due to both the complexing action of organic anions and the dissolution or exchange reactions with Al on soil solid phase by the protons released from the organic acids. The differences in concentrations of labile Al in the batch study show that the degree of Al mobilization is influenced by both the type of organic acid and its concentration (see Tables 5 and 6). Factors determining the degree of dissolution of soil Al by organic acids are the chemical affinity of the organic to chelate aluminum and the acid strength (pK_a) of the potential bonding sites (i.e. the dissociation degree) of the acidic.

Furthermore, when considering the chemical affinity of chelating agents to Al both the type and position of functional groups need to be considered. Organic acid with higher $logk_{Al}$ value leach greater amounts of Al than organic acids with lower $logk_{Al}$ values[5,16,17]. Properties of citric, oxalic and lactic acids based on the literature data are presented in Table 7. The greater ability of both citric and oxalic acid to mobilize aluminum from soils than lactic acid reflect nicely the higher $logk_{Al}$ constants of the citric and oxalic acids relative to lactic acid. On the other hand, regarding the citric acid relative to oxalic acids, the case is reverse. Apparently the use of formation-constants to evaluate the ability of organic acid to Al mobilization might be true only to a certain extent. This was also found by Pohlman and McColl[17]. Catechol and salicylic acids, both with high $logk_{Al}$ (16.6 and 12.9, respectively) mobilized much less labile Al than did citric and malic acids with lower $logk_{Al}$ (7.37 and 5.14 respectively). Thus $logk_{Al}$ might not reflect the ability of organic ligands to mobilize aluminum.

The pH value of initial extraction solution did not contribute to explain the degree of Al dissolution from soils. As shown in Figure 5, concentrations of labile Al in oxalic acids were 4 to 10 times higher than in lactic acid treatments even though the initial pH values of 1.0 mM and 5.0 mM lactic acid were 3.50, 3.00 respectively, which was nearly the same as in 0.2 mM and 1.0 mM oxalic acid (3.53 and 2.93). Furthermore, lactic acid with

Table 7. The dissociation and stability constants for Al complexes of citric, oxalic and lactic acids

Name	Molecular weight	Dissociation constants			Stability constants
		pK_1	pK_2	pK_3	$logk_{Al}$
Citric acid (H_3L)	192	3.08	4.74	5.40	7.37
Oxalic acid (H_2L)	90	1.23	4.27	—	6.53
Lactic acid (HL)	96	3.87	—	—	2.38

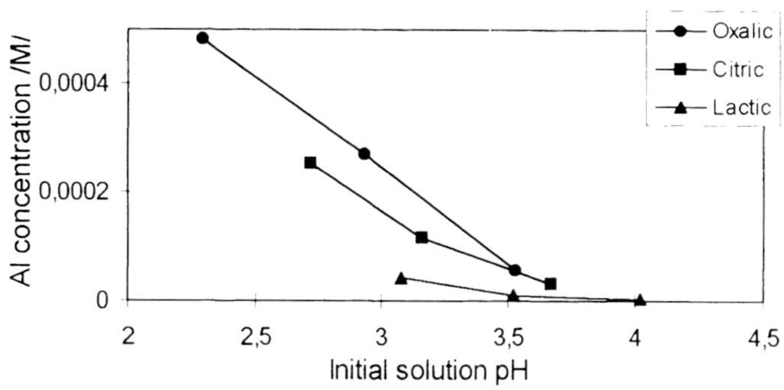

Figure 5. Al concentrations in red soil (subsoil) versus initial pH of organic acids.

initial pH values of 3.00 mobilize less Al in all the studied soils than oxalic acid with initial pH value of 3.50. The comparison of citric acid treatments with lactic acid treatments showed similar tendency.

The failure of both the initial acidity of the leachate as well as the formation constants ($logk_{Al}$) in explaining the Al mobilization suggest that the chemical structure of the organic acids must be important. At the same organic acid concentration, citric acid with three carboxylic functional groups gave rise to lower labile Al concentrations than oxalic acid with two carboxylic functional groups. This results indicate that the sterical configuration of ligand functional groups outweighed the importance of their number. The functional groups in citric acid are good potential bonding sites for Al since the complexation is favored by formation of a six membered chelate ring structure as shown in Figure 6. It was therefore assumed that organic acids possessing certain functional group arrangement, such as di- or tricarboxylic acids with a β-hydroxyl group, will dissolve more Al from soils. The interaction of β-hydroxyl and carboxyl groups with aluminum results in the formation of Al complexes with stable 5- or 6-membered chelate structure (Fig. 6). In contrast, lactic acid with only a α-hydroxyl group, could not form any chelate structure with aluminum. The dissolution of Al from soils by this kind of non-complexing organic acids can only caused by hydrogen ion exchange reaction and acid dissociation. Consequently, the amounts of labile Al in soil solutions mobilized by citric or oxalic acid are much higher than by lactic acid.

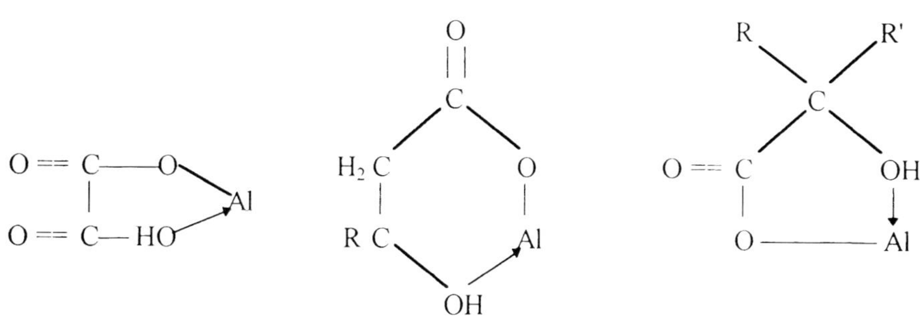

I
5-membered structure

II
6- membered structure

III
5-membered structure

Figure 6. The suggested structures involved in organo-aluminum complexes: I—oxalic acid; II, III—citric acid.

Table 8. The distribution of inorganic and organic aluminum species in oxalic acid solutions

	0.2 mM			1.0 mM			5.0 mM		
Soil type	Al-org	Al-i	Alorg/Al_L	Al-org	Al-i	Alorg/Al_L	Al-org	Al-i	Alorg/Al_L
CT-1	3.41E−05	1.60E−09	1.0	1.51E−04	4.68E−09	1.0	3.79E−04	2.00E−08	1.0
CT-2	5.80E−05	2.89E−09	1.0	2.70E−04	8.32E−09	1.0	4.84E−04	2.83E−08	1.0
F-1	6.56E−05	3.65E−09	1.0	2.41E−04	7.20E−09	1.0	4.52E−04	2.46E−08	1.0
F-2	7.67E−05	4.15E−09	1.0	2.89E−04	7.84E−09	1.0	5.90E−04	2.69E−08	1.0
UF-1	3.89E−05	2.38E−09	1.0	1.32E−04	2.26E−09	1.0	3.93E−04	1.77E−08	1.0
UF-2	4.33E−05	2.50E−09	1.0	1.44E−04	2.22E−09	1.0	4.72E−04	1.91E−08	1.0
LP-1	8.67E−06	1.08E−09	1.0	4.26E−05	5.02E−10	1.0	1.90E−04	4.17E−09	1.0
LP-2	1.53E−05	1.66E−09	1.0	8.55E−05	9.81E−10	1.0	2.63E−04	8.42E−09	1.0
LC-1	8.73E−10	2.42E−07	0.004	1.48E−06	4.18E−10	1.0	1.71E−04	5.68E−09	1.0
LC-2	8.41E−10	4.86E−07	0.002	3.33E−06	1.13E−09	1.0	2.04E−04	7.10E−09	1.0

CT—Red soil, F—Fertilized brown soil, UF—Unfertilized brown soil, LP—Peat soil, LC—Calcareous soil; 1—topsoil and 2—subsoil; Al_L—concentration of labile Al.

4. RELATIVE DISTRIBUTION OF ALUMINUM SPECIES

Concentrations of organic and inorganic Al species and their relative distribution in the batch extraction leachates are given in Tables 8–10. It may be seen that the relative concentration of organic aluminum complexes increased with the organic acid input concentrations for all the studied soils. For example, in red topsoil, the concentration of Al oxalate complexes in 0.2 mM oxalic acid is only 34.1 μM whereas in 5.0 mM oxalic acid-treated solution its concentration has increased to 379 μM. Among the studied soils, fertilized brown soil has the highest concentration of Al oxalate complexes, reaching 76.7 μM in 0.2 mM oxalic acid and 590 μM in 5.0 mM oxalic acid treatments. Although the most acid soils leachates contain the highest amounts of organic-bound aluminum, as shown in Figure 7, the concentration of organic aluminum species does not depend on the solution

Table 9. The distribution of inorganic and organic aluminum species in citric acid solutions

	0.2 mM			1.0 mM			5.0mM		
Soil type	Al-org	Al-i	Alorg/Al_L	Al-org	Al-i	Alorg/Al_L	Al-org	Al-i	Alorg/Al_L
CT-1	2.24E−05	9.72E−09	1.000	1.07E−04	6.33E−07	0.994	2.14E−04	4.45E−06	0.980
CT-2	5.19E−05	2.46E−08	1.000	1.15E−04	5.65E−07	0.995	2.48E−04	5.95E−06	0.977
F-1	6.48E−05	1.68E−07	0.997	1.41E−04	1.23E−06	0.991	2.56E−04	4.84E−06	0.981
F-2	7.39E−05	1.22E−07	0.998	2.10E−04	1.49E−06	0.993	3.37E−04	6.73E−06	0.980
UF-1	1.78E−05	1.91E−09	1.000	1.14E−04	2.51E−07	0.998	2.16E−04	2.41E−06	0.989
UF-2	2.19E−05	2.07E−09	1.000	1.26E−04	2.43E−07	0.998	2.36E−04	2.76E−06	0.988
LP-1	4.07E−06	2.06E−10	1.000	3.33E−05	1.39E−09	1.000	7.62E−05	8.34E−08	0.999
LP-2	1.07E−05	4.10E−10	1.000	5.04E−05	5.68E−09	1.000	8.72E−05	2.01E−07	0.998
LC-1	2.00E−06	1.70E−06	0.540	3.00E−05	3.07E−10	1.000	6.88E−05	1.04E−07	0.998
LC-2	1.43E−06	3.01E−06	0.323	3.41E−05	2.91E−10	1.000	7.06E−05	1.17E−07	0.998

CT—Red soil, F—Fertilized brown soil, UF—Unfertilized brown soil, LP—Peat soil, LC—Calcareous soil; 1—topsoil and 2—subsoil; Al_L—concentration of labile Al.

Table 10. The distribution of inorganic and organic aluminum species in lactic acid solutions

Soil	0.2mM			1.0 mM			5.0 mM		
type	Al-org	Al-i	Alorg/AlL	Al-org	Al-i	Alorg/AlL	Al-org	Al-i	Alorg/AlL
CT-1	3.22E–09	1.76E–06	0.002	5.25E–08	6.52E–06	0.008	3.63E–07	2.28E–05	0.016
CT-2	8.23E–09	4.34E–06	0.002	8.01E–08	1.02E–05	0.008	6.63E–07	4.22E–05	0.015
F-1	1.24E–08	6.19E–06	0.002	1.14E–07	1.72E–05	0.007	4.44E–07	3.10E–05	0.014
F-2	2.77E–08	1.32E–05	0.002	2.21E–07	3.04E–05	0.007	8.23E–07	5.32E–05	0.015
UF-1	4.40E–10	1.42E–06	0.000	4.75E–08	5.42E–06	0.009	4.76E–07	2.72E–05	0.017
UF-2	1.53E–09	3.62E–06	0.000	8.17E–08	9.45E–06	0.009	6.74E–07	3.77E–05	0.018
LP-1	8.80E–11	2.23E–07	0.000	1.67E–09	6.88E–07	0.002	7.79E–08	2.79E–06	0.027
LP-2	2.62E–10	1.43E–06	0.000	6.15E–09	2.14E–06	0.003	3.05E–07	1.15E–05	0.026
LC-1	2.12E–19	1.00E–09	0.000	1.90E–16	9.99E–10	0.000	2.14E–11	7.45E–10	0.028
LC-2	1.93E–19	1.00E–09	0.000	2.27E–16	9.99E–10	0.000	2.11E–11	7.44E–10	0.028

CT—Red soil, F—Fertilized brown soil, UF—Unfertilized brown soil, LP—Peat soil, LC—Calcareous soil; 1—topsoil and 2—subsoil; AlL—concentration of labile Al.

pH values. Driscoll et al.[18] also found that organic bound Al in natural soil solutions does not depend on soil acidity. Table 8 shows the great predominance of Al oxalate complexes in oxalic acid solutions. The organic aluminum species in oxalic acid solutions account for nearly 100% of the total labile Al species irrespective of the soil type and organic acid concentration, except for the calcareous soil. This results reflects the strong ability of oxalic acid to chelate aluminum.

The concentrations of inorganic Al species in organic acid-extracted solutions are very low, usually less than 0.03 μM. They increase with an increase of input concentrations of organic acids. The relative distribution of inorganic Al species for soils treated by 0.2 mM oxalic acid solution is shown on Figure 8. The concentration of inorganic Al species depends mainly on pH values. Fertilized soils, which have the lowest soil pH_{H_2O}, shows a predominance of trivalent Al species; i.e. up to 98%. $Al(OH)_2^+$ and $AlOH^{2+}$ are the predominant species of inorganic aluminum fraction (up to 60%–80%) in peat soil, while in calcareous soils (pH_{H_2O} ranges from 7.45 to 7.64) $Al(OH)_4^-$ dominates. The increase of organic acid input to 5.0 mM results in the predominance of trivalent Al species in all the studied soils. This was even the case in the calcareous soils (Fig. 9).

The relative distribution of aluminum species in the citric acid treatments is similar to that in the oxalic acid treatments (cf. Tables 8 & 9). The ratio of aluminum-citrate com-

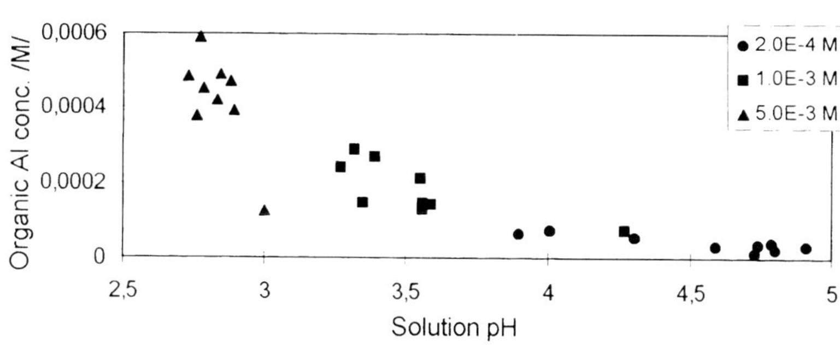

Figure 7. The changes of organic aluminum concentration with solution pH in oxalic acid treatments.

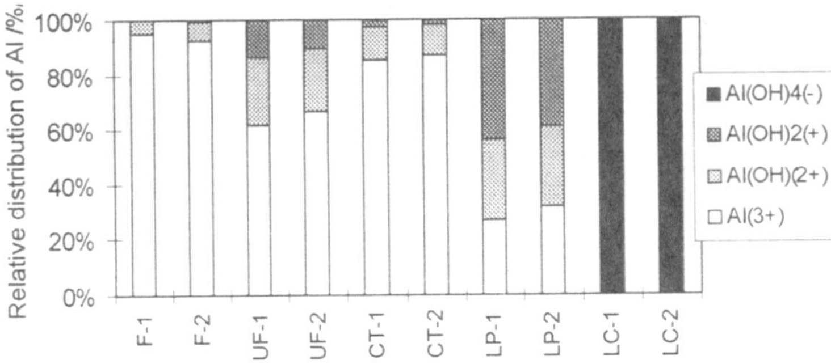

Figure 8. The relative distribution of inorganic aluminum species in 0.2 mM oxalic acid solutions. (F—fertilized brown soil; UF—unfertilized brown soil; CT—red soil; LP—peat soil; LC—calcareous soil; 1—topsoil and 2—subsoil).

plexes reaches more than 99.7%, except for the calcareous soil. In contrast to oxalic acid treatment, the distribution of aluminum-citrate complexes showed less decreasing trend with an increase of citric acid concentration, although there is still some amount of unbounded citric acids left in the solution. The distributions of inorganic aluminum species in the citric acid treatments showed the same tendency as in the oxalic solution (cf. Figs. 8 & 10).

The lactic acid treatments showed an opposite trend in the relative Al distribution compared to oxalic and citric acids (cf. Tables 8–10). In all cases, inorganic aluminum species accounted for the major contribution to aqueous aluminum. Concentrations of organic Al species in 0.2 mM lactic acid were very small, usually less than 0.03 μM, which accounts for less than 0.2% of total extracted aluminum. Even in the 5.0 mM lactic acid treatments, the organic Al contribution range only from 1.4% to 2.8% of total Al. Figure 11 shows the relative speciation of inorganic Al in 0.2 mM lactic acid treatments. Trivalent aluminum species in fertilized brown soil reach the highest content, up to 96%, while unfertilized brown soil is dominated by inorganic Al monomers ($AlOH^{2+}$ and $Al(OH)_2^+$). In calcareous soil, $Al(OH)_4^-$ is the predominant species. As lactic acid concentrations increase to 5.0 mM, trivalent aluminum become dominant like in citric and oxalic treatments.

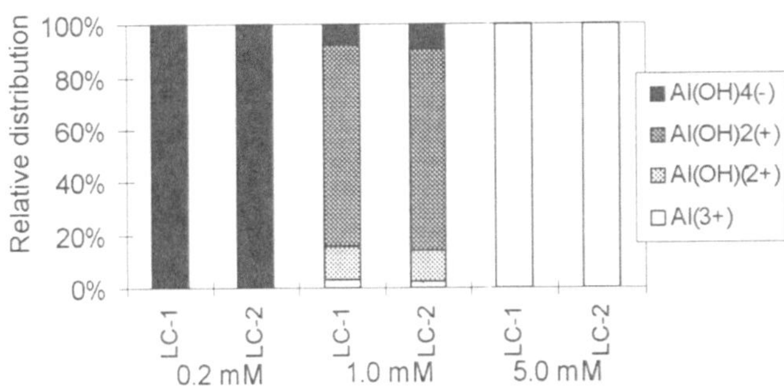

Figure 9. The relative distribution of inorganic Al species in calcareous soil treated by oxalic acid. (LC—calcareous soil; 1—topsoil and 2—subsoil).

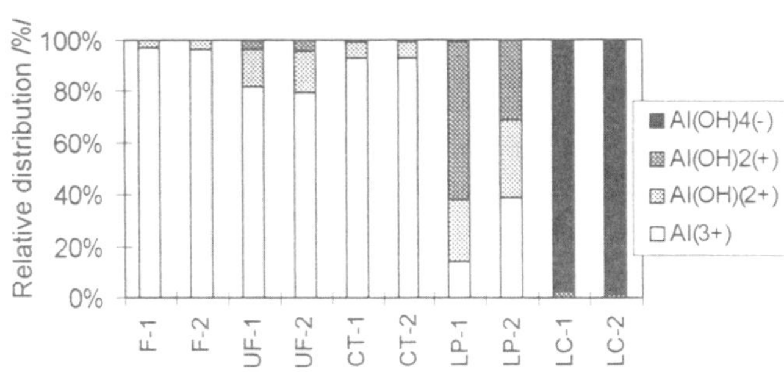

Figure 10. The relative distribution of inorganic aluminum species in 0.2 mM citric acid solutions. (F—fertilized brown soil; UF—unfertilized brown soil; CT—red soil; LP—peat soil; LC—calcareous soil; 1—topsoil and 2—subsoil).

5. CONTROL OF ALUMINUM SOLUBILITY

As depicted in Figure 12, all studied soils are highly unsaturated in citric acid and oxalic acid solutions with respect to the considered Al mineral phases such as gibbsite, imogolite, jurbanite. The degree of undersaturation increases with decreasing pH values for gibbsite and imogolite. This means that trivalent aqueous aluminum activity does not depend on the solubility of Al containing minerals in the complexing organic acid solutions and may be the complexing ability of organic acids controls the aluminum activity in solutions.

Inorganic aluminum is the predominant species in the lactic acid treatments. The soils treated by lactic acid solutions are in equilibrium with synthetic gibbsite at pH higher than 4.8. The relationship of soil Al activity to imogolite, jurbanite etc. is the same as in the citric and oxalic acid treatments.

6. CONCLUSION

The amounts of labile aluminum significantly depend on the soil type, horizon, and agricultural management. Brown soil has higher sensitivity to organic acid input than peat

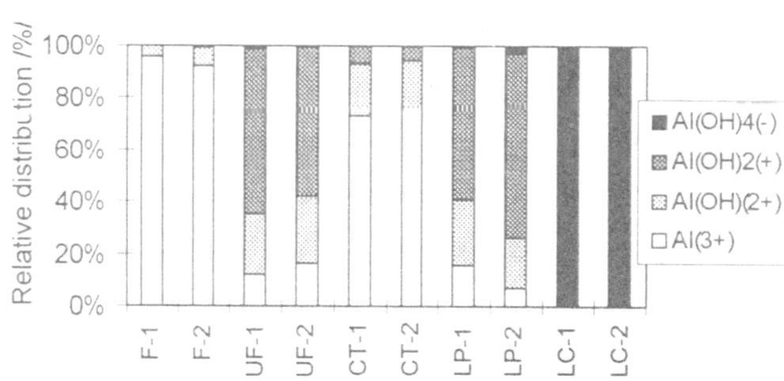

Figure 11. The relative distribution of inorganic aluminum species in 0.2 mM lactic acid solutions. (F—fertilized brown soil; UF—unfertilized brown soil; CT—red soil; LP—peat soil; LC—calcareous soil; 1—topsoil and 2—subsoil).

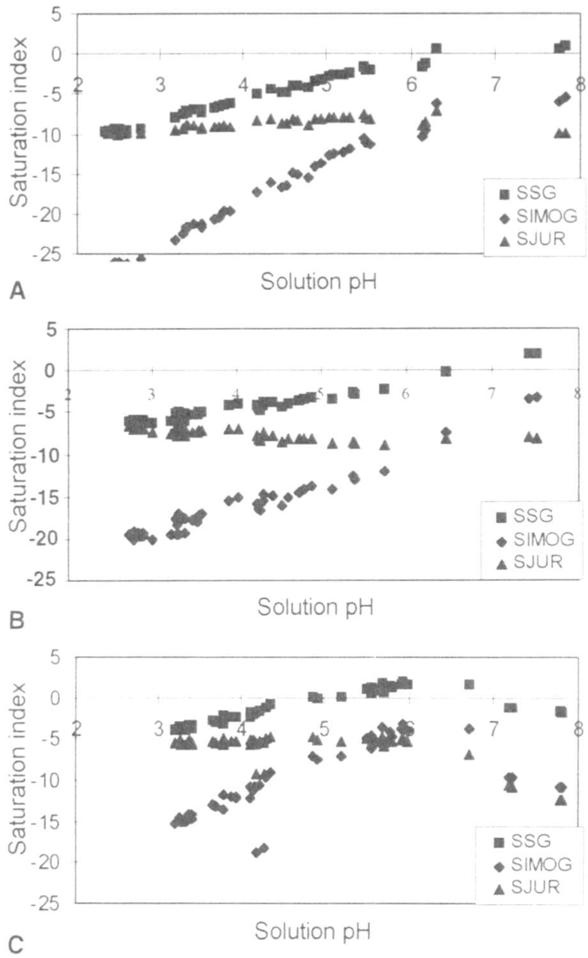

Figure 12. Saturation index versus pH for different Al mineral phases: synthetic gibbsite (SSG), imoglite (SIMOG) and jurbanite (SJUR). (A—oxalic acid, B—citric acid, C—lactic acid).

and calcareous soils. The labile aluminum concentrations increase with soil depth. Agricultural management, especially long-term ammonium fertilization, increases the soil acidity, which greatly influences aluminum mobilization. The comparison of the labile aluminum concentrations in both Chinese and Polish soils indicates that the type of nitrogen fertilizers used in soils may outweigh their amounts, i.e. NH_4NO_3 and $(NH_4)_2SO_4$ cause higher soil acidification and aluminum mobilization than NH_4HCO_3.

The ability of oxalic, citric, and lactic acid to mobilize aluminum varies due to differences in acidity, molecule structure and number and position of functional groups. Oxalic and citric acids favor the formation of the stable 5- or 6- chelate structures with aluminum, and have therefore higher ability to mobilize aluminum, while lactic acid are not able to form such complexes with aluminum. Furthermore, the released H^+ ions from the dissociation of carboxylic group play a major role in aluminum mobilization.

Generally, the ability of oxalic acid to aluminum mobilization is 1.5~2.0 times and 8.7~22 times higher in comparison to that of citric and lactic acids, respectively. Aluminum speciation in complexing organic acid treatments shows the predominance of organically bound aluminum in the extracted soil solutions. In non-complexing organic acid treatments, inorganic aluminum species account for more than 97% of labile aluminum.

ACKNOWLEDGMENT

Author would like to express his gratitude to Dr R.D. Vogt for his providing valuable comments and corrections.

REFERENCE

1. Bloom, P. R., McBride, M.B., Weaver, R.M.: Al organic matter in acid soils: Buffering and solution Al activity. *Soil Sci. Soc. Am. J.*, 1979, *143*, 488–493.
2. Cronan, C. S., Walker, W.J., Bloom, P.R.: Predicting aqueous Al concentrations in natural waters. *Nature*, 1986, *324*, 140–143.
3. Driscoll, C. T., Breeman, N., Mulder, J.: Al chemistry in a forested spodosol. *Soil Sci. Soc. Am. J.*, 1985, *49*, 437- 444.
4. Mulder, J., Stein, A.: The solubility of Al in acid forest soils: Long-term changes in acid water: the need for a reappraisal. *Geochim. Cosmochim. Acta*, 1994, *58*, 85–94.
5. Pohlman, A.A., McColl, J.G.: Soluble organics from forest litter and their role in metal dissolution. *Soil Sci. Soc. Am. J.*, 1988, *52*, 265–271.
6. Walker, W.J., Cronan, C.S., Bloom, P.R.: Al solubility in organic soil horizons from northern and southern forested watersheds. *Soil Sci. Soc. Am. J.*, 1990, *54*, 369–374.
7. James, B.R., Riha, S.J.: Al leaching by mineral acids in forest soils: I. Nitric-sulfuric acid difference. *Soil Sci. Soc. Am. J.*, 1989, *53*, 259–264.
8. James, B.R., Riha, S.J.: Al leaching by mineral acids in forest soils: II. Role of the forest floor. *Soil Sci. Soc. Am. J.*, 1989, *53*, 264–269.
9. Van Lagen, B.: *Manual for chemical soil analyses.* Department of Soil Science and Geology, Agricultural University Wageningen, 1993.
10. Driscoll, C.T., A procedure for the fraction of aqueous aluminum in dilute acidic waters. *Int. J. Environ. Anal. Chem.*, 1984, *16*, 267–283.
11. Schecher, W.D., Driscoll, C.T.: An evaluation of uncertainty associated with Al equilibrium calculations. *Water Resour. Res.*, 1987, *23*, 525–534.
12. Findlow, J.A., Duffield, J.R., Williams, D.R.: The chemical speciation of aluminum in milk. *Chem. Spec. Bioavailab.*, 1990, *2*, 3–32.
13. Porebska, G., Mulder, J.: Effect of long term nitrogen fertilization on soil Al chemistry. *J. Ecol. Chem.*, 1994, *3*, 269–280.
14. Prietzel, J., Feger, K.H.: Dynamics of aqueous aluminum species in a podzol affected by experimental $MgSO_4$ and $(NH_4)_2SO_4$ treatments. *Water Air Soil Pollut.*, 1992, *65*, 153–173.
15. Kotowski, M., Pawlowski, L., Seip, H.M., Vogt, R.D.: Mobilization of Al in soil columns exposed to acids and salt solution. *Ecol. Engin.*, 1994, *3*, 279–290.
16. McColl, J.G., Pohlman, A.A.: Soluble organic acids and their chelating influence on Al and other metal dissolution from forest soils. *Water, Air Soil Pollut.*, 1986, *31*, 917–927.
17. Pohlman, A.A., McColl, J.G.: Kinetics of metal dissolution from forest soils by soluble organic acids. *J. Envir. Qual.*, 1986, *15*, 86–92.
18. Driscoll, C.T., Schecher, W.D.: Al in the environment. In: H.Sigel and A.Sigel, (eds) *Metal Ions in Biological Systems*, Marcel Dekker, Inc. New York, 1988, *24*, 59–122.

33

WATER-SOLUBLE RARE EARTH ELEMENTS IN SOME TOP-SOILS OF CHINA

J. G. Zhu,[1] Y. L. Zhang,[1] X. M. Sun,[1] S. Yamaski,[2] and A. Tsumura[3]

[1]LMCP, Institute of Soil Science
Academia Sinica, Nanjing 210008, China
[2]Faculty of Agriculture
Tohoku University, 981 Japan
[3]National Institute of Agro-Environmental Science
Tsukuba, 305 Japan

ABSTRACT

Water soluble rare earth elements (WSREE) of sixteen typical soils in China were determined by high-resolution inductively coupled plasma mass spectrometer. Results show that the contents of WSREE of the samples ranged from 6.4–748.2 μg/kg with an average of 147.9 μg/kg. There was a positive correlation between the content of WSREE and the total content of REE in the soils ($t > t_{0.01}$). When the data of WSREE were normalized by the data of REE in chondrites, positively abnormal contents of Ce for acidic soils, negatively abnormal for alkaline soils, and negatively abnormal of Eu for all soils tested were observed.

Key words: rare earth elements, soil, water soluble, China.

INTRODUCTION

Accumulating evidence of growth-promoting effects of REE is promoting the agricultural use of REE in China. There is a strong possibility that REE's products will be disposed in the environment. It is necessary to know the background values, and the forms of REE in soils. In fact, REE are not so rare, they are ubiquitous in soils. The total amount of REE accounts for 0.01–0.02% of soil weight in eastern China[1]. REE in soil are derived from the soil parent materials, and the most of them are mainly concentrated in the minerals bearing REE and exist chiefly as fluorocarbonates, phosphates, silicates and compli-

Chemistry for the Protection of the Environment 3, edited by Pawłowski *et al.* Plenum Press, New York, 1998

cated oxides. These compounds of REE have a very low solubility in water. However, thermodynamic data show that the amounts expected to exist in soluble form are significant. The water soluble form of REE in soils is the most active one and is likely to play the most important role in both environmental behaviour and biological effect. However, owing to the complexity of soil itself, low solubility of REE compounds in soil, limited knowledge of effects of REE on ecosystems, and instrumental limit (specially in China), relatively few attempts have been made to study the occurrence of WSREE in soils. Fortunately a newly developed technique, high resolution inductively coupled plasma mass spectrometry (HR-ICP-MS), has provide useful for such studies. Recently WSREE of four typical soil profiles in China have been determined by this method[2]. In this paper we give results based on sixteen topsoil samples of China.

MATERIALS AND METHODS

Soil Samples

Sixteen typical soil profiles of China were collected, air-dried and ground to pass 60-mesh (0.28-mm) sieve. Total REE (TREE) were determined by ICP-MS (quadrupole type, PQ Plus II, VG Co.), and other soil properties were measured using standard methods[3]. Information of soils used is shown in Table 1.

Table 1. Information of soil used

Soil name						Total
Chinese name	FAO name	Code	Locality	Parent material	pH	REE
Red soil (upland)	Haplic acrisols	RE-U	Jiangxi	Q_4[1)] red clay	5.2	126.9
Red soil (wild)	Haplic acrisols	RE-W	Jiangxi	Q_4 red clay	5.5	132.5
Lateritic red earth (forest)	Haplic acrisols	LR-F	Guangdong	Granite	4.2	482.2
Lateritic red earth (upland)	Haplic alisols	LR-U	Guangdong	Granite	5.0	733.1
Paddy soil	Cambisoil Anthraquic	P(GD)	Guangdong	Granite	5.7	815.1
Calcarous purplish soil (upland)	Calcalric regosols	CP-U	Sichuan	Calcareous purplish sandstone	8.7	185.4
Calcareous purplish soil (wild)	Calcalric regosols	CP-W	Sichuan	Calcareous purplish sandstone	8.3	191.2
Paddy soil	Calcalric regosols	P(SC)	Sichuan	Calcareous purplish sandstone	8.4	169.6
Fluvoaquic soil	Calcaric fluvisols	FA	Henan	Yellow river alluvial deposit	8.8	181.9
Manured loessial soil	Cumulic anthrosols	ML	Shaanxi	Loess	8.5	187.5
Yellow cultivated soil	Calcaric cambisols	YC	Shaanxi	Loess	8.7	173.6
Dark loessial	Calcaric Kastanozems	DL	Gansu	Loess	8.7	182.6
Paddy soil	Cambisols	P(SX)	Shanxi	Diluvium	7.4	194.7
Residual slain soil	Haplic solonchaks	RS	Xingjiang	Alluvial deposit	9.0	130.7
Brown desert soil	Haplic gypsisols	BD	Xingjiang	Alluvial deposit	8.2	137.1
Grey desert soil	Calcaric cambisols	GD	Xingjiang	Alluvial deposit	9.5	172.0

[1]The Quaternary Period

Sample Preparation

One gram of soil sample was place in a plastic test tube and 40 ml deionized water was added. The suspension was then kept at 25 °C for 48 hours and shaken occasionally. The solid phase was separated from the liquid phase by centrifugation with RCF 12000g. ^{114}In was added as an internal standard. Water samples were collected in pre-cleaned polypropylene bottles with a tightly fitted screw cap, and kept in a refrigerator until analysis.

Instrumental

High Resolution ICP-MS: The measurements were carried out using a double focusing type of ICP-MS with high resolution supplied by VG Elemental, Winford, Cheshire, England. The torch box and the interface region are basically similar to those used in quadrupole type instruments. The whole interface system is electrically isolated and high voltage is applied to the sampling and skimmer cones. The double focusing mass spectrometer consists of a 70° electrostatic sector for energy focusing and 35° laminated magnetic sector for mass separation. Operating conditions and analytical parameters were essentially those recommended by the manufacturer. Ultrasonic Nebulizer: The ultrasonic nebulizer (USN) used in this study was provided by Applied Research Laboratories, En Vallaire, CH-1024, Ecublens, Switzerland. A peristaltic pump is used to deliver sample solution to the oscillating surface of quartz plate. Aerosols having smaller and more uniform particle size are produced, and transported first to the heated tube, and water in the aerosol is mostly removed by this process, and only the dry aerosol is introduced into the plasma. The temperature of the heated tube and the condenser was kept at 120°C and 1°C respectively. The sample introduction rate was adjusted to 2–4 ml/min. The HR-ICP-MS operating conditions are shown in Table 2.

Standard

Since the liner range of working curves for heavy metals is wide (from ng/l to mg/l), a single internal standard (In) with concentration 0.1 mg/l was used. Two mixed external standard solutions containing 5–500 µg/l of each REE were used. The working standards were prepared from SPEX Multi-Element Plasma Standards (XSTC-1 and XSTC-13), supplied by SPEX Industries, Inc., 3880 Park Ave., Edison, NJ., USA.

Table 2. HR-ICP-MS operating conditions

Rf power	1.2 kw	Coolant gas flow rate	13 L/min.
Auxiliary gas flow rate	0.7 L/min.	Nebulizer gas flow rate	0.8 L/min.
Solution uptake rate	2.5 ml/min.	Sample orifice diameter	1 mm
Skimmer orifice diameter	0.1 mm	Interface running pressure	1.4×10^{-4}mbal
Mass spectrometer running pressure	3.5×10^{-7}	High resolution magnet supply	2.4 A
Peak jumping dwell time	300 ms	Accelerating voltage	4.0 kv
Number of Scan over peak	2	Peak jumping Dac[1)] steps	5–40

[1]Digital to analogue convert

RESULTS AND DISCUSSION

Selection of Operation Condition

The detection limits of quadrupole type ICP-MS are good enough to analyse the REE in water solution in which concentration of REE were usually about some hundred μg/kg. But the instrument can not eliminate the interference of ^{137}BaO with ^{153}Eu in mass spectra because of low resolution. The main purpose of the development of the high resolution ICP-MS is to resolve molecular overlaps due to matrix species[4]. However, it has also become evident that the detection limits of most elements are significantly improved when the instrument is operated in the low resolution mode because of much lower background noise. A series of careful examinations of possible spectral interference, using various mixed standard solution, revealed no spectral overlap with the exception of the interference from BaO on Eu. Two methods may be used to overcome the problem. The first one was the use of high resolution mode (set the resolution at 7500). It was possible to separate both peaks completely, but this approach suffered serious loss of sensitivity. The second one is the use of a mathematical correction. First, the concentration of Sm was determined using both ^{147}Sm and ^{152}Sm, the latter overlapped with BaO. Then the magnitude of the interference of BaO on Eu was estimated from the difference of the Sm values derived from ^{152}Sm (Sm+BaO) and ^{147}Sm. It should point out that this correction was possible only when the ratio of Ba/Eu in solution was less than around 10000. The former method was used in this work. As a compensation of the loosing of sensitivity, an ultrasonic nebulizer was used which improved the sensitivity about ten times.

GSD-2 and GSD-6, which were geological reference materials provided by the Institute of Geophysical and Geochemistry Prospecting (IGGP), China, were used to check the reproducibility of the method. The relative deviation of six results obtained during three months was better than 12% for the light REE, 20% for the heavy REE.

Content of WSREE in Soils

Most of REE bearing compounds in soils, as mentioned above, are insoluble substances, and the solubility ranges from 10^{-7} to 10^{-4} mol/l. The concentration of REE in water solution should be at the mg/l level if a thermodynamic equilibrium is achieved. However, since adsorption, coordination and co-precipitation exist in a soil-water system, the concentration of REE in soil solution is usually lower than 1 mg/l. The contents of WSREE in the soil investigated varied widely. The total WSREE ranged from 6.41 to 748.15 μg/kg with an average of 147.9 μg/kg. There was a positive correlation between the content of WSREE and the total content of REE in the soils with a R^2 of 0.72 ($t > t_{0.01}$). The contents of different REE in solution also varied widely. The highest concentration was found for Ce, ranging from 1.39 to 402 μg/kg, and the lowest one was Lu, ranging from 0.03 to 1.33 μg/kg.

Among the sixteen soils, thirteen soils were collected from agricultural areas and three soils were collected from arid-desert areas. The thirteen agricultural soils can be divided into two groups: acidic soils and alkaline soils. The contents of WSREE in acidic soils were relatively higher than those in alkaline soils (see Fig. 1, A and B). The highest one (748.15 μg/kg) was from the sample of paddy soil (Guangdong) with granite as parent material. The lowest one (6.41 μg/kg) was from the sample of paddy soil (Shaanxi) with diluvium as parent material. The contents of WSREE in arid-desert soils were low except the for sample of grey desert soil (see Fig. 1, C).

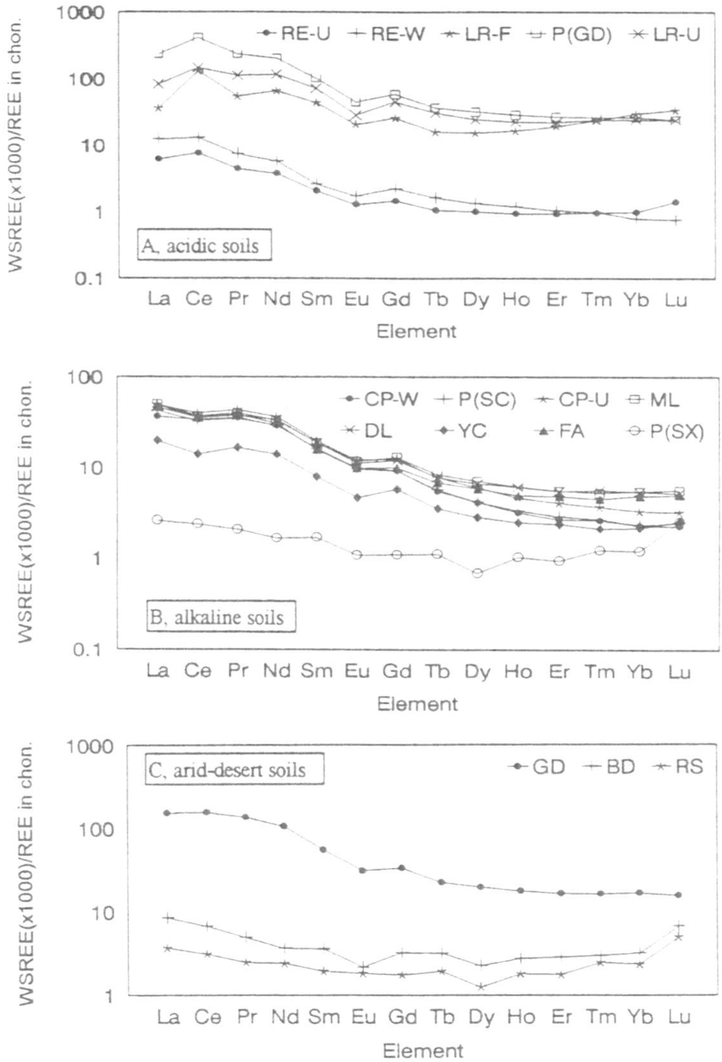

Figure 1. WSREE in some soils. For soil code, see Table 1.

Wu[5], found that 0.1 to 1 mg/l of REE in water solution would enhance the root initiation and promote ^{32}P uptake by roots of some crops. therefore, in most of the studied soils, the WSREE contents are sufficient to affect crop growth. Considering that the contents of exchangeable REE in soils usually are some hundred times higher than WSREE[6], the active ionic REE in soils should not be ignored when applying REE to the plants, i.e. some soils may already have enough available REE to crops. Since REE are poisonous heavy metals, it has been suggested that the concentration of $LaNO_3$ in terrestrial water should be controlled to less than 1 mg/l[7]. In applying REE in agriculture, one must therefore consider the delicate balance between amounts causing beneficial and amounts causing harmful effects.

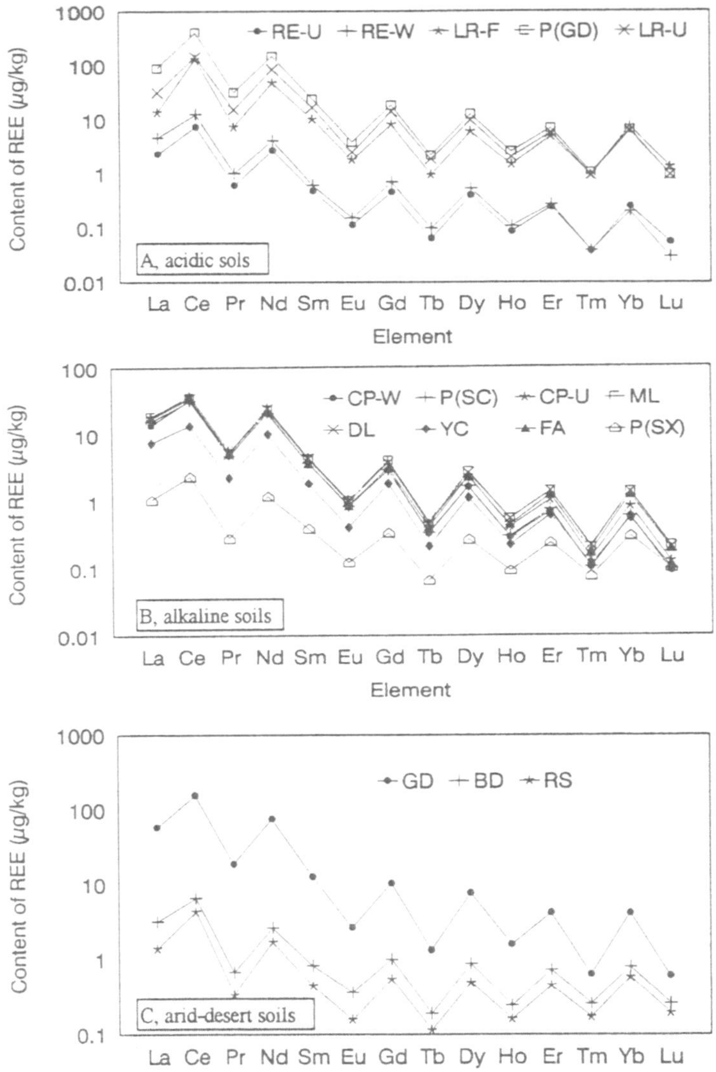

Figure 2. Chondrite-normalized curves of WSREE. For soil code , see Table 1.

Distribution of WSREE

As compared with chondrite, there is a fractionation of REE in the rock and ore forming processes that results in an enrichment or deficit for some of the REE. This characteristic of parent rocks is transferred to soils and the fractionation of REE is continued in the soil-forming process. If the contents of REE in soils were normalized by the contents of REE in chondrites, the enrichment or deficit for some of the REE would be observed to the most of soils. There have been some reports on abnormal phenomena (enrichment or deficit) of Ce and Eu in soils[8–9]. The abnormal Ce and Eu distributions varied between the soils due to different soil properties, mainly soil pH. When the data of WSREE were normalized by the data of REE in chondrite, those abnormal phenomena were also observed (see Fig. 2). As for the acidic soils , a positive abnormal content of Ce was observed (see Fig. 2, A), whereas for the alkaline soils, a negative one was observed (see Fig. 2, B). The reason is probably that Ce^{4+} is more easily reduced under acidic conditions resulting in more water soluble compounds. The negative abnormal contents of Eu for all of soils tested is probably caused by the deficit of Eu in the soils.

CONCLUSION

The combination of ICP-MS with an ultrasonic nebulizer resulted in an analytical system of excellent detection power. Although the concentration of REE in soil solutions were as low as μg/kg levels , it was determined with high precision by direct sample introduction. No sample pre-treatment, such as chemical separation and/or pre-concentration, was required. Although the WSREE in soils was low, it should have certain effects on the environment, and more attention should be paid to the possible consequences caused by applying REE in agriculture.

REFERENCES

1. Zhu, Q.Q. and Liu, Z., Rare earth elements of soils of eastern part of China. Journal of the Chinese Rare Earth Society(in Chinese) 1988, 6(4) 59–63
2. Zhu, J.G., Zhang, Y.L., Sun J., Yamasaki, S. and Tsumura, A. , Water-soluble rare earth elements in some soils of China. Pedosphere, 1997, 7 (1) 25–30
3. Li, Y.K. (editor in chief). Normal Analysis Methods of Soil and Agrochemistry (in Chinese). Science Press, Beijing. 1983
4. Bradshaw, N., Hail, E.F.H. and Sanderson, N.E., Inductively coupled plasma as an ion source for high-resolution mass spectrometer. J. Anal. At. Spectrum. 1989, 4: 801–803
5. Wu, Z.M., The effect of rare earth elements on rooting of plants cutting. Journal of the Chinese Rare Earth Society (in Chinese). 1988, 6(1) 67–70
6. Zhu, J.G. and Xing, G.X. 1992. Forms of Rare Earth Elements in Soils: I. Distribution. Pedosphere 1992 2(2) 125–134
7. Tong, S.L. and Lu, G.C., Biological effect of rare earth elements. Rare Earth (in Chinese). 1987, (4) 42–54
8. Gotoh, s and Yoshino, A.., Abundance and distribution of rare earth elements in some soil profiles in Japan. In: Trans. 14th Intern. Congr. Soil Sci. 1990, 2: 96–101
9. Zhu J.G., Xing, G.X., Forms of rare earth elements in soils: II. Differentiation of rare earth elements. Pedosphere, 1992, 2(3) 193–200

34

ION EXCHANGER COMPOSITES AS HUMUS SUBSTITUTE FOR RESTORATION OF DEGRADED SOILS

Mariola Chomczyńska, Lucjan Pawłowski, and Henryk Wasąg

Department of Environmental Protection Engineering
Technical University of Lublin
40, Nadbystrzycka str. 20-618, Lublin, Poland

ABSTRACT

Utilization of synthetic ion exchanger composites as a substitute of ion exchange materials for restoration of degraded soils has been described.

Studies showed that ion exchanger composites can be used in agriculture as fertilizers which gradually release nutrient ions. Synthetic ion exchangers show also properties similar to soil colloids which can be successfully applied in the restoration of degraded soils. In the degraded soils the ion exchanger composites are able to replace humic substances and other ion exchange colloids. Application of spent ion exchangers can significantly decrease the cost of the restoration of degraded lands.

Key words: soil recultivation, waste treatment, waste reuse, waste ion exchanger

1. INTRODUCTION

Humus being a very important soil colloid, highly influences physical, chemical and biological properties of soil. Due to their relatively high capacity and buffering properties humus, together with inorganic colloids, retains nutrient ions and adjusts pH of the soil solution. In addition, humic compounds significantly influence creation of specific soil structure (crumbs), by initiating clotting process and playing the role of an adhesive.

Thermic conditions of soil are also influenced by presence of humic compounds. Due to their dark colour, they absorb significant amount of the sun infrared radiation, raising the temperature of soil and elongating the vegetation period. Humic substances increase also the content of nutrient components in soil. During slow microbiological

Chemistry for the Protection of the Environment 3, edited by Pawłowski *et al.* Plenum Press, New York, 1998

decomposition they release additional amount of macro- and microelements. Some of humic compounds accelerate division of microbe cells and influence the nitrogen fixation and nitrification processes. By forming chelates with cations, humus blocks their bonding with PO_4^{3-} ions and prevents precipitation insoluble tricalcium phosphates[1,2].

Such different influence of humus on a soil is indubitably caused by its various properties. To find or prepare the substance of identical or even similar influence on a soil environment, seems to be a very difficult task. Experiments with synthetic ion exchangers, initiated in by Soldatovat al.[3] showed that ion exchangers may be used as a substitute for humus. Experiments showed that mixtures cation and anion exchangers can absorb macro- and microelements which are essential for plant growth. Additional advantage of the cation / anion exchanger mixtures over natural humic compounds is that this mixtures can absorb cations and anions while the humic substances can absorb only cations[4].

Ability to retain nutrient ions demonstrated by ion exchanger resins can position them as an important component of artificial soils. Results of the test on application of ion exchangers as substitutes for humic substances used for plant vegetation purposes is presented in this paper.

2. CHARACTERISTIC OF ION EXCHANGER COMPOSITES

Ion exchanger soils (called in this paper "composite") prepared by Soldatov at al.[3] are the mixtures of cation and anion exchangers saturated with ions of biogenic elements (see Table 1). Due to their high ion-exchange capacity (reaching even 10 meq/g), content of biogenes in the ion exchanger composites per mass unit, may be 10 times higher than that of the best natural soils[5]. To obtain an appropriate composite for the plant growth application it is important to assure that the proper ratios will be maintained between elements desorbed from the composite to "soil solution". To assure that, biogene ions must be sorbed by anion and cation exchangers from a solution which an ionic composition has to be maintained in the range limited by, so called, biozone.* It should be noticed that the ionic composition in the ion exchanger phase changes during the plant growing process as a result of the uptake of biogens by a plant from the ion exchanger composite. This is accompanied by changes in composition of an intergranular solution from which biogenes are uptaken directly by plants. Plant vegetation is possible as long as composition of the solution is in the range limited by biozone. Therefore, ions released from the ion exchanger phase should provide the proper composition of a intergranular solution as long as possible[4].

Mechanism of mineral nutrition of plant grown on the ion exchanger soil is, in principal, the same as for natural soil. Plants uptake macro- and microelement ions and release metabolite ions (H^+ and HCO_3^-) to the solution. Certain amount of the released H^+ cations and HCO_3^- anions is exchanged with biogenic ions from the ion exchanger composite. This way, the ion exchanger composite similarly to humus, acts as a pH buffer preventing pH fluctuations in the root zone[5,6].

Advantageous properties of the ion exchanger composites allow to prepare substrates able to produce high yields of green biomass. Theoretical productivity of the first composites IS-1 and IS-2 (Biona®-111) is equal to 70–130 kg of green biomass from 1 m^2

* Biozone is a set of ionic compositions of different nutrient media used for hydroponics plant grow. Solution which ion composition (ratio of biogenic element ions) is the range of biozone, is suitable for plant vegetation.

Table 1. The composition of some ion exchanger soils

Ion exchanger soil	Cation exchanger	Anion exchanger	Content of macroelements (meq/g)						
			K^+	NH_4^+	Ca^{2+}	Mg^{2+}	NO_3^-	$H_2PO_4^-$	SO_4^-
IS-1	KU-2	EDE-10P	0.12	—	2.01	0.27	0.77	0.15	1.00
IS-2 (Biona®-111)	KU-2	EDE-10P	0.15	—	1.45	0.26	0.67	0.24	1.11
Biona®-311	KU-2 clinoptilolite	EDE-10P	0.32	—	1.06	0.41	0.47	0.13	0.50
Biona®-122	KU -2	AN-2F	0.11	0.05	0.87	0.17	0.40	0.61	0.31
Biona®-211	clinoptilolite	EDE-10P	0.54	0.13	1.22	0.59	0.25	0.12	0.26
Biona®-221	clinoptilolite	AN-2F	0.23	0.13	1.00	0.18	0.47	0.08	0.60

of the substrate (depth of cultivated layer—20 cm)[7]. Theoretical productivity of the new generation composites[7] can reach from 200 to 300 kg of green biomass from 1 m^2

3. REVIEW OF STUDIES ON APPLICATION OF ION EXCHANGER COMPOSITES

Different aspects of application of the ion exchanger composites in agriculture have been tested. Some studies explored possibilities of use of the ion exchanger composites as substrate for plant vegetation in closed ecological systems like space ships or polar stations. Some experiments were conducted under conditions imitating a photoperiod of the moon (14 days of continuos illumination and 14 days of continuous darkness)[3]. Eleven vegetation cycles of the following species: chinese cabbage (*Brassica pekinensis* Rupr.), lettuce (*Lactuca sativa* L.) small radish (*Raphanus sativus* L. *var. sativus* Mans.), dill (*Anethum graveolens* L.) were conducted. During the experiment, the content of biogenes (including K^+) in both plant tissues and the ion exchanger composite (Biona®-111) was measured. Chemical analysis showed that 40% of the potassium absorbed in the ion exchange composite is used by plants. The portion of soil used in the experiment yield the crop equal to 160 kg of green biomass (half of the theoretical productivity). No one of the tested species after eleven vegetation cycles was malnutrited and the composite could be reused as a substrat for plant cultivation.

In another experiments the ion exchanger composites were used as substrates for the plant grow in a greenhouse. The purpose of the experiments was to evaluate productivity of the composite and possibilities of its use for growing of various species in monoculture[4]. Quality of produced vegetables and their crop were satisfying (see Table 2). However, a decrease of the green mass yield was observed for some species (carrot—*Daucus carota* L., potato—*Solanum tuberosum* L., chinese cabbage—*Brassica pekinensis* Rupr.) after several repeated vegetation cycles (see Table 3). Chemical analysis of the composite phase and samples obtained from the experiment conducted using crop rotation showed that decrease of the green mass yield was caused by potassium depletion or cumulation of plant metabolites in the substrate.

The potassium deficiency observed in those experiments leads to conclusion that conducting long cultivation cycles and obtaining high green mass yields is impossible. It was proved that shortage of potassium is not only caused by low affinity of ion exchanger to K^+, but also, to the higher degree, by sorption of metabolites on surface of ion exchanger beads blocking potassium absorbed in the deeper layers of the ion exchanger

Table 2. Characteristic of productivity (kg / m^2 of substrate) of some vegetables grown on an ion exchanger substrate[6]

Cabbage *Brassica oleracea* L. *var. capitata* L.		Radish *Raphanus sativus* L. *var. sativus Mansf.*		Potato *Solanum tuberosum* L. (after 4 cycles of vegetation)		Carrot *Daucus carrota* L. (after 3 cycles of vegetation)		Chinese cabbage *Brassica pekinensis* Rupr. (after 6 cycles of vegetation)
Cabbage cole	Inedible part	Roots	Leaves	Bulbs	Halum	Roots	Leaves	Eatable part
9.10	2.10	6.50	9.70	24.14	28.40	13.26	13.71	36.50

beads. Because of this blockade, potassium cannot diffuse from the ion exchanger beads to the solution[8].

A great number of experiments were conducted to solve the problem of potassium depletion. One of the experiments involved evaluation of different methods of plant watering[4]. One group of species (grown using crop rotation) was watered with distilled water. Another group of species was watered with 0.1% solution of potassium nitrate. Experiment was finished after twelve vegetative cycles (see Table 4) which produced 2.28 kg of green biomass from 1 kg of the substrate when watered with potassium nitrate solution. In case when distilled water was used, productivity was only half of that observed witch KNO_3 solution. Small amount of KNO_3, introduced during watering, was the additional source of K^+ ion (and *RomanNO*$_3^-$)—see Table 5.

In the next phase of studies a composition of the ion exchanger substrate was modified[8]. The used substrate was a mixture of IS-2 (Biona®-111) composite and clinoptilolite, which has high affinity to potassium. The ion exchange capacity of clinoptilolite is two time higher than that for the IS-2 composite. The crop obtained at the 30% content of the clinoptilolite in the mixture, was two times higher then that when the IS-2 composite was only used. Tests performed using four plant species (chinese cabbage—*Brassica pekinensis* Rupr., fodder beet—*Beta vulgaris* L. *ssp. cicla* Moq., radish—*Raphanus sativus* L. *var. sativus* Mans., common borage—*Borago oficinalis* L.) showed stabilizing influence of the clinoptilolite addition on crops. Successive vegetation cycles (see Table 6) showed no significant decrease in the substrate productivity. Stabilizing function of addition of clinoptilolite resulted in the additional supply of potassium. Advantageous properties of clinoptilolite are utilized in the new ion exchanger composites such as Biona®-211, Biona®-221, Biona®-311 (see Table1).

Table 3. Characteristic of productivity (kg / m^2 of substrate) of some vegetables grown on an ion exchanger composite in successive vegetation cycles[6]

Vegetation cycle	Carrot *Daucus carota* L.		Potato *Solanum tuberosum* L.		Chinese cabbage *Brassica pekinensis* L.
	Roots	Leaves	Bulbs	Halum	Eatable part
1	8.50	7.88	7.05	9.10	10.50
2	2.75	4.07	8.76	10.00	6.70
3	2.01	1.76	4.67	5.60	4.50
4	—	—	3.66	3.70	7.10
5	—	—	—	—	5.70
6	—	—	—	—	2.00

Table 4. Crop rotation of plant grown in 12 vegetation cycles[6]

Pot	Vegetation cycle 1	2	3	4	5	6	7	8	9	10	11	12
I	c	f	r	l	m	b	c	f	r	l	m	b
II	f	r	l	m	b	c	f	r	l	m	b	c
III	r	l	m	b	c	f	r	l	m	b	c	f
IV	l	m	b	c	f	r	l	m	b	c	f	r
V	m	b	c	f	r	l	m	b	c	f	r	l
VI	b	c	f	r	l	m	b	c	f	r	l	m

Where:
c—chinese cabbage (*Brassica pekinensis* Rupr.)
f—foliage beet (*Beta vulgaris* L. *ssp. cicla* Moq.)
r—radish (*Raphanus sativus* L. *var. sativus* Mans.)
l—lettuce (*Lactuca sativa* L.)
m—mustard (*Sinapis sp.* L.)
b—common borage (*Borago officinalis* L.)

Table 5. Changes of K^+ and NO^- content in the IS-1 composite after 12 vegetation cycles[6]

Ions	Initial content in composite (meq/kg)	Initial content in pot (meq)	Amount delivered with watering (meq)	Removal with crop (meq)	Use of initial content (%)
K^+	0.27	2.11	2.40	1.93	42.70
NO_3^-	0.35	2.73	2.40	2.66	52.00

Table 6. Green biomass (kg/m^2) of vegetables produced during 13 successive vegetation cycles[3]

Vegetation cycles 1	2	3	4	5	6	7	8	9	10	11	12	13
18.956	14.069	17.413	13.363	12.920	14.356	11.607	12.351	10.286	11.610	9.121	8.737	8.743

In the experiments discussed above, the ion exchanger composites were used as an artificial soil alone. In other experiments composites were used as fertilizer additives mixed with materials such as sand, peat, perlite etc. Such experiments were also conducted by Kloc and Szwed[9]. Objective of their studies was to evaluate the influence of three ion exchanger composites on some parameters characterizing the plant vegetation process (number of germinated plants, plant height, green and dry biomass of shoots, dry biomass of roots). The standard polyvalent composite Biona®-111, the powdered composite Biona®-111, and the composite prepared from monoionic form of ion exchangers were used. These composites were chosen to evaluate substrates prepared from the physically degraded ion exchangers (powdered form of Biona®-111 composite) and from the monovalent ion exchanger forms. Nine mixtures of garden soil with ion exchanger composites and one control sample containing only garden soil were prepared (see Table 7). Seeds of corn (*Zea mays* L. *var. sacharata* Korn.) were sown on the above substrates. Results of this experiment are presented in Table 7. The results show, that addition of any type of the ion exchanger composite (apart from 100% M*) increases the yield and accelerates the

* Composite is a mixture of monoionic ion exchanger forms (ion exchanger grains sorb only one kind of ion i. e. one grain sorbs only Mg^{2+} the other - only Ca^{2+} ions etc.)

Table 7. Results of experiment with corn—*Zea mays* L. *var. sacharata* Korn[9]

Substrate series	Average amount of germinated plants: 3rd day	Average amount of germinated plants: 11th day	Average height of plants after 5 weeks of vegetation (cm)	Average green mass of shoots (g)	Average dry mass of shoots (g)	Average dry mass of roots (g)
100% B	0.670	3.000	25.500	23.510	8.097	4.097
50% B + 50% s	0.670	3.000	17.200	18.760	6.247	1.941
25% B + 75% s	0.330	4.000	14.000	10.040	3.910	1.303
100% Bp	2.000	4.670	16.700	18.430	9.657	4.714
50% Bp + 50% s	1.000	5.000	16.000	12.250	6.803	2.269
25%Bp + 75% s	0.670	3.670	16.900	14.780	3.743	0.849
100% M	0.330	3.000	3.200	0.230	0.230	0.142
50% M + 50% s	0.330	2.670	15.800	5.820	1.657	0.576
25% M + 75% s	0.330	4.000	12.600	3.230	1.527	0.721
100% s	0.000	3.330	7.800	0.230	0.790	0.628

Where:
B—standard Biona
Bp—powdered Biona
M—substrate composed of monoionic forms of ion exchangers
s—garden soil (control substrate)

plant growth. Influence of the tested composites on germination process (after 11 days) could not be strictly estimated. It was, however, observed that an addition of the ion exchange composite accelerates germination at the beginning of the vegetation period.

Positive results of experiments with the ion exchanger composites inspired their use for recultivation of degraded soils. Such soils, often depleted of humic compounds, has small ion exchange capacity for biogenic ions. Therefore, next series of experiments tested a possibility to substitute humus with ion exchanger composites[10]. The main object of the studies was to find a minimal dose of the Biona®-111 ion exchanger composite required to maintain the normal plant grow on degraded soil (a sand was a model of the degraded soil). The other task was to determine the amount of the ion exchanger composite, which is necessary to be added to sand to obtain substrate of similar productivity as the garden soil. Nine samples of substrates with increasing content of the ion exchanger composite and one control sample of a sand were prepared. Orchard grass (*Dactylis glomerata* L.) was used as a tested plant. The experiment results are summarized in Table 8. As it can be seen the addition of 1% of the ion exchanger composite (by volume) can be considered as a minimal dose for the 5 week vegetation period.

In the second part of the experiment five substrates of different composition (1% Biona +99 % sand, 2% Biona + 98% sand, 4% Biona + 96% sand, 8% Biona + 92% sand, 16 % Biona + 84 % sand) and one control sample (100% of natural soil) were used. Perennial ryegrass (*Lolium perenne* L.) was used as a tested plant. Plant growing were monitored on the prepared substrates during four months of vegetation. The test showed that the addition of 2% of the composite to sand, was sufficient to achieve effects comparable with control substrate of garden soil during first two weeks of the experiment. However, for the experiment lasting 4 weeks the same results were obtained with addition of only 1% of ion exchanger composite.

4. CONCLUSIONS

Results of the experiments showed that the ion exchanger composites can be successfully used as substrates for plant cultivation. These composites can also be used as ad-

Table 8. Results of experiment with ryegrass—*Lolium perenne* L[10]

Substrate series	Average shoots growth (cm)		Average shoots green mass (g)		Average shoots dry mass (g)		Average roots dry mass (g)	
	2 weeks	5 weeks	2 weeks	5 weeks	2 weeks	5 weeks	2 weeks	5 weeks
1% B + 99% s	6.00	13.50	0.71	2.58	0.11	0.72	0.11	0.65
2% B + 98% s	8.00	15.50	1.65	3.77	0.29	0.98	0.42	1.13
3% B + 97% s	9.00	16.50	2.01	7.29	0.31	2.10	0.34	1.45
4%B + 96% s	11.50	18.00	2.73	8.88	0.38	2.06	0.43	1.26
5% B + 95% s	11.00	20.50	1.57	12.14	0.24	2.88	0.21	2.15
10% B + 90% s	11.50	25.50	3.56	12.73	0.46	2.83	0.39	2.55
20% B + 80% s	13.50	18.50	2.76	10.32	0.36	2.18	0.19	1.43
100% s	4.50	6.00	0.25	0.51	0.07	0.17	0.07	0.27

Where:
B-Biona[R]-111 composite
s-sand

ditives acting as soil colloids, gradually releasing ions of biogenic elements at the rate depending on plant demand. However, application of the ion exchanger composite require further investigation and research work. One of problems which have to be solved is the lack of success in growing plants on the substrate prepared by mixing of the monovalent ion exchanger forms. Solution of this problem is important for static method of a preparation of the ion exchanger composites. This method is simpler and cheaper in comparison to methods of preparation of polyvalent composites.

Our interest in an use of synthetic ion exchangers for preparation of artificial soil substrates is connected with environment protection aspects. Water purification plants discharge thousands of cubic meters of spent ion exchange resins as a waste material. The main reason for their discharge is decrease in their ion exchange capacity caused by an irreversible sorption of humic compounds from water. Incineration is one of methods used for neutralisation of such resins. However, during incineration of cation exchangers SO_2 is emitted and during combustion of anion exchangers - NO_x is emitted. An use of ion exchangers for recultivation of degraded soils seems to be economically and ecologically attractive method for their reuse and a viable alternative to their disposal as a waste material.

On the other hand there is a problem of degraded soils deprived of humic substances. It was found, that the waste ion exchangers can be used for restoration of the degraded soil.

REFERENCES

1. Dobrzański, B. and S. Zawadzki, *Gleboznawstwo*, PWRiL, Warszawa, 1981, pp. 614.
2. Lityński T. and H. Jurkowska, *Żyzność gleby i odżywianie się roślin*, PWN, Warszawa, 1982, pp. 643.
3. Soldatov V.S., Terent'ev V.M. and N.G. Perishkina, Iskustvennye pochvy na osnove ionoobmennykh materialov, *Dokladi AN BSSR*, 1968, *12*, 357–359.
4. Soldatov V. S., Perishkina N., G. and R. P. Horoshko, *Ionitnye pochvy*, Nauka i Tekhnika, Minsk, 1978, pp. 272.
5. Soldatov V. S. Ion exchange nutrient substrates for plants, typescript.
6. Soldatov V. S. Ion exchanger mixtures used as artificial nutrient media for plants, in: *Ion exchange for industry*, ed. M. Streat, Ellis Horwood, London, England, 1988, pp. 652–658.
7. Soldatov V.S. and N. G. Perishkina, *Iskusstvennye pochvy dla rastenii*, Nauka i Tekhnika, Minsk, 1985, pp. 63.

8. Myt'ko L.V., Lukasevich L. I., Hirsanova I. F., Verbitskya H. A. and A. L Ol'shanikova, Ispol'zovanie ionoobmennykh substratov v dlitel'noi kul'ture ovoshchnykh rastenii. *Agrochimia*, 1989, *7*, 51–58.
9. Kloc E. and R. Szwed, *Wymieniacze jonowe jako nocniki bioelementów. Badania nad możliwościami wykorzystania wymieniaczy jonowych do polepszania właściwości gleb*, M. Sc. Thesis, Politechnika Lubelska, Lublin, 1995, pp. 132.
10. Lesowska B. and A. Plucisz, *Symulacje w warunkach laboratoryjnych rekultywacji zdegradowanych i zdewastowanych gleb z zastosowaniem substratu jonitowego Biona® 111*, M. Sc. Thesis, Politechnika Lubelska, Lublin, 1996, pp. 125.

35

EFFECT OF CONCENTRATION AND DURATION OF ACID TREATMENT ON WATER ADSORPTION AND TITRATION BEHAVIOUR OF SMECTITE, ILLITE AND KAOLIN

G. Józefaciuk,[1] A. Szatanik-Kloc,[1] and Jae-Sung Shin[2]

[1]Institute of Agrophysics
Lublin, Poland
[2]National Agricultural Science and Technology Institute
Suweon, Republic of Korea

1. ABSTRACT

Three clay minerals: bentonite, illite and kaolin were treated with acid at different concentration and time. The changes of surface areas, water vapor adsorption energies, variable charge and apparent surface dissociation constants were observed depending on the kind of the mineral and acidification conditions. After acid treatment, the surface area increased for kaolin and illite, but decreased for smectite and the opposite tendency occured for the adsorption energy. For all minerals the amount of variable charge increased and the apparent surface dissociation constant decreased indicating the decrease of the surface acidity strength.

2. INTRODUCTION

Soil acidification became recently a very important problem in many countries[14]. The change in soil reaction influences microbial activity, heavy metals and nutrient status, acid-base equilibria and many others[12]. Acidification effects both soil solution and solid phase chemistry. The alteration of soil mineralogical features is frequently reported: mainly destruction of clay minerals and/or accumulation of iron and silica oxides. It seems that the surface properties of soil minerals should be good indicators of soil acidificationas well as of its initial stages. To gain a better insight into such changes, the investigations of surface adsorption and charging properties under influence of protons were undertaken.

Chemistry for the Protection of the Environment 3, edited by Pawłowski *et al.* Plenum Press, New York, 1998

3. MATERIALS AND METHODS

Three minerals: bentonite (Chmielnik, Poland), illite (known as "Hungarian illite") and kaolin (Valencia, Spain) were used. These minerals were acidified (1:50 mineral:acid ratio) with HCl of different concentrations: 0.05, 0.1, 0.5, 1 and 5 mole/liter during different periods: 1, 3, 9 and 27 days. After completing each acidification step, the minerals were converted to homoionic calcium form by triplicate equilibration with 1N $CaCl_2$, washed with distilled water to the lack of chlorides ($AgNO_3$ reaction) and air dried. The homoionic Ca forms of nonacidified minerals were prepared as controls.

For the treated minerals and the control samples the experimental adsorption isotherms were taken in triplicate using a vacuum chamber method. The samples were placed in the chamber and equilibrated with water vapor of different (increasing) pressures over sulfuric acid solutions of different (decreasing) concentrations. The amount of adsorbed vapor was measured after 48h by weighing. The differences between the replicates did not exceed 4%.

The surface area and the adsorption energy were calculated using the linear form of the Aranovich[1] equation:

$$x/[a(1-x)^{1/2}] = 1/(a_m C) + x/a_m, \tag{1}$$

where x is the relative water vapor pressure at the temperature of the measurement, p/p_o, a_m [mg/g] is the monolayer capacity and $C=\exp[E_a - E_c]/RT$ is the thermodynamic equilibrium constant , E_a [J/mole] is the adsorption energy, E_c[J/mole] is the condensation energy of water vapor and R and T have their usual meaning. By fitting eq. (2) to the experimental data in the range of 0.05 to 0.55 p/p_o the a_m and C values were found. The surface area S [m^2kg^{-1}] was calculated as:

$$S = L\sigma a_m/M, \tag{2}$$

where $\sigma = 10.8*10^{-19} m^2$ is the area occupied by a single water molecule, L [mol^{-1}] is the Avogadro number and M is the molecular mass of water. From the value of C, the adsorption energy was calculateded in dimensionless units i.e. as $[E_a - E_c]/RT$. The experimental methods and calculations are described in details elsewhere[6].

For the treated minerals and the control samples the back titration curves[4] were taken using the modified back-titration procedure as described by Józefaciuk and Shin[7]. After removal of exchange aluminum, the suspensions of the minerals in 1N NaCl were titrated with 0.1 M NaOH from pH=3 upward to pH=9 and from these curves the titration curves of the equilibrium solutions (blank) were subtracted to obtain the titration curves of solid phases alone. The amount of variable charge present at any pH on the mineral surfaces, Q(pH) [mole], was assumed to be equal to the amount of base [mole] used to titrate the solid phase. As the total amount of variable charge, the base consumed with the rise of pH from 3 to 9 was taken. The average value of the negative logarithm of the apparent surface dissociation constant, pK_{app}, was calculated also. The latter value was established using site heterogeneity approach[9,11] using a method presented by Józefaciuk and Shin[8]. From titration curves of solid phases, the distribution function of the apparent surface dissociation constant, $F(pK_{app})$, was derived:

$$F(pK_{app}) = dQ(pH)/dpK_{app};\ pK_{app} = pH \tag{3}$$

The average value of pK_{app} was then calculated as:

$$pK_{app,\,average} = \int pK_{app} F(pK_{app}) dpK_{app}. \tag{4}$$

4. RESULTS AND DISCUSSION

The dependencies of surface areas, water vapor adsorption energies, variable charge and apparent surface dissociation constants on the concentration and time for the studied minerals are presented in Figures 1 to5.

For the bentonite the surface area (Fig.1) seems to increase slightly after moderate time and concentration treatments and high time-concentration regimes lead to a significant drop of the surface area. The maximal changes of the surface area accounts in our experiments for about 20% of the initial value. However, much shorter treatment with 2M acid (but at the temperature 348K) was reported to markedly increase (more then three

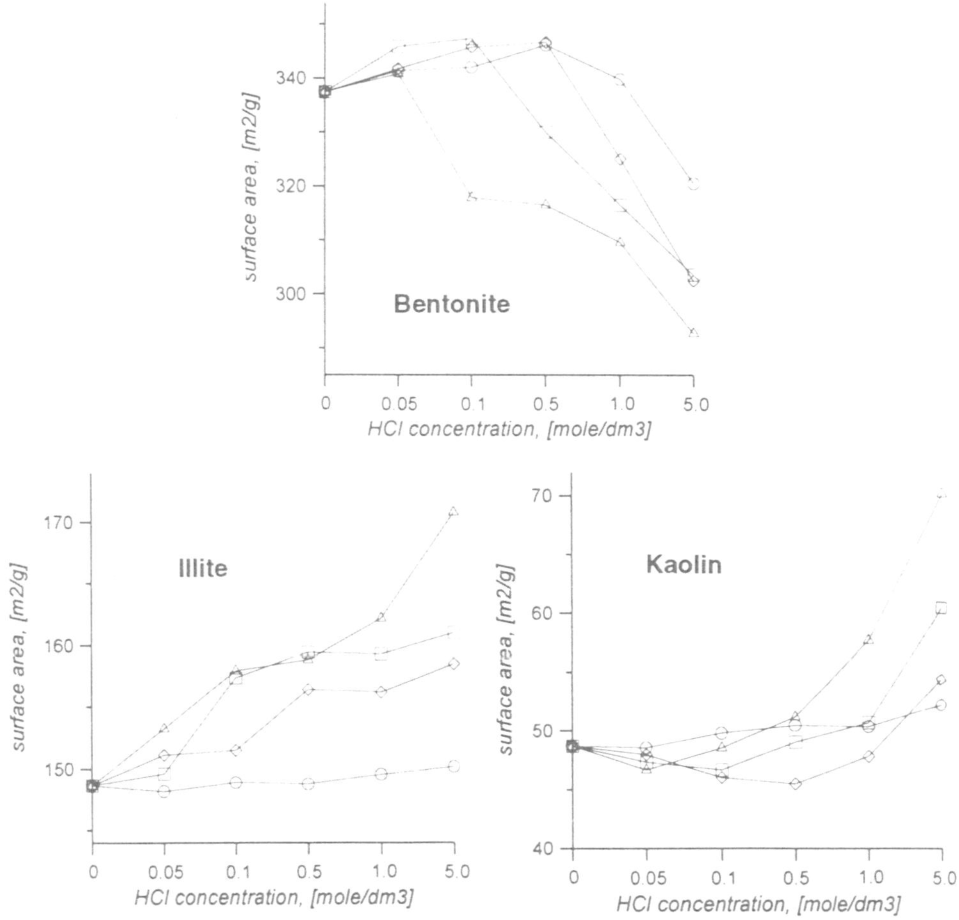

Figure 1. Surface areas of the studied minerals under influence of acidification. Abbreviations: circles: 1 day; rhombs: 3 days; squares: 9 days and triangles: 27 days treatment.

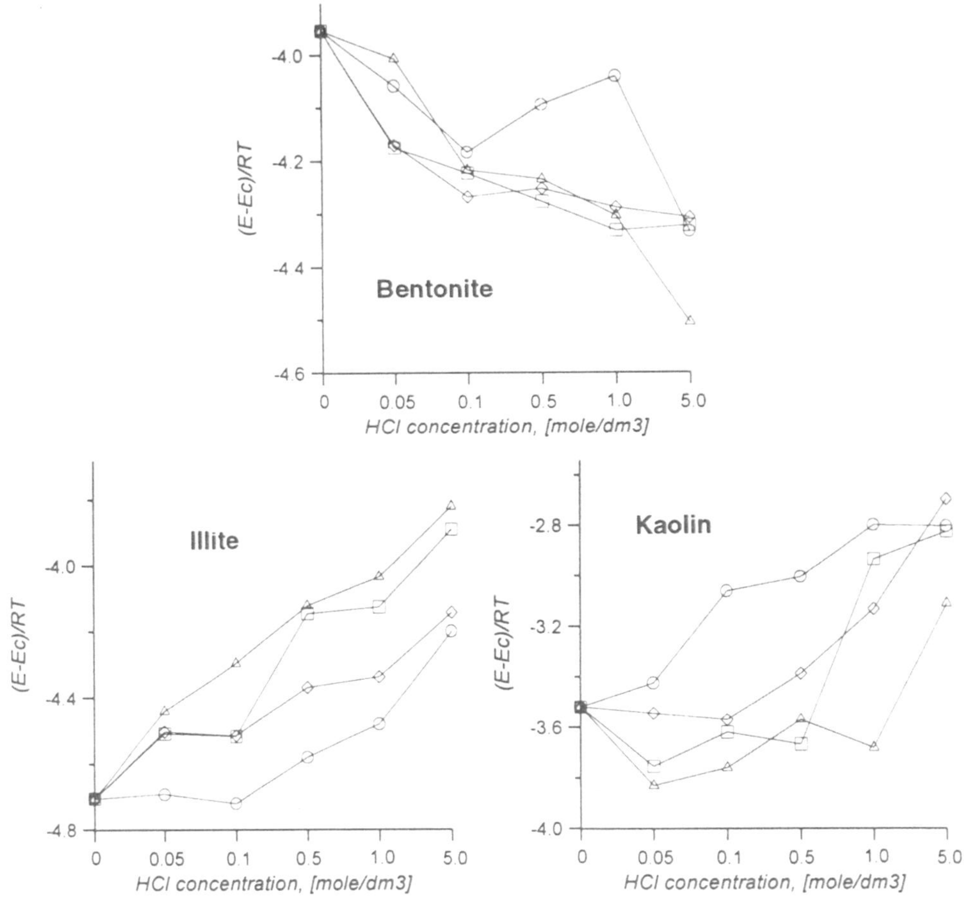

Figure 2. Adsorption energies of the studied minerals under influence of acidification. Abbreviations as in Figure 1.

times) the Serbian bentonite surface area measured by nitrogen adsorption[5]. Because for the minerals the nitrogen and water adsorption are positively correlated, comparing the latter results with our data, one can conclude either that different bentonites can react in different ways or that the mechanism of the reaction of acid with the bentonite surface depends strongly on temperature. The surface area of the illite seems not to be altered by 1day treatment with all acid concentrations applied. However, the consecutive increase of the surface area with acid concentration increase during longer treatments occurs, up to about 25%. The kaolin surface area remains approximately constant up to 0.5M–1M acid treatment. Only very high time-concentration regimes lead to the rise of kaolin surface area. For kaolin, the relative changes of the surface area are the highest—up to 50 % of the initial value. The surface area of the acidified minerals may be connected with the altered mineral surfaces and/or with the surfaces of some residues coming from minerals destruction. Yeoh and Oades[16] reported that such residues (after exhaustive phosphoric acid treatment) for bentonite have less surface area and for the illite and kaolin, higher surface area than the parrent mineral. This is important to mention that surface areas calculated in this paper using Aranovich approach are 1.2 to 1.5 times higher then these resulting from the standard BET method for that same materials.

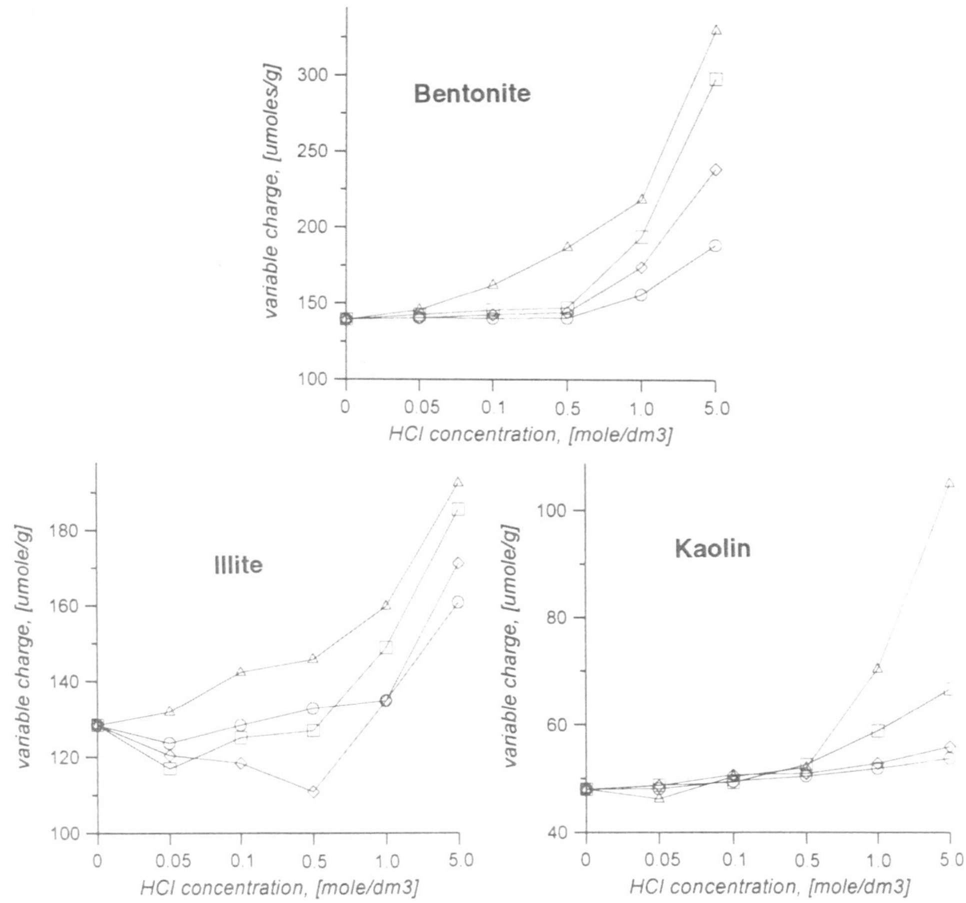

Figure 3. The total amount of variable charge of the studied minerals under influence of acidification. Abbreviations as in Figure 1.

The adsorption energy (Fig. 2.) for the bentonite decreases tendentially with increasing time and concentration of acidification for all but one day treatments. However, for the latter case the final decrease of adsorption energy is evident also. In the notation used the decrease of the (negative) adsorption energy means the increase of adsorption forces. Thus despite the drop in surface area, the adsorption is stronger after acid treatment. In the case of the bentonite this may be connected with the increase of surface porosity caused by the face surfaces being attacked with protons in addition to the attacked edges[3]. The latter are attacked mainly in the case of illite and kaolin[10]. For the illite only the lowest acid concentrations in 1day treatment do not alter adsorption energy. For the other cases the adsorption energy increases with the increase of concentration-time regime. For the kaolin, adsorption energy increases regurarly with concentration in 1day treatments, but in the longer treatments the initial drop followed by the rise of the adsorption energy with the acid concentration occurs. Thus despite the increase of the surface area, the acid treatment of bentonite and illite decreases their surface activity.

The total amount of surface variable charge (Fig. 3.) for the bentonite is practically not altered by acid treatments up to 9 days and 0.5N concentration conditions. More severe acidification regimes lead to the marked increase of the variable surface charge. Similar situation occurs for the kaolin, but for this mineral the higher time-concentration regime is necessary to change the total amount of variable charge. However, for the illite at low acidification regimes the total amount of variable charge decreases and this rises again at higher time-concentration conditions. Under acid treatment the accompanying decrease of the layer charge of the minerals can occur also[15]. This should be stressed that the total variable charge is taken in the method used as the difference between the initial charge at pH=3 and the final charge at pH=9 including charge sign. So, if the variable charge surfaces of the studied minerals have a point of zero charge, the variable charge calculated is equal to the sum of the amount of the positive charges (below PZC) and the amount of negative variable charges above the PZC. The edge surfaces of clay minerals are constant potential surfaces of the definite PZC values[13]. On kaolinite surfaces some amount of positive charge at pH values as high as 9 and 10 was measured by ions adsorption[2].

The values of pK_{app} for all studied minerals increase with the increase of acidification regime (Fig. 4.). This indicates, that the studied surfaces have consecutively weaker acidic character with increasing acidification. For the illite this increase starts practically

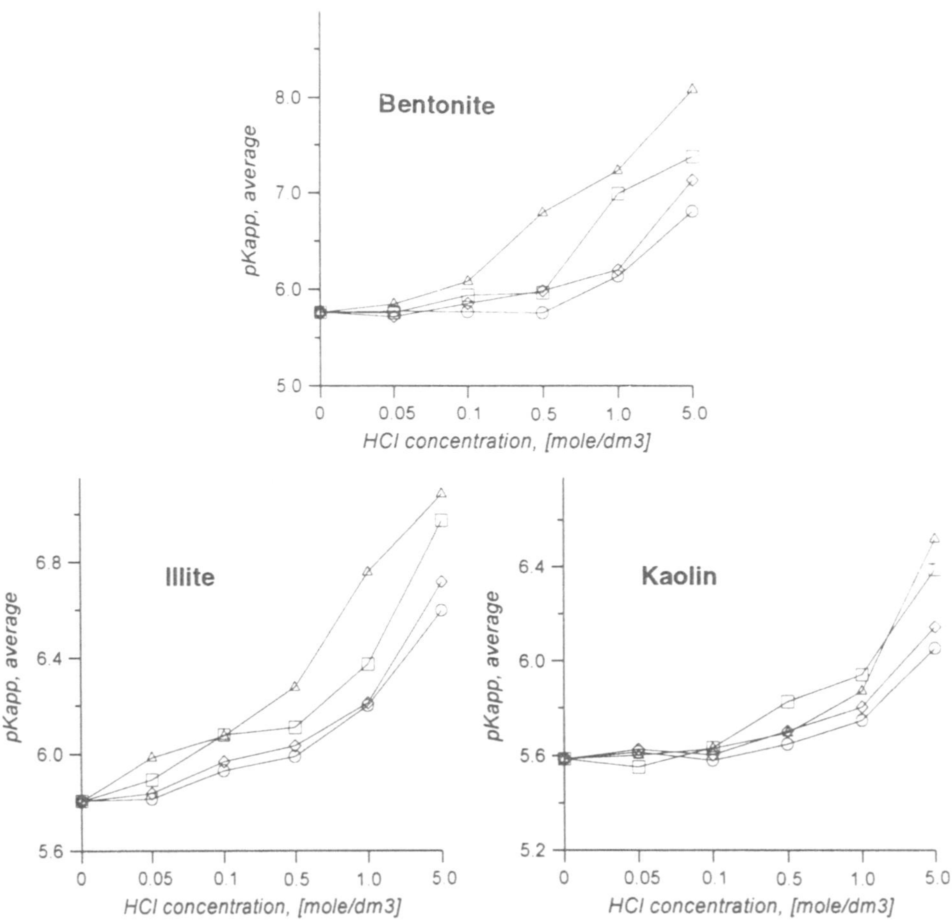

Figure 4. The value of the average apparent surface dissociation constant of the studied minerals under influence of acidification. Abbreviations as in Figure 1.

for the weakest treatment, for the bentonite at acid concentrations about 0.1M and for the kaolin at about 0.5M.

5. CONCLUSIONS

From the data obtained one can conclude that the kaolin is most resistant against acid treatment. All its properties studied but the adsorption energy change very slightly until the highest acid concentrations are applied. The observed changes of the surface properties of the clay minerals studied may come from the surfaces of these minerals altered in different ways by acid attack as well as from the residues lasting after acidic destruction of the finest mineral particles. Also, the overall picture may be influenced by some amounts of impurities present in the parrent materials. Taking into account the different directions of the changes of surface areas and adsorption energies for different minerals, these two properties may not be good indicators of soil acidification. However the amount of variable charge and especially the average apparent dissociation constant value can possibly be used for acidification diagnosis.

REFERENCES

1. AranovichG.L., 1992: The theory of polymolecular adsorption. Langmuir, 8, 736–739;
2. Bolland M.D.A., Posner A.M. and Quirk J.P., 1976: Surface charge on kaolinites in aqueous suspension. Aust. J. Soil Res., 14, 197–216;
3. Davis L.E., 1961: The instability of neutralised H, Al -bentonites. Soil Sci. Soc. Am. Proc. 25, 25–27;
4. Duquette M. and Hendershot W., 1993: Soil surface charge evaluation by back-titration: I. Theory and method development. Soil Sci. Soc. Am. J., 57, 1222–1228;
5. Jovanoviæ N. and Janackoviæ J., 1991: Pore structure and adsorption properties of an acid-activated bentonite. Applied Clay Science, 6, 59–68;
6. Józefaciuk G. and Jae-Sung Shin, 1996a: Water vapor adsorption on soils. I. Surface areas and adsorption energies as calculated by the BET and a new Aranovich theories. J. Korean Soc.Soil Sci. Fertilizer, 29/2, 86–91;
7. Józefaciuk G. and Jae-Sung Shin, 1996b: A modified back-titration method to measure soil titration curves minimizing exchange acidity and dilution effects. J. Korean Soc.Soil Sci. Fertilizer, 29/4, 321–327;
8. Józefaciuk G. and Jae-Sung Shin, 1996c: Distribution of apparent surface dissociation constants of some Korean soils as determined from back titration curves. J. Korean Soc.Soil Sci. Fertilizer, 29/2, 86–91;
9. Koopal L.K., van Riemsdijk W.K. and Roffey M.G., 1987: Surface ionization and complexation models: a comparison of methods for detrmining model parameters. J. Coll. Interface Sci., 118, 117–136;
10. Miller R.J., 1965: Mechanisms for hydrogen to aluminum transformation in clays. Soil Sci. Soc. Am. Proc. 29, 36–39.
11. Nederlof M.M., De Witt J.C., van Riemsdijk W.K. and Koopal L.K., 1993: Determination of proton affinity distributions for humic substances. Environ. Sci. Technol., 27, 846 856;
12. Nilsson I., 1983. Effects on soil chemistry as a consequence of proton input. In: B. Ulrich and J. Pankrath (eds.), Effect of Accumulation of Air Pollutants in Forest Ecosystems, D. Reidel Publ. Comp. 105–111;
13. Parks G.A., 1967. Aqueous surface chemistry of oxides and complex oxide minerals. Isoelectric points and zero point of charge. In: Advances in Chemistry Series. R.F. Gould (ed.), 67, 121–160;
14. Ulrich B., 1986. Natural and anthropogenic components of soil acidification. Z. Pfl. Bodenkunde. 149, 702–717;
15. Warren C.J., Dudas M.J. and Abboud S.A., 1992. Effects of acidification on the chemical composition and layer charge of smectite from calcareous till. Clays Clay Minerals, 40, 6, 731–739;
16. Yeoh N.S. and Oades J.M., 1960. Properties of soils and clays after acid treatment. I. Clay minerals. Aust. J. Soil Res., 19, 147–158.

ABOUT THE EDITORS

LUCJAN PAWŁOWSKI

Lucjan Pawłowski, was born in Poland, 1946. He studied chemistry at the Marie Curie-Skłodowska University and got his Ph.D. in 1976. And D.Sc. (habilitation) in 1980 both at the Technical University of Wrocław. In 1986 the President of Poland nominated L. Pawłowski to a full professor. He started research on application of ion exchange for water and wastewater treatment. As a result he together with B. Bolto from CSIRO Australia, has published a book "Wastewater Treatment by Ion Exchange" in which they summarized their own results and experience of ion exchange area. In 1980 L. Pawłowski was elected President of International Committee "Chemistry for Protection of the Environment". In 1980–1984 has was Chairman of the Environmental Chemistry Division of the Polish Chemical Society. In 1994 he was elected the Deputy President of the Polish Chemical Society and in the same year, the Deputy President of the Presidium Polish Academy of Science Committee "Men and Biosphere". In 1997 he was awarded a title of Honorary Professor of China Academy of Science and in 1998 was elected a member of the European Academy of Science and Arts. He is member of editorial board of international journals: "Reactive Polymers", "The Science of the Total Environment", "Environment International", and "Journal of Ecological Chemistry". In 1990 he served as a adviser to President Lech Wałęsa. In 1991 he was elected as the Deputy Rector of the Technical University of Lublin, and this post he has hold for two terms (1991–1996). He has published 19 books, 128 papers, and authored 68 patents.

MARJORIE A. GONZALEZ

Marjorie A. Gonzalez is the Environmental Assurance Manager for the Defense and Nuclear Technologies Directorate at Lawrence Livermore National Laboratory (LLNL). She currently is responsible for providing environmental technical and management advice to researchers, with her primary focus being pollution prevention efforts. She has also

been a principal co-investigator on an environmental projects for the Middle Urals' region of Russia, collaborating with the environmental scientists from the Institute of Industrial Ecology of the Russian Academy of Sciences located in Yekaterinberg, Russia and scientists from the Russian federal Nuclear Center Institute of Technical Physics located in Snezhinks, Russia.

An LLNL employee for 20 years, Ms. Gonzalez's work has covered a broad spectrum of environmental issues, but has always focused on cost-effective integration of programmatic goals with technical, policy, and regulatory requirements. She has authored numerous articles related to on-line detection of pollutants in wastewater and the development and implementation of source reduction and recycling technologies. Ms. Gonzalez received her bachelor of Science degree in Mathematics from California State Polytechnic University, San Luis Obispo, California, USA and her master of Science degree in Environmental Engineering form the University of Cincinnati, Ohio, USA. She is Registered Environmental Assessor in the State of California and is a Qualified Environmental Assessor certified by Institute of Professional Environmental Practice.

MARZENNA R. DUDZIŃSKA

Marzenna R. Dudzińska received M.Sc. in chemistry in 1983 from Marie Curie-Skłodowska University (UMCS) in Lublin, Poland. In 1983–1988 she was employed as research worker and teacher at Physical Chemistry Department at UMSC. In 1987 she worked in Institute of Spectroscopy and Applied Spectrochemistry in Dortmund, FRG on the project combining chromatography and spectroscopy. In 1988 she was hired by Technical University of Lublin, Poland, and worked for Water and Wastewater Treatment Department She got a Fulbright Scholarship in 1989, and spent 1989/1990 at University of Houston, Texas, in Civil and Environmental Engineering Department. She received Ph.D. in environmental chemistry from Marie Curie-Skłodowska University in 1992. Now she works for Department of Environmental Protection Engineering at Technical University of Lublin. She has authored numerous papers related to adsorption and ion exchange processes, complexation mechanism in water systems and lately waste incineration. She was awarded by Polish Chemical Society for her master thesis in 1983, by Polish Ministry of Science and Education in 1987 and in 1992 and in 1994 and 1997 by Rector of Technical University of Lublin. She is a member of Polish Chemical Society, and served as a President of Student Section of the Society in 1984–1988.

WILLIAM J. LACY

Dr. William J. Lacy, a registered professional chemical engineer, is President of LACY & CO., an international environmental consulting firm to industries and governments. He received his B.Sc. from the University of Connecticut, completed his Masters degree at New York University College of Engineering, and studied at the Oak Ridge In-

stitute for Nuclear Studies, University of Michigan and Michigan State University. The School for Advanced Chemistry at Paul Sabatier University, France, awarded him the University Medal in 1983. He has an honorary Doctorate of Science in environmental engineering.

Dr. Lacy has authored 198 technical publications, including 14 textbooks. He has three patents and serves on the editorial advisory board of five technical journals. He has chaired or co-chaired 34 National International Conferences.

For his efforts, Dr. Lacy has received numerous awards and medals from Thailand, India, Egypt, Russia, Belgium, France, Poland and Italy. Other honors include American Defense Preparedness Association award, Secretary of Defense Special Service Award, Presidential Recognition Award, EPA, Bronze Medal, and the U.S. Governmental Distinguished Service Medal.

He is a fellow in AIChE, Life Member International Ozone Association, ASTM, Life Member Washington Academy of Science, American Academy of Environmental Engineers, WEF, ACS, Res. Soc. of America and Alpha Phi Omega.

AUTHOR INDEX

SUBJECT INDEX

GPSR Compliance

The European Union's (EU) General Product Safety Regulation (GPSR) is a set of rules that requires consumer products to be safe and our obligations to ensure this.

If you have any concerns about our products, you can contact us on ProductSafety@springernature.com

In case Publisher is established outside the EU, the EU authorized representative is:

Springer Nature Customer Service Center GmbH
Europaplatz 3
69115 Heidelberg, Germany

FSC
www.fsc.org
MIX
Papier | Fördert
gute Waldnutzung
FSC® C083411

Kolibri 360 GmbH
Ferdinand-Jühlke-Straße 7
99095 Erfurt, Deutschland
produktsicherheit@kolibri360.de